AIRBORNE AND ALLERGENIC POLLEN OF NORTH AMERICA

WALTER H. LEWIS
   Professor of Biology, Washington University
   Senior Botanist, Missouri Botanical Garden

PRATHIBHA VINAY
   Research Associate of Biology, Washington University

VINCENT E. ZENGER
   Allergy Trainee, Washington University School of Medicine

Airborne and Allergenic POLLEN of North America

WALTER H. LEWIS • PRATHIBHA VINAY • VINCENT E. ZENGER

THE JOHNS HOPKINS UNIVERSITY PRESS
BALTIMORE AND LONDON

This book has been brought to publication with the generous
assistance of the Commonwealth Fund and the Monsanto Company.

© 1983 by THE JOHNS HOPKINS UNIVERSITY PRESS
All rights reserved
Printed in the United States of America

THE JOHNS HOPKINS UNIVERSITY PRESS, Baltimore, Maryland 21218
THE JOHNS HOPKINS PRESS LTD., London

Library of Congress Cataloging in Publication Data
Lewis, Walter Hepworth.
  Airborne and allergenic pollen of North America.

  Bibliography: p.
  Includes index.
  1. Pollen—North America—Atlases. 2. Allergens—North America—Atlases. 3. Hay-fever plants—North America—Identification. 4. Air—Microbiology. 5. Botany—North America—Classification. 6. Phytogeography—North America—Maps. 7. Plants, Flowering of—North America—Flowering time. I. Vinay, Prathibha. II. Zenger, Vincent E. III. Title. [DNLM; 1. Pollen.
QK 658 L677a]
QK658.L48 1983   582.097   82-21183
ISBN 0-8018-2940-2

To Gunnar Erdtman[†] and Associates,
    Palynological Laboratory,
    Swedish Museum of Natural History, Stockholm,
    whose training in palynology was instrumental
    to the development of this book.

Contents

Preface   ix

Acknowledgments   xi

Color Plates

    Trees and Shrubs   I
    Grasses and Grasslike Plants   VII
    Weeds and Herbs   IX
    Guide to Pollen Types   XIII

Introduction   1

Chapter 1. Trees and Shrubs   7

Chapter 2. Grasses and Grasslike Plants   105

Chapter 3. Weeds and Herbs   129

Appendix 1. Distribution Maps of Species   177

Appendix 2. Methodology   237

Glossary   239

Literature Cited   243

Index   249

Preface

This is a book about plants, specifically their pollen. It is not a medical volume, although it was written primarily for allergists and others in the medical profession and for aerobiologists. We in the botanical sciences have been remiss in not providing a concise, visually oriented reference on pollen that could aid the allergists in their quest to understand better the major sources of offenders in inhalant allergy that adversely affect the lives of millions.

Many antibody-mediated allergies occur because humans intrude inadvertently in the plant transmissional process of transferring the pollen from the anther to a receptive stigma of the same species. This process of pollination is essential because the pollen is the male gametophyte, which at maturity possesses the two sperm necessary for sexual reproduction. Immediately following pollination, or accidental contact with human mucosal surfaces, pollen releases proteins. Often these serve as species-specific recognition compounds in plants or, among atopic humans, the same proteins may act as allergens and, following sensitization, promote allergic reactions.

Because there is no comprehensive North American flora, either of indigenous or cultivated plants, basic botanical data involving species distribution, flowering times, and taxonomy are not readily available. In fact, for some of the most important allergenic plants, such as *Ambrosia* (ragweeds), no available publication provides fundamental information on flowering times and distributions of the 17 or more species native to the continent. Furthermore, there is no survey of pollen morphology. As allergists and aerobiologists are particularly interested in those plants whose pollen becomes airborne, our synthesis brings together a large amount of divergent, yet relevant, information about such plants. This volume includes descriptions of plants and pollen, with an updated nomenclature and current phylogenetic classification, as well as information on species distribution and flowering times given by regions of North America. Also included are aerobiological occurrences of pollen and levels and frequencies of allergenicity among human populations. Visual presentation is emphasized, thereby allowing the user an opportunity to assimilate rapidly botanical facts on plant morphology, geography, phenology, and palynology from among the great array of native and introduced vegetation found in North America.

A number of aspects of the book are of particular interest to the user. In the case of pollen, no major treatment has presented photomicrographs using both scanning electron and light microscopy that include acetolyzed and stained whole grains. This uncommon but important approach is of relevance, for example, to botanists and palynologists who often fail to examine whole stained pollen and thus overlook the variable morphology of the intine as a potentially valuable character in comparative morphology and evolutionary studies.

The pollen illustrated includes samples from 67 Angiosperm families, as well as a few Gymnosperm and fern families, that represent the majority of plant groups shedding pollen into the ambient atmosphere of the continent and contributing to the modern pollen rain. The absence of a comprehensive treatment of airborne pollen of the native and cultivated vegetation of North America seriously handicaps accurate diagnoses of modern and historic pollen needed in many biomedical and geologic fields. For example, pollen of plants like *Acalypha, Boehmeria, Garrya, Laportea, Myriophyllum, Thalictrum,* and *Urtica,* that release wind-dispersed pollen as their primary mode of pollination, in addition to many genera that are secondarily anemophilous or incidentally release pollen, have never, or only incompletely, been illustrated and hence remain virtually unknown. Yet each may be important to the health of allergic persons, and all are biological contaminants of the atmosphere.

For purposes of plant identification and for understanding the morphology of staminate flowers, high quality photographs are important. Unfortunately, only infrequently are photographs available of wind-pollinated plants. There are a number of reasons for this: catkins of many trees mature very early in the spring, often maturing and abscissing long before observation and collection; grass inflorescences are small, seemingly lacking in diversity, and consequently often ignored by the nonspecialist; and flowers of many weeds are small and not showy and generally are considered unattractive, thus many are only infrequently collected and even fewer are photographed.

Many of the approximately eight hundred species distribution maps illustrating potentially wind-dispersed pollen are original contributions. The medical and scientific communities interested in where such plants occur are hard-pressed to find even generalized maps of these species. In order to obtain an overview of these distributions, composite maps of species densities of 26 genera (excluding grasses)

PREFACE

having allergenic/aerobiologic significance are also provided. By examining these maps the user can observe those regions of the continent free of an offending aeroallergen or, at the other extreme, those regions heavily populated by species eliciting inhalant allergies. In a society as mobile as that of North America, the geography of potential offenders must always be considered.

Too many botanical studies ignore flowering times. Although approximate because of varying years and seasons, times of flowering have here been summarized from many sources for all major, wind-dispersed species in North America. These are given by month for six regions. Monthly frequencies of major pollen types captured by aerosampling are also summarized for certain families and genera. They show by region when pollen can be anticipated atmospherically and in what frequency, data that usefully corroborate flowering periods obtained from plant collections.

There is no comprehensive modern discussion of plants known to or suspected of eliciting immediate, antibody-mediated hypersensitivities in North America. Therefore, an important aspect of this project was to consider those plants previously thought of minor allergenic importance but not known to have serious implications in pollinosis. For example, *Myrica* (bayberry) pollen is responsible for many cases of severe allergic reactions in the spring, but only a few years ago this source of allergens was largely unrecognized. There are many similar examples among the plants included in this book. A secondary, but nevertheless cognizant, aspect has been the recognition that many allergenic plants, while of minor importance continentally or regionally, may be of relevance at the microenvironmental level causing acute pollinosis among a few individuals exposed to those plants having limited wind-dispersal.

In the progress of allergy research, an important consideration is the existence of common and disparate allergens in pollen of different species. When conducting skin tests, hyposensitizations, and research with allergens, the wise use of information involving plant relationships can serve only to augment the quality of patient treatment and the value of research results. Such botanical data are often difficult to extract from the taxonomic literature, but wherever they are of potential importance they have been included.

Perhaps the most important aspect of this project has not been these individual aspects, but rather the bringing together of all facets into a single cohesive treatment. The concept of the whole book is, therefore, what is unique. It provides a treatment of a large spectrum of indigenous, adventive, and cultivated North American plants important to the allergist, aerobiologist, palynologist, ecologist, evolutionary and population biologist, and geologist concerned with wind-dispersed pollen from either modern or historic periods. Nowhere have data involving distributions, flowering times, pollen dispersal including aerobiological sampling and pollination data, pollen morphology, plant relationships, and allergic potential been brought to focus on the health of so large an atopic population and on the pollen aerobiology of so vast an area.

Acknowledgments

We wish to acknowledge major support from the Commonwealth Fund Book Program, National Institutes of Health Biomedical Research Grant Program, and Washington University. Without their generous support and interest, as well as grants-in-aid from the Center Laboratories and Hollister-Steir Laboratories, this book would never have been completed.

Over the years of preparation we have had assistance from many persons interested in the project. In particular we are grateful to those who contributed figures: J.W. Nowicke, Smithsonian Institution, Washington, D.C. (Figs. 93b, 93d, 93e, 122b, 163), J. Miller (Fig. 148d), W. Wagner (Figs. 126, 150d, 150e, 182), and W. Zenger (Figs. 51a, 51b, 66a, 66b); and determined specimens of several complex taxonomic groups: G.W. Argus (*Salix*), W.H. Blackwell, Jr. (Chenopodiaceae), T.B. Croat (*Solidago*), and G. Davidse (Cyperaceae, Poaceae). Likewise, many helped in a variety of ways, and although our list falls short of recognizing all those who assisted, we thank L.C. Anderson, W.E. Black, M.P.F. Elvin-Lewis, E. Haber, S.F. Hampton, B. Hansen, P. Hock, G.B. Johnson, K.C. Lewis and staff, M.F.R.M. Lewis, W.H. Lewis, Jr., R.F. Lockey, V. Kumar Mohan, V. Muehlenbach, E. Murray, W.H. Outlaw, Jr., J.L. Reveal, R.G. Slavin, S.S. Trott, G.M. Veith, P. Muthiah Vinay, S. Vinjamuri, H.J. Wedner, E.D. White, and R.P. Wunderlin.

We also recognize the generous assistance of allergists from throughout the United States who completed the questionnaire on wind-pollinated allergic plants, viz., A.T. Baugh, Jr., M.L. Brandon, V.V. Chambers, J.T. Chiu, C.C. Cohrs, M.S. Conte, R.L. Cutter, W.C. Deamer, R.L. Don, C. Dubovy, T.W. Green, R. Hale, P. Huffman, F. Kessler, N.M. Kudelko, R.F. Lockey, E.J. Luippold, W.R. MacLaren, G.M. McCaskey, A.W. Neilson, Jr., J.E. Newland, J.S. O'Toole, W.C. Sawyer, J.M. Steffey, J.D. Teigland, W.L. Venning, and A.D. Wert. The help of undergraduate students at Washington University who worked in the laboratory during all phases of the project is gratefully recognized, and we thank in particular J.A. Acosta, R.J. Boos, T.P. Dyer, L.R. Frank, S.L. Murray, M.B. Perlman, W.D. Phipps, C.B. Stevens, L.A. Sunn, L.A. Vawter, J.J. Viscardi, A.L. Watterman, B.T. Woodson, and K.A. Zalumsky.

Finally, we thank K. Persons and A. Papian who typed the manuscript so skillfully.

AIRBORNE AND ALLERGENIC POLLEN OF NORTH AMERICA

Color Plates

Trees and Shrubs

Acer negundo (box-elder), a = × 1⅓, b = × ⅔ (p. 12).

Acer negundo (box-elder), a = × 1⅓, b = × ⅔ (p. 12).

Alnus rubra (red alder) × 1 (p. 23).

Betula populifolia (gray birch) × 1 (p. 23).

Quercus alba (white oak) × 1⅓ (p. 51).

Quercus palustris (pin oak) × 1½ (p. 51).

Quercus prinoides (dwarf chestnut oak) × 1 (p. 51).

Quercus phellos (willow oak) × 1 (p. 51).

Quercus virginiana (live oak) × 1 (p. 51).

Fagus sylvatica (European beech) × 1 (p. 51).

Juglans nigra (black walnut) × 1½ (p. 58).

Carya x laneyi (hybrid hickory) × ½ (p. 58).

III

Morus alba (white mulberry) × 2 (p. 65).

Myrica cerifera (southern wax-myrtle) × 1½ (p. 66).

Populus deltoides (cottonwood) × 2 (p. 85).

Salix bebbiana (long-beaked willow) × 3 (p. 85).

Fraxinus pensylvanica (green ash) × 2 (p. 73).

Celtis occidentalis (hackberry) × 1 (p. 92).

Ulmus parvifolia (Chinese elm) × 1 (p. 92).

Grasses and Grasslike Plants

FIGURE 128

Morphology of representative Poaceae (*Bromus-Lolium*), p. 106.

Bromus inermis (smooth brome) × ¹⁄₁₀.

Dactylis glomerata (orchard grass) × ⅛.

Dactylis glomerata (orchard grass) × ½.

Festuca elatior (meadow fescue) × ⅑.

Bromus inermis (smooth brome) × 1.

Festuca elatior (meadow fescue) × 1.

Hordeum jubatum (foxtail barley) × ⅛.

Hordeum jubatum (foxtail barley) × 1.

Lolium perenne (English ryegrass) × ¹⁄₁₅.

Lolium perenne (English ryegrass) × 1.

FIGURE 129

Morphology of representative Poaceae (*Phleum-Sorghum*), Cyperaceae (*Carex-Scirpus*), and Juncaceae (*Luzula*), p. 106.

Phleum pratense (timothy) × ⅛.

Phleum pratense (timothy) × 1½.

Poa pratensis (Kentucky bluegrass) × ⅓.

Sorghum halepense (Johnson grass) × ½.

Poa pratensis (Kentucky bluegrass) × ¾.

Scirpus maritimus (saltmarsh bulrush) × ⅗.

Sorghum halepense (Johnson grass) × ½.

Carex crinita (fringed sedge) × ⅓.

Cyperus esculentus (yellow nutgrass) × ½.

Luzula multiflora (many-flowered woodrush), from dried herbarium sheet, × ⅓.

Weeds and Herbs

Ambrosia trifida (giant ragweed) × ⅓ (p. 139).

Artemisia ludoviciana (western mugwort) × ½ (p. 139).

Iva annua (rough marsh-elder) × ⅓ (p. 139).

Xanthium strumarium (cocklebur) × 2 (p. 139).

Acalypha rhomboidea (three-seeded mercury) × 1 (p. 158).

Chenopodium hybridum (maple-leaved goosefoot) × ½ (p. 152).

Humulus lupulus (hops) × 1 (p. 150).

Salsola kali (Russian thistle) × ⅒ (p. 152).

Plantago lanceolata (English plantain) × 1 (p. 164).

Urtica dioica (stinging nettle) × ½ (p. 173).

Guide to Pollen Types

Pollen stained ×425 (except as noted)

Polyads, Tetrads

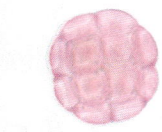
Acacia (Fabaceae), p. 45.

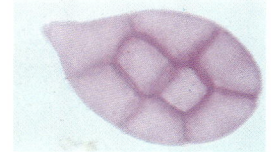
Calliandra ×175 (Fabaceae), p. 45.

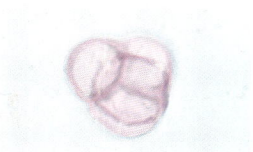
Calluna (Ericaceae), p. 40.

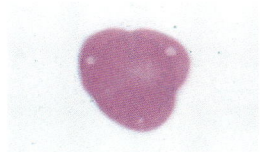
Kalmia (Ericaceae), p. 40.

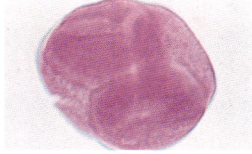
Luzula (Juncaceae), p. 123.

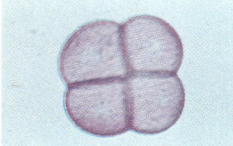
Typha (Typhaceae), p. 125.

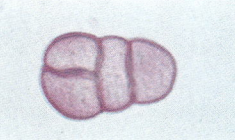
Typha (Typhaceae), p. 125.

Monads—
Shape Different at Poles, Winged

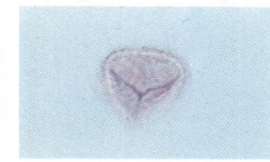
Lycopodium (fern ally), p. 129.

Monads—
Shape ± Similar at Poles
No Apertures

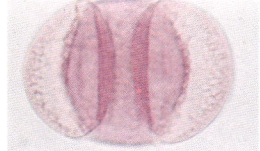
Pinus (Pinaceae), p. 9.

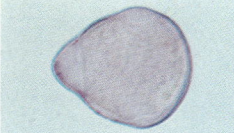
Carex (Cyperaceae), p. 121.

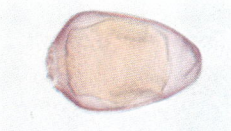
Scirpus (Cyperaceae), p. 121.

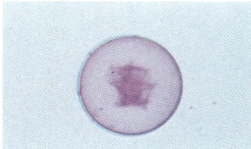
Chamaecyparis (Cupressaceae), p. 7.

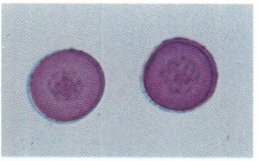
Cupressus (Cupressaceae), p. 7.

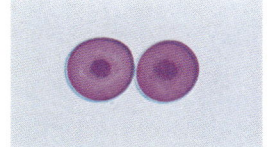
Juniperus (Cupressaceae), p. 7.

Larix (Pinaceae), p. 9.

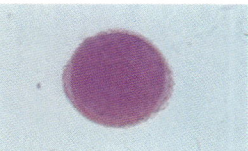
Tsuga ×175 (Pinaceae), p. 9.

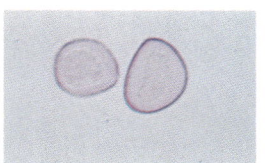
Taxus (Taxaceae), p. 11.

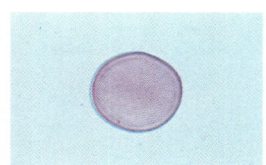
Populus (Salicaceae), p. 85.

Apertures with Pores
1-Porate

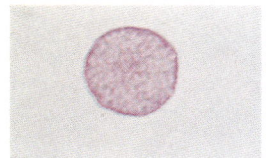
Potamogeton (Potamogetonaceae), p. 125.

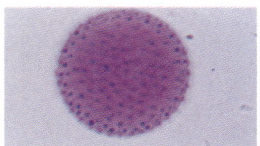
Sassafras (Lauraceae), p. 60.

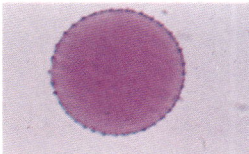
Sassafras (Lauraceae), p. 60.

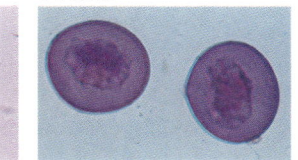
Sequoia (Taxodiaceae), p. 11.

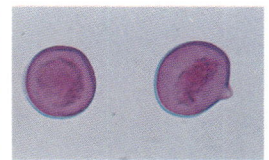
Taxodium (Taxodiaceae), p. 11.

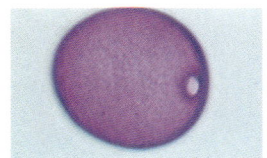
Avena (Poaceae), p. 106.

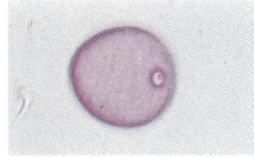
Dactylis (Poaceae), p. 106.

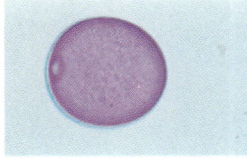
Festuca (Poaceae), p. 106.

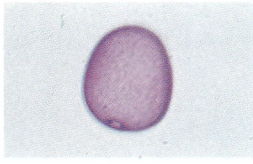
Lolium (Poaceae), p. 106.

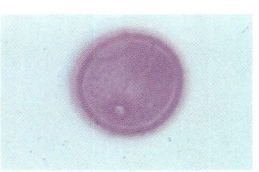
Sorghum (Poaceae), p. 106.

2–3(–4)-Porate—
Not Aspidate, Small Pollen

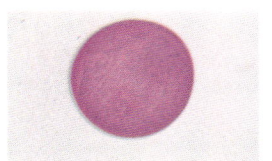
Sorghum (Poaceae), p. 106.

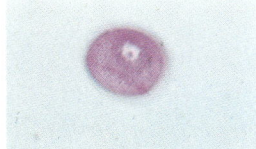
Typha (Typhaceae), p. 125.

Sparganium (Sparganiaceae), p. 127.

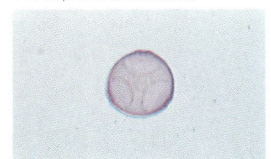
Humulus (Cannabaceae), p. 150.

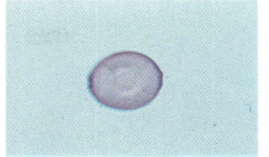
Humulus (Cannabaceae), p. 150.

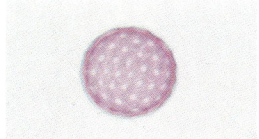 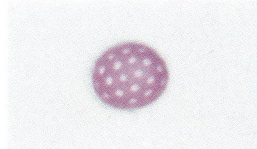 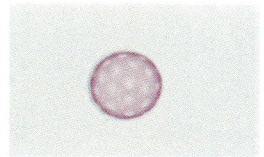 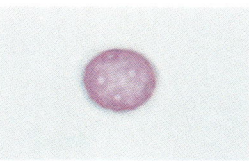 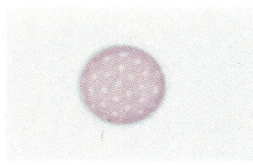

Kochia (Chenopodiaceae), p. 151.   *Salicornia* (Chenopodiaceae), p. 151.   *Salsola* (Chenopodiaceae), p. 151.   *Sarcobatus* (Chenopodiaceae), p. 151.   *Suaeda* (Chenopodiaceae), p. 151.

Apertures with Colpi (Furrows)
1-Colpate

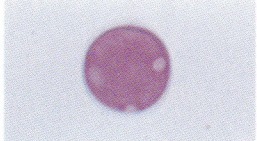 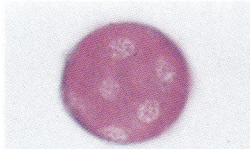 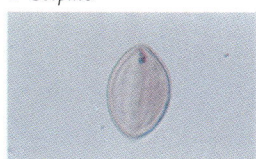 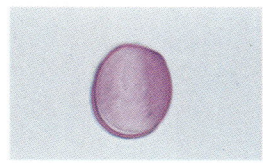 

Pistacia (Anacardiaceae), p. 15.   *Liquidambar* (Hamamelidaceae), p. 58.   *Ginkgo* (Ginkgoaceae), p. 11.   *Chamaedora* (Arecaceae), p. 23.   *Cocos* (Arecaceae), p. 22.

3–4-Colpate

   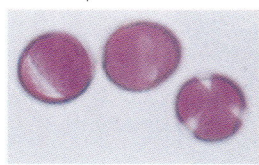 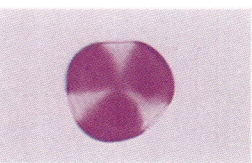

Serenoa (Arecaceae), p. 23.   *Serenoa* (Arecaceae), p. 23.   *Magnolia* (Magnoliaceae), p. 62.   *Acer* (Aceraceae), p. 12.   *Acer* (Aceraceae), p. 12.

   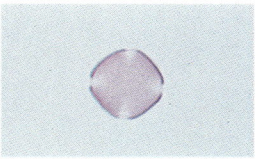 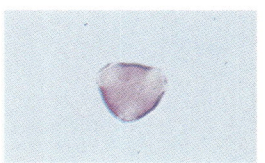

Acer (Aceraceae), p. 12.   *Quercus* (Fagaceae), p. 51.   *Quercus* (Fagaceae), p. 51.   *Fraxinus* (Oleaceae), p. 73.   *Platanus* (Platanaceae), p. 80.

5+-Colpate, Syncolpate

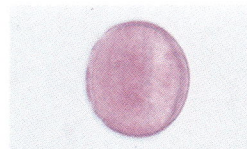 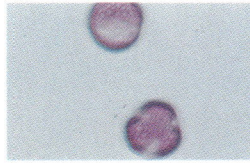 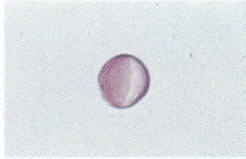 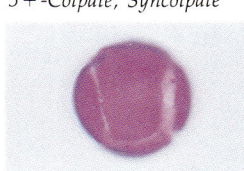 

Cercocarpus (Rosaceae), p. 81.   *Salix* (Salicaceae), p. 85.   *Tamarix* (Tamaricaceae), p. 89.   *Berberis* (Berberidaceae), p. 23.   *Eschscholzia* (Papaveraceae), p. 162.

Apertures with Pores and Colpi (Colporate)
Mostly 3-Colporate, No Spines

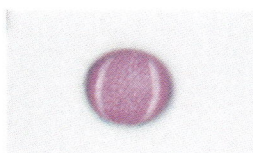  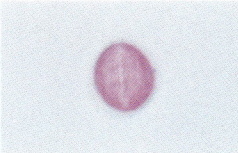 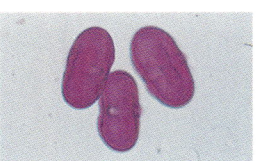 

Eschscholzia (Papaveraceae), p. 162.   *Rhus* (Anacardiaceae), p. 17.   *Schinus* (Anacardiaceae), p. 17.   *Daucus* (Apiaceae), p. 135.   *Artemisia* (Asteraceae), p. 139.

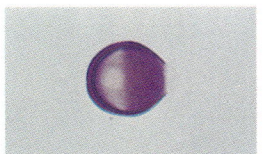 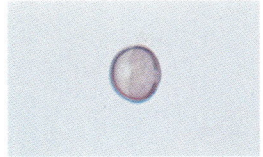  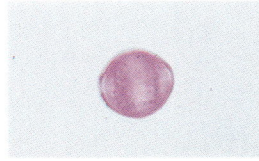 

Artemisia (Asteraceae), p. 139.   *Sambucus* (Caprifoliaceae), p. 31.   *Cornus* (Cornaceae), p. 39.   *Cornus* (Cornaceae), p. 39.   *Ricinus* (Euphorbiaceae), p. 45.

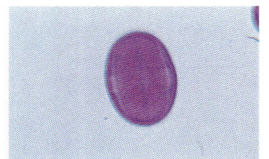 *Melilotus* (Fabaceae), p. 159.
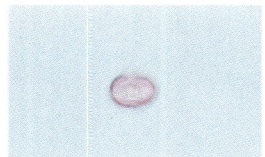 *Castanea* (Fagaceae), p. 51.
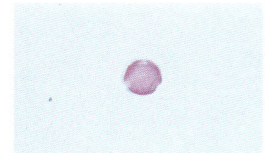 *Castanopsis* (Fagaceae), p. 51.
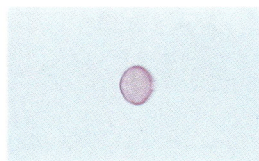 *Castanopsis* (Fagaceae), p. 51.
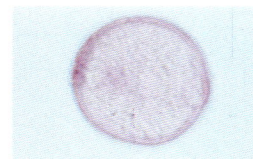 *Fagus* (Fagaceae), p. 51.

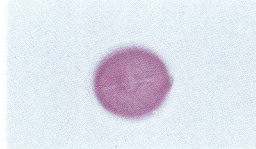 *Garrya* (Garryaceae), p. 57.
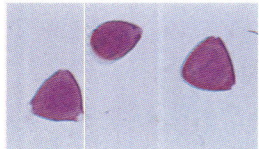 *Callistemon* (Myrtaceae), p. 69.
 *Eucalyptus* (Myrtaceae), p. 69.
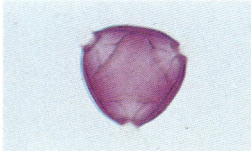 *Nyssa* (Nyssaceae), p. 71.
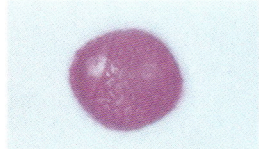 *Ligustrum* (Oleaceae), p. 73.

 *Olea* (Oleaceae), p. 73.
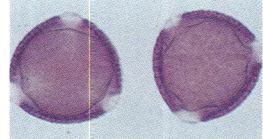 *Syringa* (Oleaceae), p. 73.
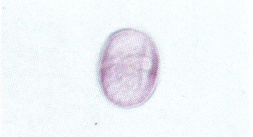 *Polygonum* (Polygonaceae), p. 169.
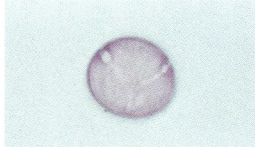 *Rumex* (Polygonaceae), p. 167.
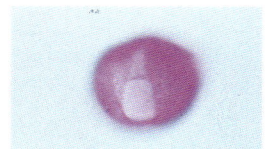 *Crataegus* (Rosaceae), p. 82.

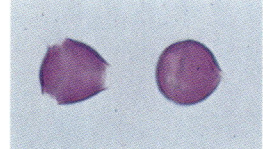 *Prunus* (Rosaceae), p. 82.
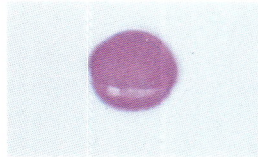 *Phellodendron* (Rutaceae), p. 84.
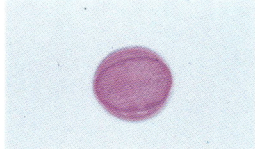 *Ailanthus* (Simaroubaceae), p. 87.
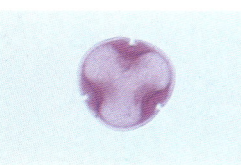 *Tilia* (Tiliaceae), p. 90.
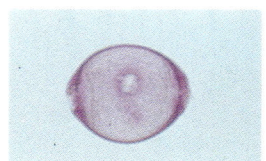 *Tilia* (Tiliaceae), p. 90.

Mostly 3-Colporate, with Spines

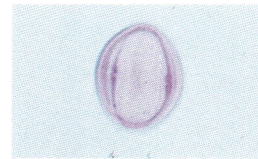 *Parthenocissus* (Vitaceae), p. 98.
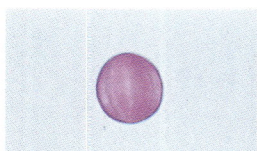 *Larrea* (Zygophyllaceae), p. 98.
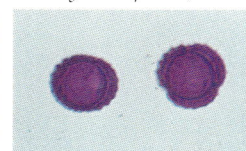 *Ambrosia* (Asteraceae), p. 139.
 *Arctium* (Asteraceae), p. 147.
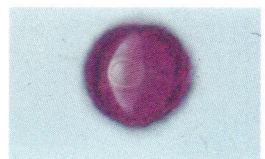 *Arctium* (Asteraceae), p. 147.

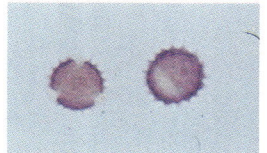 *Baccharis* (Asteraceae), p. 139.
 *Eupatorium* (Asteraceae), p. 139.
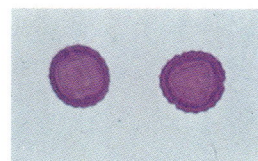 *Iva* (Asteraceae), p. 139.
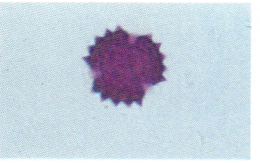 *Tanacetum* (Asteraceae), p. 139.
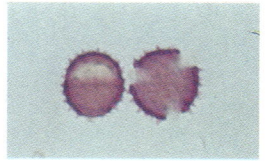 *Solidago* (Asteraceae), p. 139.

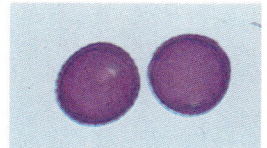 *Xanthium* (Asteraceae), p. 139.
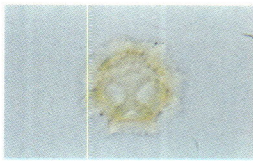 *Taraxacum* acetolyzed (Asteraceae), p. 139.
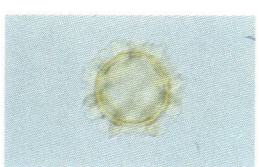 *Taraxacum* acetolyzed (Asteraceae), p. 139.

Introduction

Visual presentation of a large amount of botanical data that could be found quickly has been the focus of this publication. The relatively simple format and the initial determination of pollen types through the four-page color key also enable those without extensive training in the field to have ready access to information that may be needed in the management of patients with allergic problems and in recognizing the biological components of the ambient atmosphere.

For each plant family known to elicit major allergic reactions(*), illustrative material includes one or more photographs of representative plants, usually several pollen photomicrographs using different techniques of preparation, and maps of North American distributions of important species including composite species maps for the most significant genera. Flowering data are presented in a series of histograms (grasses), in tabular form, or in the text. Aerobiological data based on the Statistical Report of the Pollen and Mold Committee of the American Academy of Allergy (1974-78) are tabulated where available. Brief descriptions of plant morphology, flowering and pollen aerobiology, pollen morphology, and allergenicity accompany the visual presentation.

For those families eliciting only minor or localized allergic reactions concomitant with releasing only limited windborne pollen, a briefer format is typical, although data on pollen morphology, flowering, and allergenicity are always included if known.

Plant Names

Each species has a legitimate binomial of two Latin words, a generic name followed by a specific epithet. Sometimes more than one Latin binomial refers to the same species. These synonyms (syn.) are given with the correct binomial in Appendix 1. Also in this appendix is listed the authority for each species, an important reference for understanding origin and nomenclatural history. If in current use, one or more common (vernacular) names is also given. These are often a source of confusion unless a common name is clearly understood to apply to only one species. Names below the rank of species (subspecies, varieties) are not used.

Names of genera and families occasionally possess synonyms or misapplied names and their nomenclature has been updated.

Morphology

Each family of major allergenic importance is described in sufficient morphological detail to aid the nonspecialist in identifying the offending plants. A brief reference to economically relevant plants is also included. These descriptions supplement the photographs of representative genera. If families are only minor offenders in inhalant allergy, morphological diagnoses are not given, but a photograph of the plant is usually included.

Classification

The use of plant taxonomy to infer cross-reactivity of pollen antigens/allergens depends on two premises, as aptly described by Weber (1981). First, more closely related plants will have greater similarities and more shared antigens/allergens than more distantly related plants. Second, the current botanical classification truly reflects phylogeny, whereby two species in the same genus evolved from a recent common ancestor and would share almost all of their antigens, and two species of different genera of the same family evolved from a more distant common ancestor and would share fewer common antigens. Eventually, shared antigens might be few, particularly if the families were classified in different Phylogenetic Groups, that is, two species of different Phylogenetic Groups would have evolved from a common ancestor a very long time ago and would share few, if any, antigens. Thus, there is essentially no cross-reactivity between pollen extracts of the Betulaceae (alder, birch, hazel) classified in Group 2 and the Poaceae (grass) found in Group 14 (Dalen & Voorhors 1981), and they undoubtedly share no major and perhaps no minor allergens. There is no cross-reactivity between Gymnosperms and Angiosperms (Weber 1981), as might be expected, for Angiosperms appeared from gymnospermlike or fernlike ancestors 65-135 million years ago.

Those Angiosperm families known or suspected of eliciting pollinosis are shown in a schematic phylogenetic tree (Fig. 1). This diagram reflects the current botanical classification and may serve as a useful working model for broad research concerning pollen allergens. There is great diversity among flowering plants, accumulating since their explosive evolution during the Cretaceous period, yet two families are responsible for more inhalant allergy from pollen than

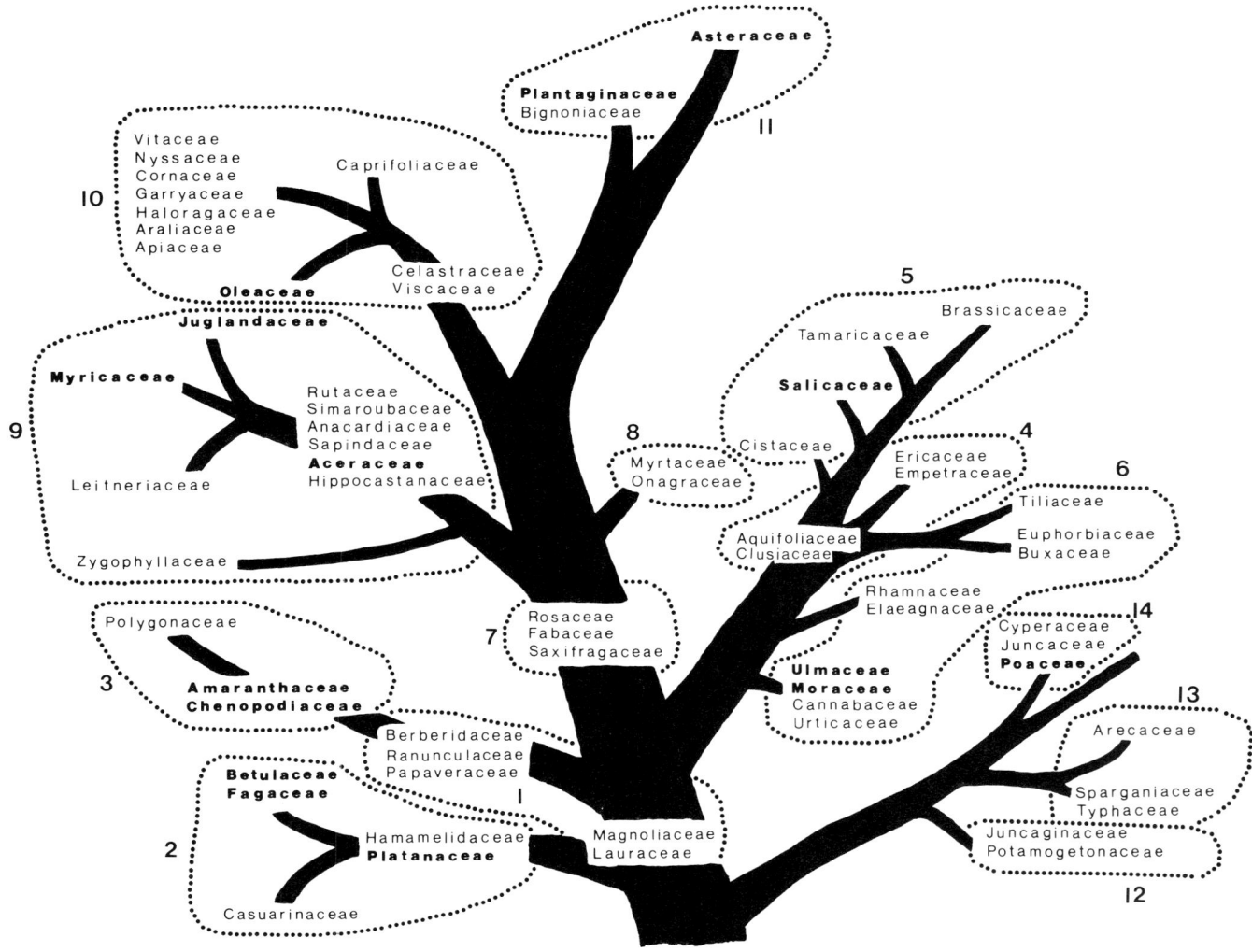

Fig. 1. Diagrammatic phylogenetic tree of flowering plants (Angiosperms) showing families with offending species in inhalant allergy. Important families in pollinosis are in boldface. Phylogenetic groups (or superorders) showing relationships of families are numbered (1) to (14) (Lewis & Elvin-Lewis 1977).

Trees and Shrubs

Aceraceae (maple) 9
Anacardiaceae (cashew) 9
Aquifoliaceae (holly) 4
Araliaceae (ginseng) 10
Betulaceae (birch) 2
Bignoniaceae (bignonia) 11
Buxaceae (boxwood) 6
Caprifoliaceae (honeysuckle) 10
Casuarinaceae (Australian pine) 2
Celastraceae (staff-tree) 10
Cistaceae (rockrose) 6
Clusiaceae (St. John's wort) 4
Cornaceae (dogwood) 10
Elaeagnaceae (oleaster) 6
Empetraceae (crowberry) 7
Ericaceae (heath) 4
Euphorbiaceae (spurge) 6
Fabaceae (legume) 7
Fagaceae (beech) 2
Garryaceae (silk-tassel) 10
Hamamelidaceae (witch-hazel) 2
Hippocastanaceae (horse-chestnut) 9
Juglandaceae (walnut) 9
Lauraceae (laurel) 1
Leitneriaceae (corkwood) 9
Magnoliaceae (magnolia) 1
Moraceae (mulberry) 6
Myricaceae (wax-myrtle) 9
Myrtaceae (myrtle) 8
Nyssaceae (tupelo) 10
Oleaceae (olive) 10
Platanaceae (planetree) 2
Rhamnaceae (buckthorn) 6
Rosaceae (rose) 7
Rutaceae (citrus) 9
Salicaceae (willow) 5
Sapindaceae (soapberry) 9
Saxifragaceae (saxifrage) 7
Simaroubaceae (quassia) 9
Tamaricaceae (tamarisk) 5
Tiliaceae (linden) 6
Ulmaceae (elm) 6
Viscaceae (mistletoe) 10
Vitaceae (grape) 10
Zygophyllaceae (lignum vitae) 9

Grasses and Grasslike Plants

Cyperaceae (sedge) 14
Juncaceae (rush) 14
Juncaginaceae (arrowgrass) 12
Poaceae (grass) 14
Potamogetonaceae (pondweed) 12
Sparganiaceae (bur-reed) 13
Typhaceae (cattail) 13

Weeds and Herbs

Amaranthaceae (amaranth) 3
Apiaceae (carrot) 10
Asteraceae (aster) 11
Brassicaceae (mustard) 5
Cannabaceae (hemp) 6
Chenopodiaceae (goosefoot) 3
Euphorbiaceae (6)
Fabaceae (legume) 7
Haloragaceae (milfoil) 10
Onagraceae (evening-primrose) 8
Papaveraceae (poppy) 1
Plantaginaceae (plantain) 11
Polygonaceae (buckwheat) 3
Ranunculaceae (buttercup) 1
Urticaceae (nettle) 6

all others combined. These very large families, the Asteraceae (Compositae) and Poaceae (Gramineae), are necessarily subdivided into subfamilies (or groups) and tribes. The most current infrafamilial classification that may be important for comprehending antigenic/allergenic diversity inherent within each has been provided. Also in the text, subgroupings of large genera, such as *Quercus* (oak) and *Ambrosia* (ragweed), may be meaningful, particularly when allergenic research extends to many species of these genera.

The initial separation of the volume follows, with minor modification, the traditional grouping of allergenic plants into "trees," "grasses," and "weeds." Although in part botanically unnatural, it is a useful distinction, because it follows a generalized flowering sequence from spring to autumn in temperate areas. The phylogenetic tree does not take into account this superdivision of plants but the legend for Fig. 1 does group the families accordingly.

Distribution

Distribution maps of some of the most important indigenous species, and also of introduced grasses, are based on a number of general sources besides specific monographic treatments (Appendix 1). For the trees, an *Atlas of United States Trees* (Little 1971–78) was indispensable and was supplemented by *The Complete Trees of North America* (Elias 1980) and *Native Trees of Canada*, 4th ed. (1949). For the grasses, the *Manual of the Grasses of the United States*, 2d ed. (Hitchcock 1971) was fundamental. For plants in general, and particularly for the shrubs and herbs, the *Atlas of the Flora of the Great Plains* (Barkley 1977) was used freely as were all the floras listed under "Flowering."

For introduced species of trees and shrubs, perennial weeds, and herbs, distributions are based on vegetational hardiness zones (*Hortus Third* 1976) from 1 (extreme northern Canada) to 10 (subtropical Florida, Texas, and California) (Fig. 2). Appendix 1 gives their origin and hardiness zone, the number representing the northern limit of cold tolerance of the species. The species may be planted south (or west) of this zone.

For the most significant allergenic taxa (usually genera or, as for oaks, subgeneric groups), composite distribution maps of indigenous species are also provided. These maps show species densities and are based on the maps of individual species provided in Appendix 1. They are intended to give a rapid overview of species distributions and frequencies if important in pollinosis.

Flowering

It is difficult to obtain reliable data on flowering times of North American plants. Many botanical mono-

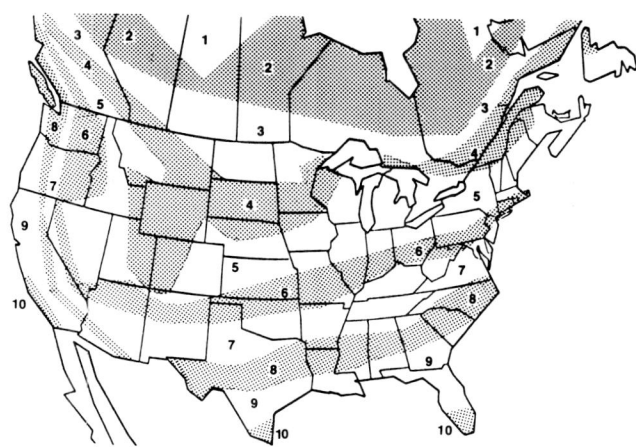

Fig. 2. Map showing hardiness zones from northern Canada (1) to subtropical Florida, Texas, and California (10). A plant recorded for Zone 6, for example, will survive in that zone and to south and west, but not north of it. Most data are for cultivated plants (*Hortus Third* 1976).

graphs and floras, for example, ignore times of flowering, and even some of the best do not include flowering for all genera. Nevertheless, we have prepared data from a number of modern source volumes, one for each major region of the continent: nNE (northern Northeast)—*The Flora of New England* (Seymour 1969), sNE (southern Northeast)—*Flora of Missouri* (Steyermark 1963), SE (Southeast)—*Manual of the Vascular Flora of the Carolinas* (Radford et al. 1968), sFL (southern Florida)—*A Flora of Tropical Florida* (Long & Lakela 1976), NC (northcentral)—*Vascular Plants of South Dakota* (Bruggen 1976), SC (southcentral)—*Manual of the Vascular Plants of Texas* (Correll & Johnson 1970), NW (Northwest)—*Vascular Plants of the Pacific Northwest* (Hitchcock et al. 1955–69), and SW (Southwest)—*A California Flora & Supplement* (Munz 1959, 1968). In addition, secondary sources were sometimes needed, viz., *The Flora of Canada* (Scoggan 1978–79), *Gray's Manual of Botany*, 8th ed. (Fernald 1950), *Flora of West Virginia*, 2d ed. (Strausbaugh & Core 1978), *A Flora of New Mexico* (Martin and Hutchins 1980–81), and *Arizona Flora and Supplement* (Kearney & Peebles 1951, 1960). When flowering times for particular taxa did not exist in these volumes or when they appeared incomplete, herbarium specimens found in several major collections were also consulted.

Often, plants important to the allergist and aerobiologist are not included in botanical floras, for the latter include only "wild" species (indigenous and introduced but only when naturalized). For species of horticultural and agricultural significance we consulted *Hortus Third* (1976), the *Manual of Cultivated Trees & Shrubs*, 2d ed. (Rehder 1947), and also herbarium collections.

Flowering is summarized by meterological months and seasons: December-February = winter, March-May = spring, June-August = summer, and Sep-

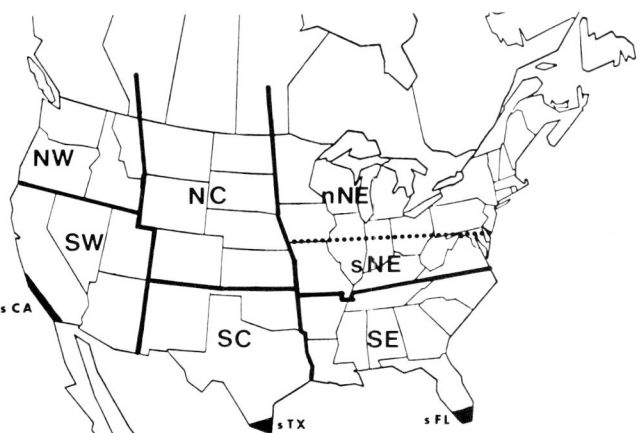

Fig. 3. North America divided into regions for grouping plant flowering and aeropollen data.

tember-November = autumn. During an exceptional year flowering might occur during a greater or shorter period than that given.

For many wide-ranging species it is inadequate to provide a flowering time that encompasses the total distribution. Therefore, we divided the continent into eight regions, which considers the importance of latitude across the continent and of major vegetational differences that separate the eastern deciduous forests from the grassy prairies and plains of the central portion and the highly variable flora from the Rocky Mountains to the Pacific Coast (Fig. 3). The northeast as a whole is similar, but differences in flowering times between New England and Missouri, for example, necessitated the refinement of this highly populous region into a northern and southern part. The three discontinuous, ±subtropical areas are in southern Florida (sFL), Texas (sTX), and California (sCA). Unfortunately, flowering data are usually available only for the adjacent major region, except for southern Florida where a separate flora exists. The remaining regions are self-explanatory: southeast, northcentral, southcentral, northwest, and southwest. The system, albeit crude, gives a generalized view of flowering according to region.

Pollen Aerobiology

At maturity the anthers release pollen that transmits the male genetic material in sexual reproduction. The pollen may be wind-pollinated (anemophilous) and, depending on bouyancy, sculpturing, shape, stickiness, and other inherent characteristics as well as the ambient environment (wind, humidity, time, temperature), may be carried some distance from the immediate vicinity of the parent. Species that are wind-pollinated are the source of the vast majority of pollen allergens. Some species may also be browsed by insects seeking pollen for food and may be secondarily or facultatively insect-pollinated, or they may be only incidentally visited by vectors.

Wind-pollination is typical of many temperate zone plants. In the tropics grasses abound in savannas and other areas and are a major source of allergens as in the temperate areas. Even in the lowland tropical rain forest, however, wind-pollination occurs (Bawa & Crisp 1980). Plants that have evolved wind-pollination as the predominant mode of pollination, generally as a specialization from insect-pollination, share a set of adaptive features. These include a reduction in the size and number of perianth parts; little or no nectar or aromatic compounds; small, dry, and smooth pollen grains (20–40 μm in diameter) with slow terminal velocities (2–6 cm/sec); a complex feathery stigma with a relatively large surface area; and the production and release of a large number of grains per ovule (Adams et al. 1981). Exceptions exist for each of these major adaptations, however, an important one being the spiny pollen sculpturing of the wind-pollinated and highly allergenic members of the subtribe Ambrosiinae (Asteraceae).

The majority of flowering plants have not evolved anemophily but have maintained a vector-mediated pollination system. Such plants are of little general concern in pollinosis, not because their pollen lacks antigens, but because exposures to their pollen are so limited that sensitizations are prevented from occurring. Yet some animal-pollinated (zoophilous) species produce abundant pollen, have an open floral morphology with exposed anthers, produce few lipids and have less sticky pollen, and may release at least moderate amounts of pollen that is wind-dispersed (Lewis & Vinay 1979). These imperfectly entomophilous or, more broadly, zoophilous species (amphiphilous, Wodehouse 1971) are sometimes partially adapted to wind-dispersal as a secondary mode of pollination and are facultatively anemophilous, while for other species pollen released to the atmosphere is incidental to pollination and is an accidental event related to local environmental conditions. Because of either facultative anemophily or incidental release of pollen it is fallacious to ignore all zoophilous species as potential offenders in localized cases of inhalant allergy.

Aeropollen data have been summarized from the Statistical Report of the Pollen and Mold Committee (1974–78) for important genera or families where available. These summaries, selected from 22 stations in the United States and Canada over a period of five years, provide frequencies, but not volumetric counts, of total pollen captured by Durham samples (gravitational method only) per month by regions of the continent.

A number of recent aerobiological treatises are important references. For the morphology of ambient pollen in Canada, see Bassett et al. (1978), and in Europe, Ciampolini & Cresti (1981), Charpin et al.

(1974), Hyde & Adams (1958), and Nilsson et al. (1977). For a broad reference on aerobiology, the best is Edmonds (1979).

Pollen Morphology

The pollen wall consists of three layers: an outer layer resistant to acetolysis known as the exine, which is separable into an external sexine and an internal impervious nexine, and an inner nonresistant intine (Fig. 4). It is normally the sexine, the outer surface of which is usually permeated by thousands of micropores, and the intine that have direct roles in pollinosis. A second important feature of the pollen wall is the aperture, either a pore or furrow or both, from which the pollen tube emerges following successful pollination. It is through these large gaps in the wall that compounds held in the intine are able to escape the inner confines of the pollen. These compounds, as well as those released from the sexinous micropores on the surface, are the pollen allergens that elicit IgE production and give rise in sensitized individuals to symptoms of inhalant allergy (Lewis & Vinay 1980).

Pollen was examined using three techniques from either fresh material that was air dried (ad) or acetolyzed specimens (a) of various ages. Without further preparation, pollen was mounted on studs, sputter-coated with gold, and examined using the scanning electron microscope (SEM). Pollen of a few species having a thin exine was acetolyzed and critical point dried in order to prevent wall collapse before proceeding with the sputter coating. The SEM photomicrographs give a three-dimensional view of the pollen and show well the sexine surface and often the aperture. A second lot of untreated pollen was stained with basic fuchsin and photographed using light microscopy (LMs). Such pollen shows the protoplasm intact and the thickness of the innermost wall, the intine; these are illustrated in black-and-white throughout the text and in color on Pls. XIII–XVI, which are the visual key to major pollen types. The third technique involved an acid-heat treatment called acetolysis (Erdtman 1952) that removed all material from the pollen except the resistant exine. Photomicrographs of acetolyzed pollen (LMa) are excellent for showing details of the sexine's internal architecture and its relation to the nexine and of the anatomy of the aperture. These methodologies and the equipment used are described in Appendix 2.

By using the results of one or all of these techniques, identification of the pollen is usually possible without serious qualification. Many pollen grains can be identified to the level of genus, but only rarely to that of species. Sometimes a cluster of genera forming a tribe or family possesses very similar pollen, and it may be impossible to distinguish them even after a very careful and detailed micromorphological study. Indeed, in some instances, there is no certain way using conventional light microscopy of distinguishing pollen of all members of related families, as, for example, Amaranthaceae and Chenopodiaceae, and it is prudent to lump them into a single pollen morphological group even though a few distinct pollen types typical of genera may easily be recognized. The vast majority of pollen types found in the ambient atmosphere of North America are illustrated, however, and can be identified with some degree of certainty.

A brief palynological description accompanies the photomicrographs. Terminology has been reduced to a minimum, although all terms are defined in the Glossary. Descriptions of most grains include shape, size (if radially symmetrical, measurements made between poles [the polar axis P] and/or of the equatorial diameter [E] as seen in polar view), aperture number and type (colpus or furrow, pore or os), and wall stratification (including sexine, nexine, and intine).

For more detailed SEM of pollen for a few regions of the continent, the reader is referred to Adams & Morton (1972–79) for woody plants of eastern Canada, Lieux (1980–82) for Louisiana and the Southeast, and Martin & Drew (1969–70) for the Southwest.

Allergenicity

This volume includes only type I, antibody-mediated hypersensitivity in which pollen allergens lead to allergic reactions in a sequence approximately as follows. Pollen enters the nose or eyes and lands on mucous membranes of the upper respiratory tract; mucous liquid solubilizes the pollen allergens released rapidly from the micropores on the surface and more gradually from the intine through the apertures, which then penetrate the mucous tissues; mucous membranes of allergic patients contain mast cells that have many IgE antibodies attached to their membranes; pollen allergens quickly complex with IgE antibodies; the complex activates enzymes that cause the release of mediators from specialized organelles in these cells; chemical mediators, e.g., his-

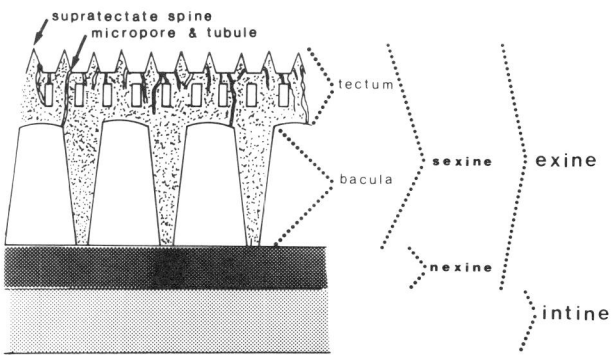

Fig. 4. Diagrammatic pollen wall cross-section showing exine components and intine.

tamine, induce the allergic symptoms by the dilation of blood capillaries, contraction of nasal and bronchial muscles, constriction of nasal or bronchial passages, hypersecretion of watery nasal fluids, and edema of mucous membranes (Stanley & Linskens 1974).

Allergic reactions are discussed briefly for each family or genus as a consequence of exposure to pollen allergens. Data relevant to cross-reactivity are also summarized. Unfortunately, only a few allergens from pollen of *Ambrosia* and its allies, some grasses, *Betula*, and *Alnus* have been characterized, too few to obtain an overview of allergenic homology, or lack of it, among any major group of flowering plants. It is clearly an important area of future research using pollen material from vouchered collections of known origin which have been correctly determined. Nevertheless, we have presented current botanical classifications for major genera of trees, in particular, and some weeds so that patients allergic to oak pollen, for example, who may be sensitive to allergens found in one or several oak taxonomic groups, may be skin tested and desensitized using extracts from the major classes of oaks.

We have added as appropriate unpublished data on pollen hypersensitivity among atopic patients suffering from pollinosis sent to us by many allergists from across the United States. These results were based on skin tests using prick or scratch testing at a usual concentration of 1:20 w/v in 50% glycerine. Supplemental intradermal (intracutaneous) tests were occasionally performed when the initial skin test reactions to suspected allergens were negative or weakly positive. Aqueous (nonglycerinated) concentrations were 1:500 w/v, commonly 1:1000 w/v, or occasionally more dilute. Controls were typically buffered saline with phenol (negative) and/or histamine acid phosphate (positive).

Chapter 1

Trees and Shrubs

Gymnosperms (*Conifers*)

Gymnosperms are cone-bearing trees and shrubs with scalelike or needlelike leaves which are dominant in northern North America. They are differentiated from Angiosperms by their exposed ovules (not enclosed in an ovary), single fertilization, haploid endosperm, and so forth. The assemblage included here are very diverse and include many distantly related lines of phylogeny, though two lines are usually distinguished, the Cycad and the Conifer-Ginkgo. The latter is common, consisting of the closely related Cupressaceae, Pinaceae, Taxaceae, and Taxodiaceae, and the distantly allied Ginkgoaceae. All are pollinated by wind, but only the Cupressaceae are important in pollinosis.

CUPRESSACEAE*
Cypress Family

Leaves of the Cupressaceae are small and scalelike, opposite or whorled, and their cones are usually smaller than those of other Gymnosperms, the cone scale being opposite or whorled. Mature cones are small, dry, and woody, or fleshy and berrylike.

The important genera are *Chamaecyparis* (false cypress), *Cupressus* (cypress), particularly common in the Southwest, *Juniperus* (red cedar, juniper) (Fig. 5), *Libocedrus* (incense cedar), and *Thuja* (arborvitae). The majority of pollinosis has been attributed to *Juniperus* pollen, but all may be involved.

Juniperus (syn. *Oxycedrus*, *Sabina*) is widely distributed in North America, but the number of species found at one locality is not great, with the exception of a few areas in the Rocky Mountains, North Dakota, Texas, and Wisconsin (Fig. 6).

Flowering and Pollen Aerobiology

Most species shed enormous quantities of pollen that can be carried great distances by the wind. Shedding is largely in the winter and early spring months, although *Juniperus* pollen may also mature in the autumn in Texas. *Libocedrus* pollen is dispersed late, usually in April and May.

Pollen Morphology (Fig. 7)

Grains are usually spheroidal, 25–36 μm in diameter, and 1-aperturate with a faintly demarked circular pore. The exine is thin; the sexine generally microverrucate or microtuberculate and sometimes granular; the nexine thinner than the sexine; and the intine markedly thick, to 6.5 μm, usually corresponding to the cytoplasmic contents, which are either spheroidal or distinctly stellate, as in *Juniperus*. The pollen is uniform in the family and different genera are difficult to distinguish.

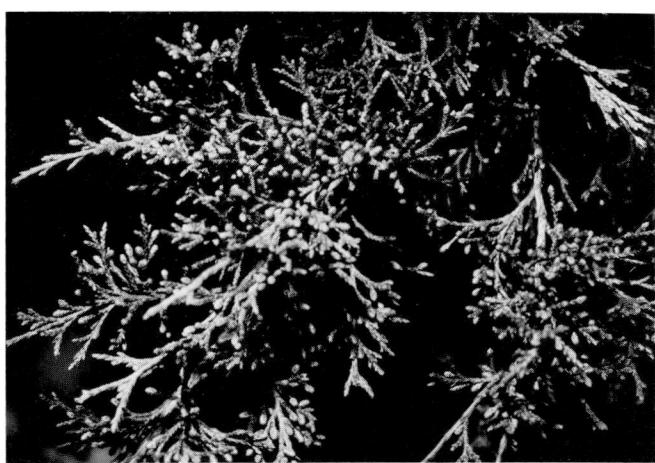

Fig. 5. *Juniperus virginiana* (red cedar, juniper) branch showing staminate cones, ×½.

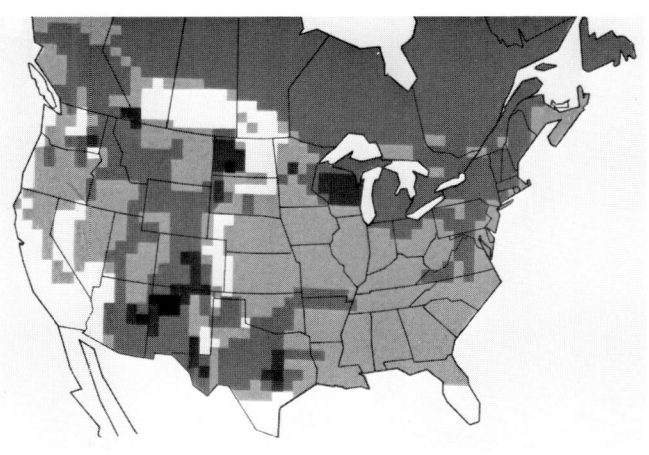

Fig. 6. Generalized composite distribution of 10 indigenous species of *Juniperus* (☐ 1 sp., ▨ 2 spp., ▨ 3 spp., ■ 4 spp.)

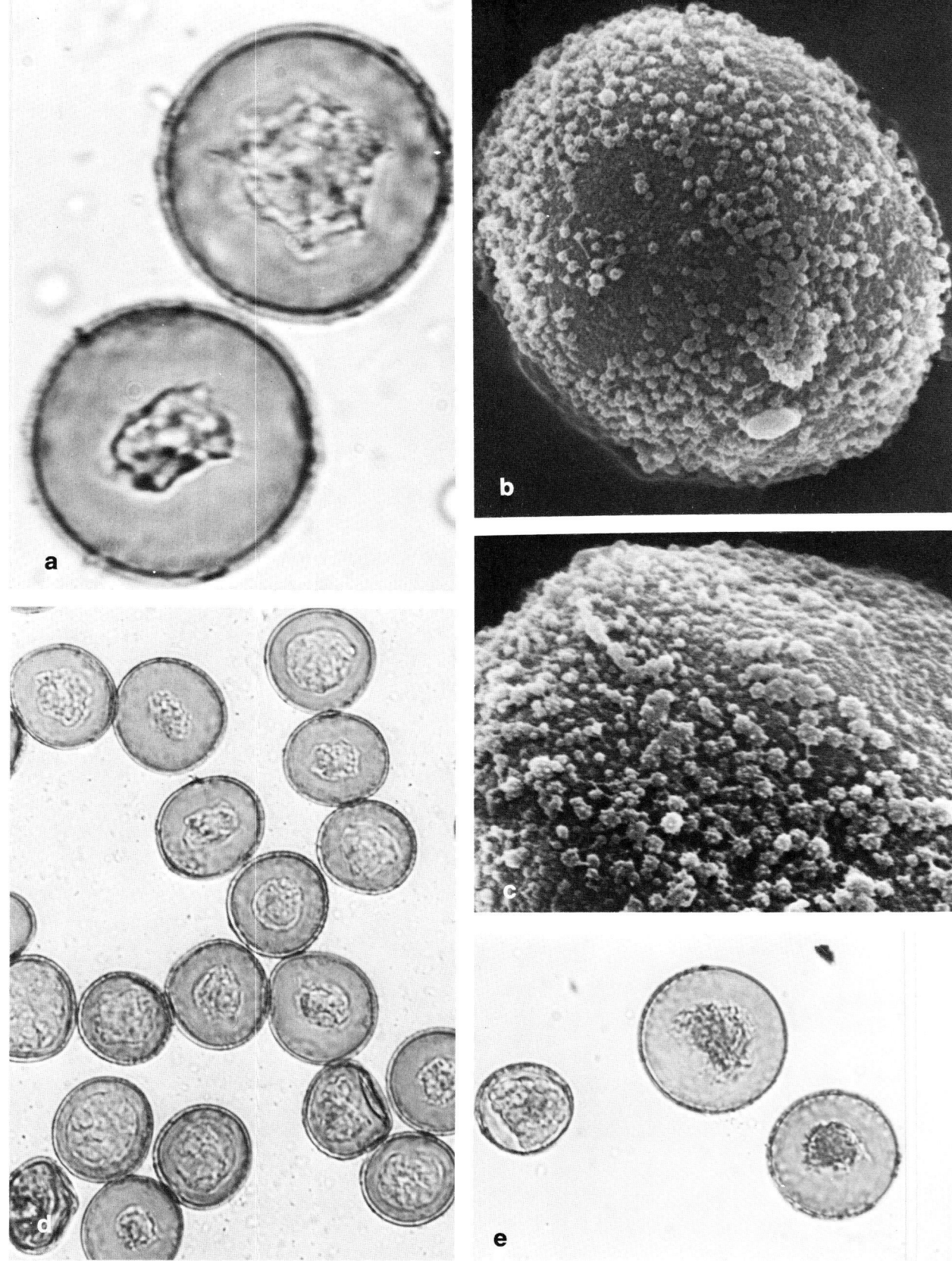

Fig. 7. Pollen of representative Cupressaceae. a–c: *Juniperus scopulorum* (Rocky Mountain juniper), a = LM(s) showing thick intine and cytoplasmic contents ×1830, b = SEM(a) ×5600, c = SEM(a) showing verrucate surface × 10000; d: *Cupressus arizonica* (Arizona cypress), LM(s) ×720; e: *Chamaecyparis lawsoniana* (Lawson's false cypress), LM(s) ×720.

Allergenicity

A small group of atopic patients in the San Francisco area proved skin sensitive to 10 species of Cupressaceae, including *Chamaecyparis, Cupressus, Juniperus, Libocedrus,* and *Thuja.* There was cross-reactivity between most of these, but not between them and members of the Taxodiaceae (Yoo et al. 1974). All have been implicated in pollinosis, but most troublesome in inhalant allergy are the numerous endemic species of *Cupressus* found in California, in addition to *C. arizonica* (Ordman 1945) and the introduced *C. sempervirens* (Tas 1964), and the many species of *Juniperus,* particularly those occurring in the western and southcentral regions. Significant are the allergenic properties of *J. ashei* (syn. *J. mexicana, J. sabinoides*) (Black 1929; French 1930; Sellers 1935) from Texas, *J. occidentalis* from central Oregon (R.L. Cutter, personal communication), *J. pinchotii* from Texas (Wolf 1948), and undoubtedly others in these areas. Additional species, such as *J. virginiana,* probably including *J. silicicola,* have also been implicated in winter and early spring pollinosis in the southern states (Kahn & Grothaus 1931) and St. Louis (Lewis & Imber 1975b), but either this widely distributed species in eastern North America is not as allergenic as the others noted, it is not identified as an offender at this time of year, or exposure to its pollen is limited, since few reports of allergenicity exist. Likewise, only very occasionally are the widespread and northern *J. communis* and *J. horizontalis* implicated as offenders.

PINACEAE
Pine Family

The largest and most important family of Gymnosperms, the Pinaceae, has evergreen, needlelike, or linear leaves arranged in spirals or fascicles with flattened cone scales distinct from the subtending bract. The mature cones are woody, the seeds winged or wingless.

The principal genera in North America are *Abies* (fir), *Larix* (larch, tamarack), *Picea* (spruce), *Pinus* (pine) (Fig. 8), *Pseudotsuga* (Douglas fir), and *Tsuga* (hemlock). An immense tonnage of airborne pollen is dispersed annually by these conifers, but much of it is of little or no clinical importance when it appears in the spring and early summer (Solomon & Durham 1967).

Pollen Morphology (Fig. 9)

Grains are either winged, nonwinged, or with rudimentary wings or bladders (sacci), heteropolar, bilateral, and 1-aperturate or inaperturate. Aperturate grains are generally with a longitudinal colpus ventrally and with a thin operculum. The exine is thicker at the proximal face and thinner at the distal face.

Fig. 8. *Pinus sylvestris* (Scotch pine) branch showing staminate cones, ×1.

Among winged grains, the body (corpus) is subspheroidal to broadly ellipsoidal, often with a marginal frill near the proximal face of the bladder. The sexine is either granular, finely reticulate, or vermiculate dorsally, and either smooth, granular, or verrucate ventrally, particularly near the colpus. Among nonwinged grains, the sexine is rugulate, insulate, or microverrucate. Bladders are generally reticulate or occasionally smooth. The sexine varies greatly in thickness, from 1–2 μm in *Pinus* to 4–6 μm in *Tsuga.* The nexine is thinner than the sexine, and the intine is up to 3 μm thick.

Of the biwinged grains, those of *Abies* and *Picea* are the largest (up to 160 μm [including bladders]), and those of *Pinus* are the smallest (up to 85 μm). Grains of *Tsuga* are inaperturate, often disc-shaped or folded, and (excepting the winged *T. mertensiana*) nonwinged or with a rudimentary bladder of fingerlike projections (frilled pollen).

Allergenicity

Infrequent reports of pine pollen allergenicity generally at low levels of sensitivity are known from California (Rowe 1939; W.R. MacLaren, personal communication), Colorado (Newmark & Itkin 1967), and also from Alabama (A.T. Baugh, Jr., personal communication), Connecticut (F. Kessler, personal communication), and Florida (R.F. Lockey and E.J. Luippold, personal communications). Like that of the Cupressaceae, pine pollen is light and may be carried great distances, but unlike the Cupressaceae it is insignificant in pollinosis. For other genera in the Pinaceae in this continent and elsewhere, the pattern is the same, i.e., very limited toxicity and no relevance in pollinosis. It is very interesting to speculate why there should be such a great disparity in pollen allergens between the two major families of Conifers. What adaptations, for example, have been effective

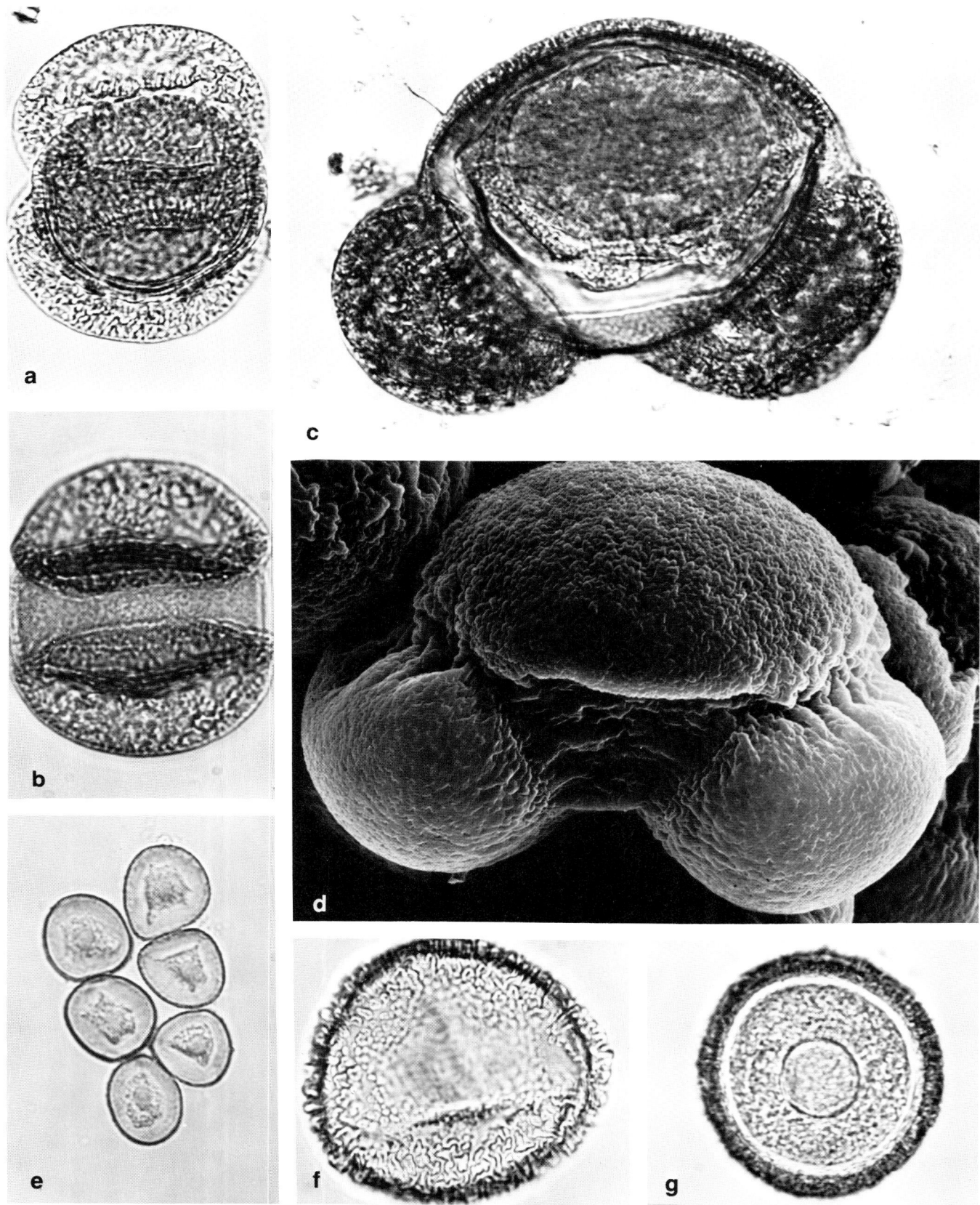

Fig. 9. Pollen of representative Pinaceae and Taxaceae. a: *Abies balsamea* (balsam fir), LM(s) showing biwinged pollen ×720; b: *Picea sitchensis* (Sitka spruce), LM(s) showing biwinged pollen ×720; c: *Pinus strobis* (white pine), LM(a) showing biwinged pollen ×1830; d: *Pinus mugo* (mountain pine), SEM(a) showing biwinged pollen ×2100; e: *Taxus cuspidata* (Japanese yew), LM(s) without wings ×720; f–g: *Tsuga canadensis* (eastern hemlock) without wings, f = LM(s) surface view ×720, g = LM(s) optical view showing intine (left ring) and cytoplasmic content ×720.

in their evolution in order to possess pollen so different?

TAXACEAE
Yew Family

Well-known because the evergreen shrubs are commonly planted around the home, yews (*Taxus*) (Fig. 10) have two-ranked, linear leaves with three light-green and two dark-green bands beneath. The solitary seed is surrounded by a fleshy aril that turns bright red at maturity. A second genus, *Torreya*, is represented by two species, but only the Californian species can be found readily.

Pollen Morphology

Pollen of *Taxus* (see Fig. 9) closely resembles that of *Juniperus* (see Fig. 7) in its sexine pattern, thick intine to 6.5 µm, and stellate cytoplasmic contents. Aerosamples with small amounts of yew pollen have been obtained from British Columbia (Bassett et al. 1978).

Fig. 10. *Taxus cuspidata* (Japanese yew) branch showing staminate cones, ×2.

Allergenicity

Allergenicity is not known, but the increased use of yews as bedding plants makes this possibility a matter of particular clinical importance (Solomon & Durham 1967).

TAXODIACEAE
Bald Cypress Family

The bald cypress family has two well-known genera, *Sequoia* (redwood) and *Taxodium* (bald cypress), recognized by their scalelike or needlelike leaves, deciduous or persistent, and by their flat or peltate cone scales lacking distinct bracts with each scale producing two to nine seeds.

Pollen Morphology (Fig. 11)

Grains are generally spheroidal, to 36 µm in diameter, and 1-aperturate, the pore ±circular and papillate. The sexine is microverrucate, and the intine is markedly thickened to 6 µm, corresponding to the stellate or spheroidal cytoplasm. Grains of *Sequoia* are smaller (22–25 µm) than those of *Taxodium* (28–36 µm) and have a thinner sexine (ca. 0.5 µm vs. 1 µm). Taxodiaceous grains also resemble those of the Cupressaceae, except that they are slightly larger and have papillate apertures.

Allergenicity

Pollen allergenicity is not well understood. *Taxodium* pollen is considered a minor offender during the winter months in the Tampa, Florida, area (R.F. Lockey and R.P. Wunderlin, personal communications), suggesting that further research is warranted.

GINKGOACEAE
Ginkgo Family

The family possesses a single species, *Ginkgo biloba* (maidenhair tree) (Fig. 12), distinguished by its deciduous, flabellate, often bifid leaves that are dichotomously veined. The seed (appearing as a fruit) is plumlike and putridly fetid, with the odor of rancid butter at maturity.

Pollen Morphology

Grains (see Fig. 11) are oblate or cymbiform, bilateral, 1-colpoidate; the exine thin (to 1.6 µm), with an undulating rugulate sexinous surface; and the intine thin (to 1 µm). The morphology is distinct from other Gymnosperms described.

Allergenicity

Ginkgo sheds large quantities of light pollen in the spring which have occasionally been suspected of causing hay fever (Wodehouse 1971).

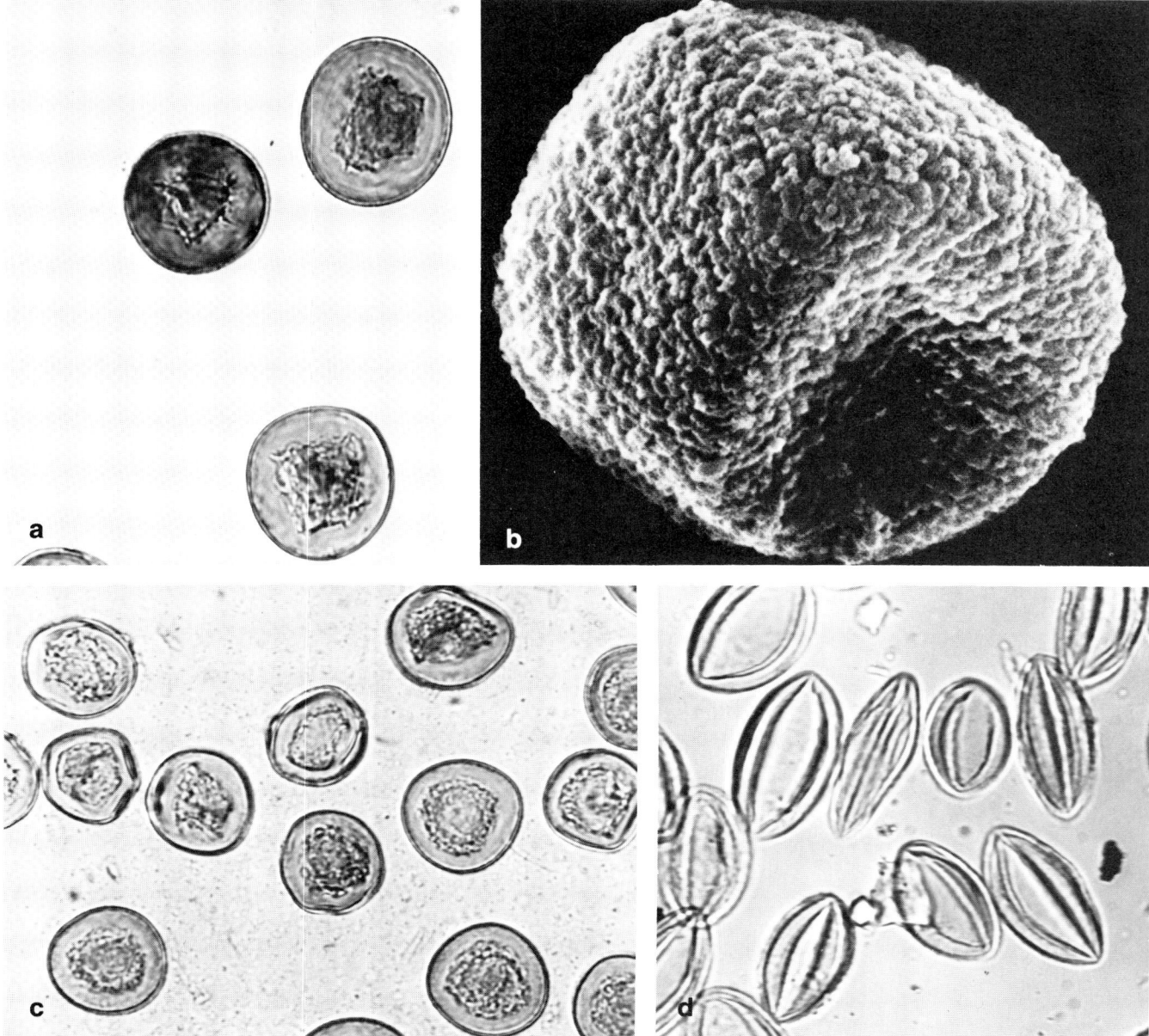

Fig. 11. Pollen of representative Taxodiaceae and Ginkgoaceae. a–b: *Sequoia sempervirens* (coastal redwood), a = LM(s) showing thick intine and ±stellate cytoplasm ×720, b = SEM(a) ×4100; c: *Taxodium distichum* (bald cypress), LM(s) showing intine and cytoplasm similar to that of *Sequoia* ×720; d: *Ginkgo biloba*, LM(a) ×720.

Angiosperms (*Flowering Plants*)

ACERACEAE*

Maple Family

A family of up to 200 species centered in China, the maples (*Acer*) (Fig. 13) are predominantly medium to small deciduous trees of northern temperate regions. There are 10 native North American species that occur predominantly in the east and secondarily in the Pacific Coast region (Fig. 14), but others are widely planted as ornamentals.

Leaves are opposite, simple, usually palmately lobed and veined, but sometimes pinnately compound (as *Acer negundo*). Flowers are bisexual or unisexual with four to five sepals, four to five petals or none, four to ten stamens, and a compound pistil of two carpels with a superior ovary and two styles. Fruit is a 2-winged samara.

Classification

Because maples are closely related, with the exception of the box-elder, it is probably unnecessary to distinguish them for potentially diverse allergens. Nevertheless, the following separation of the native

Fig. 12. *Ginkgo biloba* (ginkgo) branch showing staminate cones, ×1½.

species may be useful for future research: (1) box-elder (*Acer negundo*); (2) soft maples: vine maple (*A. circinatum*), Rocky Mountain maple (*A. glabrum*), striped maple (*A. pensylvanicum*), red maple (*A. rubrum*), silver maple (*A. saccharinum*), and mountain maple (*A. spicatum*); (3) hard maples: big-toothed maple (*A. grandidentatum*), a big-leaved maple (*A. macrophyllum*), and sugar maple (*A. saccharum*).

Pollination

Maple pollination is variable. The European *Acer platanoides* and *A. pseudoplatanus* are both entomophilous, but they may disperse pollen incidentally (Ogden et al. 1974). The majority of North American species are pollinated partly by insects and by wind (amphiphilous), but as their flowers are sweet scented and eagerly sought by bees, their principal mode of pollination is undoubtedly entomophilous. Since their pollen may be dispersed by the wind and commonly found in aerosamples, however, they are facultatively anemophilous. However, the weedy and ubiquitous *A. negundo* is characteristically anemophilous and sheds abundant pollen in the early spring in both urban and rural areas throughout much of the continent.

Flowering and Pollen Aerobiology

Maples flower predominantly in the early spring, but some species, like the earliest-flowering *Acer saccharinum*, begin during the winter months in January and February, and a few western species and *A. spicatum* extend flowering to June or even July. Consequently, pollen aerosamples show high frequencies in the northeastern region from February to April where maples dominate, but somewhat later in April and May in the northwestern region where they are found frequently (Table 1).

Pollen Morphology (Figs. 15–17)

Grains vary in shape from prolate (30–51 × 20–36 μm) to ±spheroidal (22–26 μm); the amb rounded-triangular with slightly convex sides, and 3-colpate or 3-colporate; long colpi having tapering, blunt, or rounded ends (18–48 × 1–8 μm); when present, circular to lalongate ora, 2–3 μm in diameter; and granular or smooth opercula. The sexine is 0.8 μm thick, striate to striato-reticulate or finely reticulate to rugulate; the nexine 0.6–0.8 μm thick; and the intine 0.8–1.0 μm thick.

Maple pollen is divisible into two groups based on sexine patterns. The anemophilous *Acer negundo* pollen possesses a distinctly rugulate-undulating sexine and the amphiphilous *A. macrophyllum*, *A. rubrum*, and *A. saccharum* pollen is finely striate, distinctly striate, and pertectate-reticulate sexine, respectively. *A. macrophyllum* pollen is comparatively large (40–50 × 24–36 μm) and that of *A. circinatum* (27–30 × 27 μm) and *A. tataricum* (ca. 26 × 24 μm), relatively small.

Allergenicity

All species of maple are potentially allergenic, but different modes of pollination vary the extent of pollen exposure and therefore sensitizations. Botanically, the most distinct species is *Acer negundo* (box-elder) and it should possess more unique pollen antigens/allergens than the other species of *Acer*; furthermore, it is wind-pollinated and the greatest numbers of atopic individuals are sensitized by its pollen. Based on skin test reactions in St. Louis, Missouri, results showed that box-elder pollen extracts were more allergenic on the average than the other maple species tested (*A. rubrum*, *A. saccharinum*, *A. saccharum*), which were all low in levels of allergenicity. Moreover, box-elder pollen proved the most allergenic of all tree pollen tested among patients allergic to spring pollen (Lewis & Imber 1975b). In Manitoba box-elder proved a cause of early spring pollinosis (Walton & Dudley 1947). Based on our survey this species and a few other maples were considered common incitants of pollinosis (skin test data averaged) in Alabama (3.0+), California (3.0+), Connecticut

Fig. 13. Morphology of representative Aceraceae. a–b: *Acer negundo* (box-elder) having 3-parted leaves showing flowers on long peduncles (a, ×1½) and fruit (b, ×1⅓); c–d: *Acer saccharinum* (silver maple) showing typical leaves (c, ×⅓) and staminate flowers with anthers (d, ×3) before leaf formation; e: *Acer macrophyllum* (big-leaved maple) branch with inflorescences, ×¼; f–g: *Acer saccharum* (sugar maple) with long pedunculate flowers and emerging leaves (f, ×1) and developing fruit (g, ×2); h: *Acer platanoides* (Norway maple) branch showing clumps of flowers; ×1; i: *Acer rubrum* (red maple) branch with flowers and exposed stamens ×2½; j: *Acer circinatum* (vine maple) flowers, ×2.

TREES AND SHRUBS

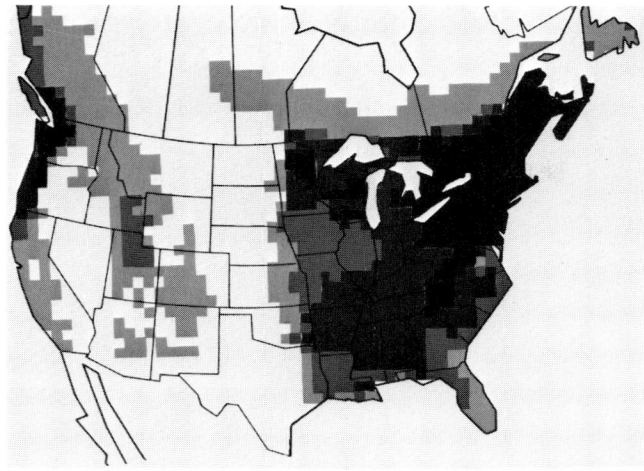

Fig. 14. Generalized composite distribution of nine indigenous species of the typical maples (*Acer*), excluding the widespread and weedy box-elder (*A. negundo*) (☐ 1 sp., ▨ 2 spp., ▨ 3 spp., ■ 4–5 spp.).

TABLE 1. *Acer* (maple) aeropollen frequency in regions of North America based on percentage of pollen captured, by month (1974–78).

REGION	MONTH							
	1	2	3	4	5	6	7	8
nNE		13.9	11.8	25.4	4.5	0.2		
sNE		12.7	9.2	13.4	5.2	3.1	3.1	
SE		0.5	8.7	11.4	4.6			
NC			11.4	3.4	2.9	0.1		
SC			0.6	2.7	?	0.3	0.3	
NW			0.5	20.0	14.0	0.4		
SW			4.4	3.1	3.4	1.0		
sCA	0.7							

(3.0+), Iowa (2.5+), New Jersey (3.0+), Missouri (3.0+), Oregon (3.0+), and Texas (3.0+), and less commonly in Florida (3.5+).

ANACARDIACEAE
Cashew Family

The Anacardiaceae contain about 57 genera and 600 species of trees and shrubs mostly of tropical distribution but also growing in temperate areas of the Mediterranean basin, eastern Asia, and North America. Although some are popular ornamentals and other species produce commercially valuable fruit (mango) and nuts (cashew, pistachio), the family is better known medically for its resinous exudate that causes severe contact dermatitis, as in poison ivy, poison oak, poison sumac, and pepper tree.

Genera that occur in North America include *Anacardium* (cashew cultivated in subtropical Florida), *Cotinus* (smokebush or smoketree), *Mangifera* (mango cultivated in subtropical Florida), *Pistacia* (pistachio, one species native to southern Texas, but *P. vera*

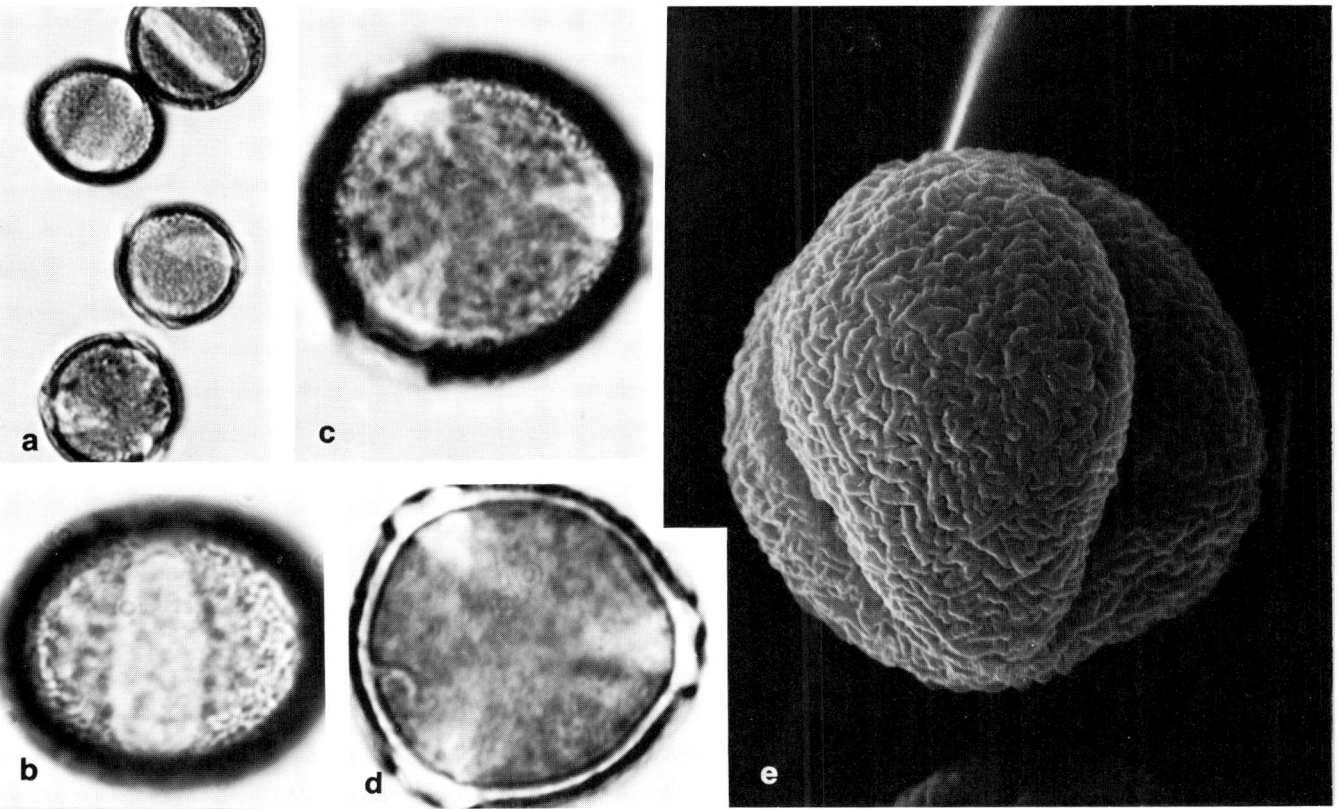

Fig. 15. Pollen of box-elder (*Acer negundo*). a = LM(s) ×720; b = LM(s) equatorial view ×1830; c–d = LM(s) equatorial views ×1840; e = SEM(a) equatorial view showing 2 apertures ×3460.

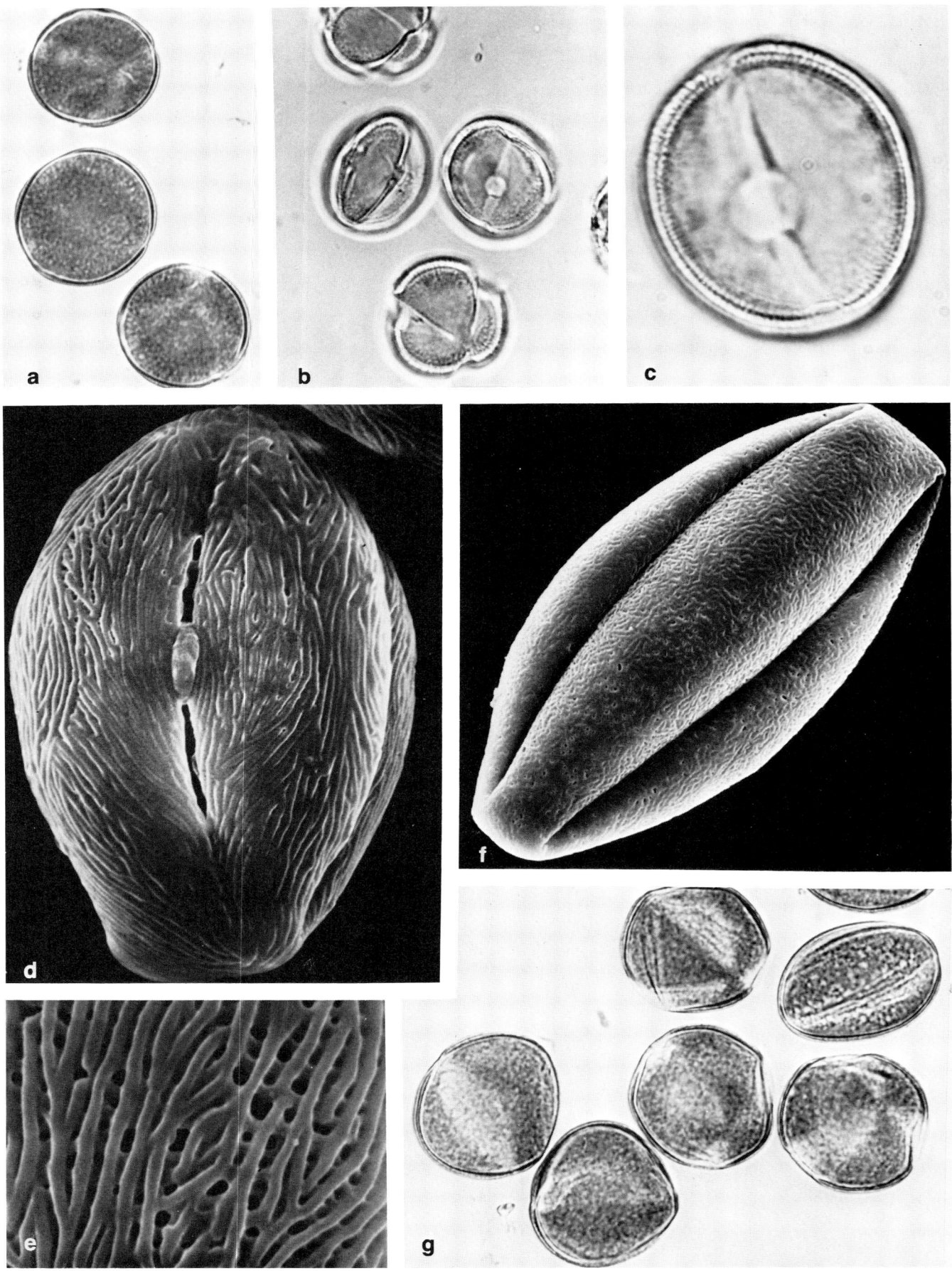

grown commercially, especially in California), *Rhus* (sumac) (Fig. 18), *Schinus* (pepper tree) (Fig. 18), and *Toxicodendron* (poison ivy, oak, and sumac) (Fig. 18). Many elicit type IV (delayed) hypersensitivity on contact. The majority are entomophilous with only incidental dispersal of pollen, excepting *Pistacia*, which is wind-pollinated.

Pollen Morphology (Fig. 19)

Grains are prolate (26–40 × 18–34 µm) to occasionally spheroidal (24–28 µm); the amb triangular with convex sides and 3-colporate or occasionally 4–5(–8)-brevicolpate to porate (*Pistacia*); the typical colpi long and narrow, sometimes slightly constricted equatorially, 20–32 × 0.5–3.5 µm; the ora generally lalongate to 10 × 1 µm; and the opercula granular. The sexine is often striate, less commonly striato-reticulate, reticulate, or scabrate, 0.7 (*Pistacia*) to 1.2 (*Mangifera*) µm thick; the nexine slightly thinner or as thick as the sexine; and the intine 0.8–1.0 µm thick, but much thicker below the apertures where 2.0–3.2 µm thick.

The wind-pollinated pollen of *Pistacia* differs markedly from that of the the other genera in that it is spheroidal, 4–5(–8)-brevicolpate to porate; and the reticulate sexine has minute spinules projecting from the muri. Pollen of the others is mostly prolate, 3-colporate, and variously striated.

MANGIFERA   *mango*

As for many entomophilous plants, close proximity during anthesis may result in sensitization and allergic symptoms following repeated exposures, which for mango is from June to August. Pollen allergenicity was reported as questionable in Florida (Lewis & Vinay 1979) and as moderate in Puerto Rico (Marchand 1948).

PISTACIA   *pistachio*

There is no report of allergenicity of wind-pollinated species of *Pistacia*.

RHUS   *sumac*

Sumacs are common in eastern North America, where they flower from spring to July (occasionally later), and also in southern California where flowering is year-round. Airborne pollen is found occasionally and is considered a mild cause of hay fever (Scheppegrell 1922).

Fig. 16. Pollen of soft maples. a–e: *Acer circinatum* (vine maple), a = LM(s) ×720, b = LM(a) ×720, c = LM(a) ×1830, d = SEM(ad) equatorial view showing 1 central aperture ×3800, e = SEM(ad) showing sexine surface ×8800; f = *Acer saccharinum* (silver maple), SEM(ad) equatorial view showing 2 apertures ×2600; g: *Acer rubrum* (red maple), LM(s) ×720.

SCHINUS   *pepper tree*

Two species of *Schinus* are cultivated frequently in the southern United States. *S. molle* (Peruvian pepper tree) has become naturalized in Texas and California, whereas *S. terebinthifolius* (Brazilian pepper tree) is an aggressive colonizer in peninsular Florida and Hawaii. Airborne pollen has been reported for each species (Lewis & Vinay 1979) during anthesis from June to August and from March to April and September to November, respectively. Some patients react positively to skin tests (averaging 3.4+) during flowering (Luippold 1974; Lewis & Vinay 1979), and although the number is not great, those that are sensitive are very much so (E.J. Luippold, personal communication). Morton (1978) also reported that flowering pepper trees are a source of widespread respiratory and skin irritations, including rashes, eye inflammations, and facial swelling for people who handle the plant, indicating delayed (type IV) hypersensitivity, as would be expected from a relative of poison ivy (*Toxicodendron*).

AQUIFOLIACEAE

Holly Family

Hollies (Fig. 20) are a large genus of about 400 species of trees and shrubs widespread in tropical and temperate regions which are valued as ornamentals. Consequently, incidental exposure to the largely insect-dispersed pollen around the home and elsewhere is possible (Hyde 1950; Markgraf 1980) and although pollinosis due to *Ilex* pollen has not been reported, it should be considered of local significance during flowering in the spring and early summer in most regions and year-round in southern Florida. The genus is particularly common in the Southeast.

Pollen Morphology (Fig. 21)

Grains are prolate to prolate-spheroidal; the amb rounded or triangular and 3-colporoidate; the colpi long, 27–28 × 0.5–3.0 µm, narrowed equatorially, with slightly thickened margins and ends tapering, blunt, or rounded; the faintly demarked ora lolongate (ca. 5.0 × 3.2 µm) to lalongate; and the opercula densely granular. The sexine is thick (ca. 2.5 µm), warty (either dense or with few warts), clavate, or pilate; the nexine 1.2–1.5 µm thick; and the intine generally 1.0–1.5 µm thick, but much thicker to ca. 3 µm below the apertures.

Pollen of *Ilex opaca* is 28–34 × 22–30 µm in size with few scattered excrescences; that of *I. aquifolium* is slightly larger with dense large excrescences ca. 3.2 × 4.0 µm in size and with free pilae ca. 2 µm high.

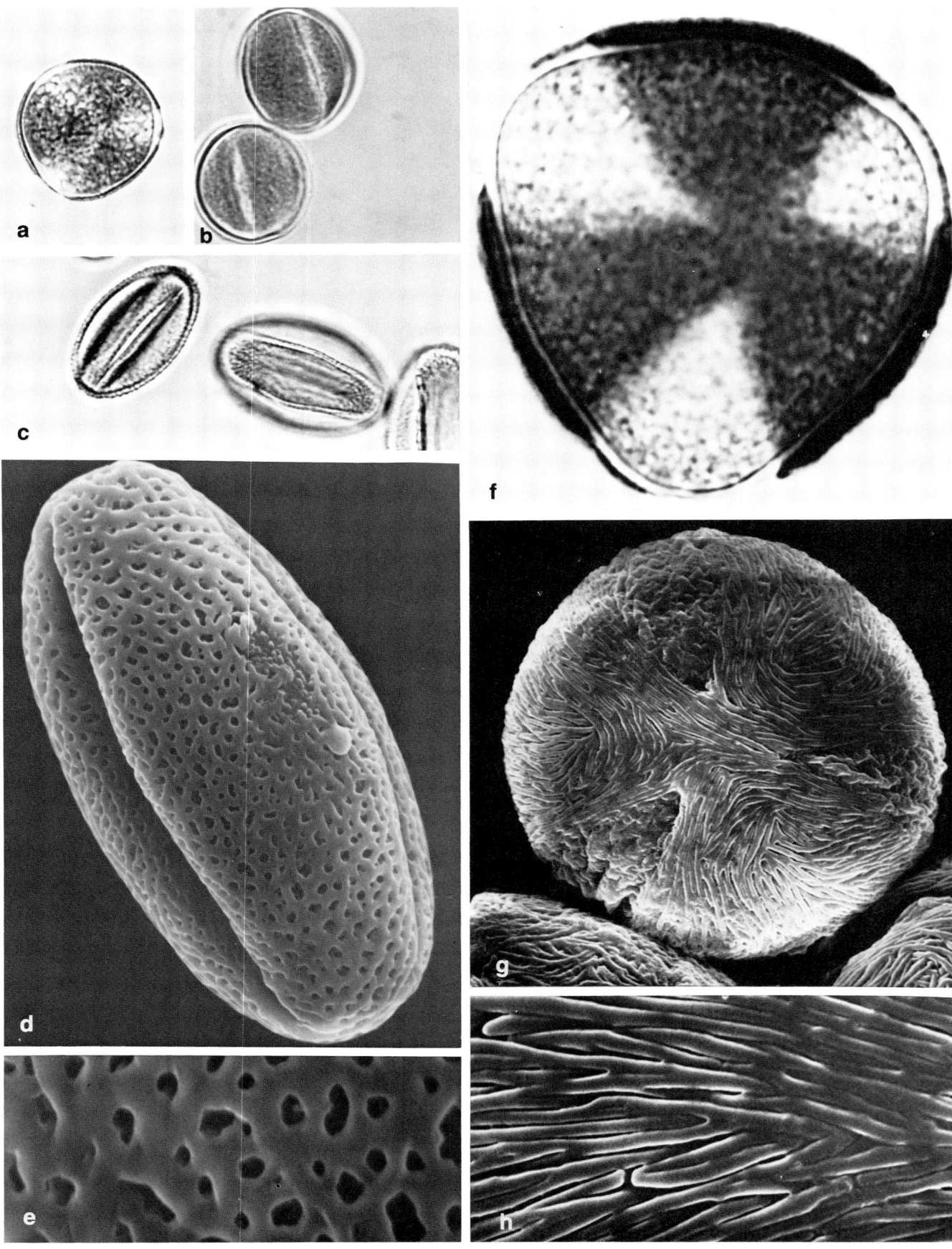

Fig. 18. Morphology of representative Anacardiaceae. a–b: *Toxicodendron diversilobum* (western poison oak) branch (a, ×½) and flowers (b, ×4); c: *Rhus glabra* (scarlet or smooth sumac) branch with inflorescence, ×⅕; d: *Schinus terebinthifolius* (Brazilian pepper tree) branch with fruit, ×⅔.

ARALIACEAE
Ginseng Family

Like the preceding genus, the largely insect-pollinated members of the Araliaceae are not known to elicit pollinosis. Yet English ivy (*Hedera*) (Fig. 22) is widely cultivated near homes and buildings and its exposed anthers could shed pollen incidentally when in anthesis during September and October. *Hedera* pollen has been identified from aerosamplers in Switzerland (Markgraf 1980).

Pollen Morphology (Fig. 23)

Grains of *Hedera* are prolate to prolate-spheroidal, 32–36 × 25–32 μm; the amb triangular with convex sides or circular and 3–4-colporate, rarely 6-colpate; the colpi long and narrow, 24 × 1–2 μm, with slightly thickened margins and ends tapering or blunt; and when present the ora faint and lalongate, ca. 3.2 × 10.0 μm. The sexine is ca. 0.9 μm thick, reticulate; the lumina polygonal to elongate, 0.5–1.8 μm, but smaller toward the apertures; the muri broad, 0.5–

Fig. 17. Pollen of hard maples. a–e: *Acer saccharum* (sugar maple), a = LM(s) polar view ×720, b = LM(s) equatorial views ×720, c = LM(a) equatorial views ×720, d = SEM(ad) equatorial view ×2600, e = SEM(ad) showing sexine surface ×8200; f–h: *Acer macrophyllum* (big-leaved maple), f = LM(s) polar view ×1830, g = SEM(ad) polar view ×2800, h = SEM(ad) showing sexine surface ×6800.

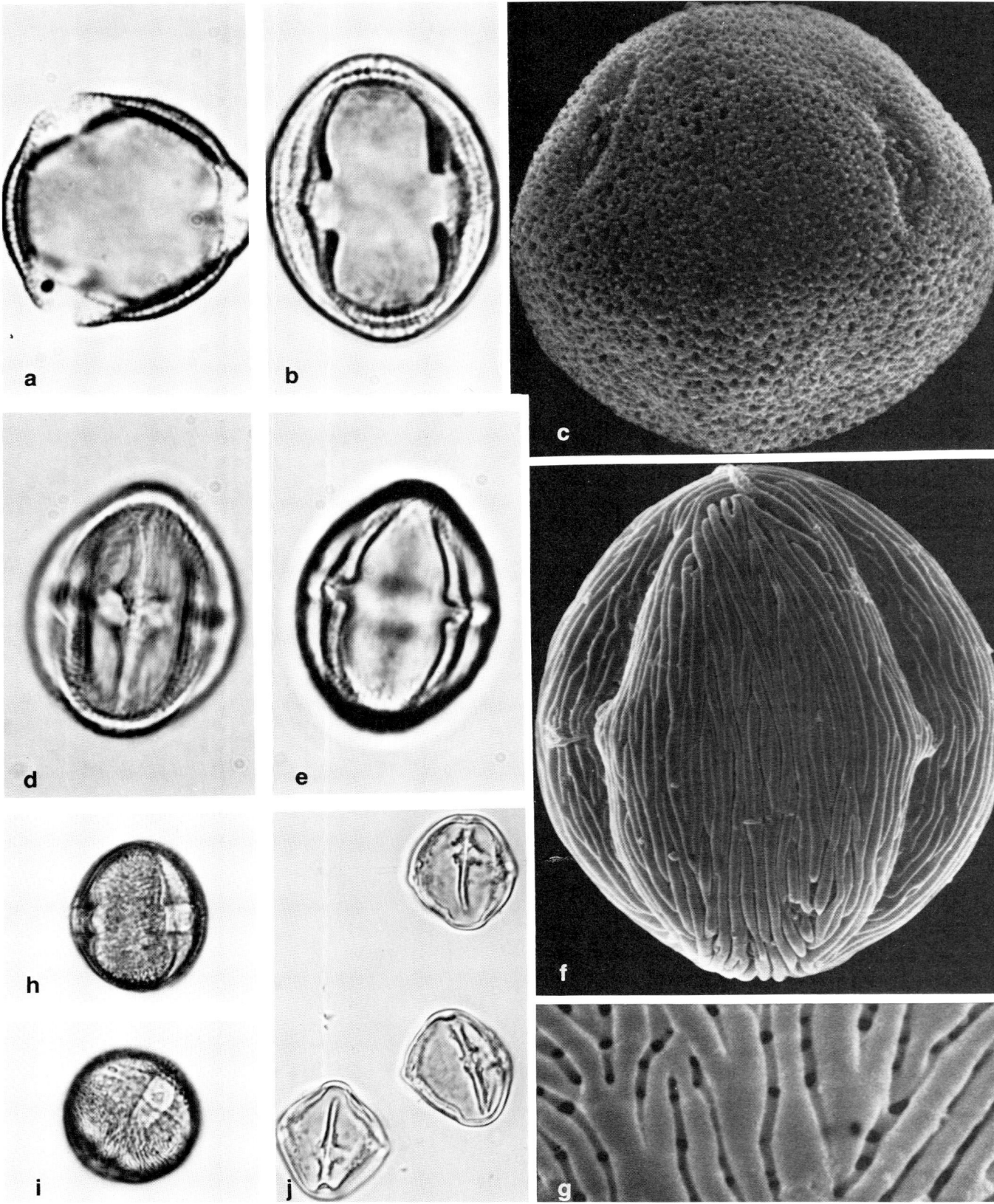

Fig. 19. Pollen of representative Anacardiaceae. a–b: *Mangifera indica* (mango), a = LM(a) polar view ×1830, b = LM(a) equatorial view ×1830; c = *Pistacia chinensis* (Chinese pistachio), SEM(a) oblique-equatorial view showing 2 apertures ×3800; d–g: *Schinus terebinthifolius* (Brazilian pepper tree), d = LM(a) surface equatorial view ×1830, e = optical equatorial view ×1830, f = SEM(a) equatorial view showing 2 apertures ×4600, g = SEM(a) showing sexine surface ×12000; h–j: *Rhus glabra* (scarlet or smooth sumac), h = LM(s) equatorial view ×720, i = LM(s) oblique-equatorial view ×720, j = LM(a) equatorial views ×720.

TREES AND SHRUBS

Fig. 20. *Ilex aquifolium* (English holly) cultivar showing flowers and variegated leaves, ×½.

1.0 μm wide; the nexine about as thick as the sexine, somewhat thicker adjacent the apertures; and the intine 0.8–1.0 μm thick, considerably thickened below the apertures.

ARECACEAE (Palmae)*   *Palm Family*

Palms are largely tropical, with only a few temperate outliers, such as *Sabal* (palmetto), which occurs naturally as far north as North Carolina. A number of other palms are found in the southeastern and southwestern regions, mostly in Florida and southern California as introductions.

All palms have a single apical bud, and if that is destroyed the stem dies. Stems are woody, forming unbranched trees or shrubs to over 30 m tall. The few leaves are fan-shaped or pinnately compound, formed one at a time from the stem tip, each with a broad sheathing base, and sometimes enormously large as in *Raphia* (raffia palm), which reaches 20 m long. Inflorescences are usually borne in the crown of leaves; they are large, paniculate, and subtended by one or more spathes. Floral parts are in threes and flowers are bisexual or unisexual, the sexes being borne either on the same (monoecious) or separate (dioecious) plants. Fruit is mostly a 1-seeded berry or drupe, the seed among the largest, as in coconuts.

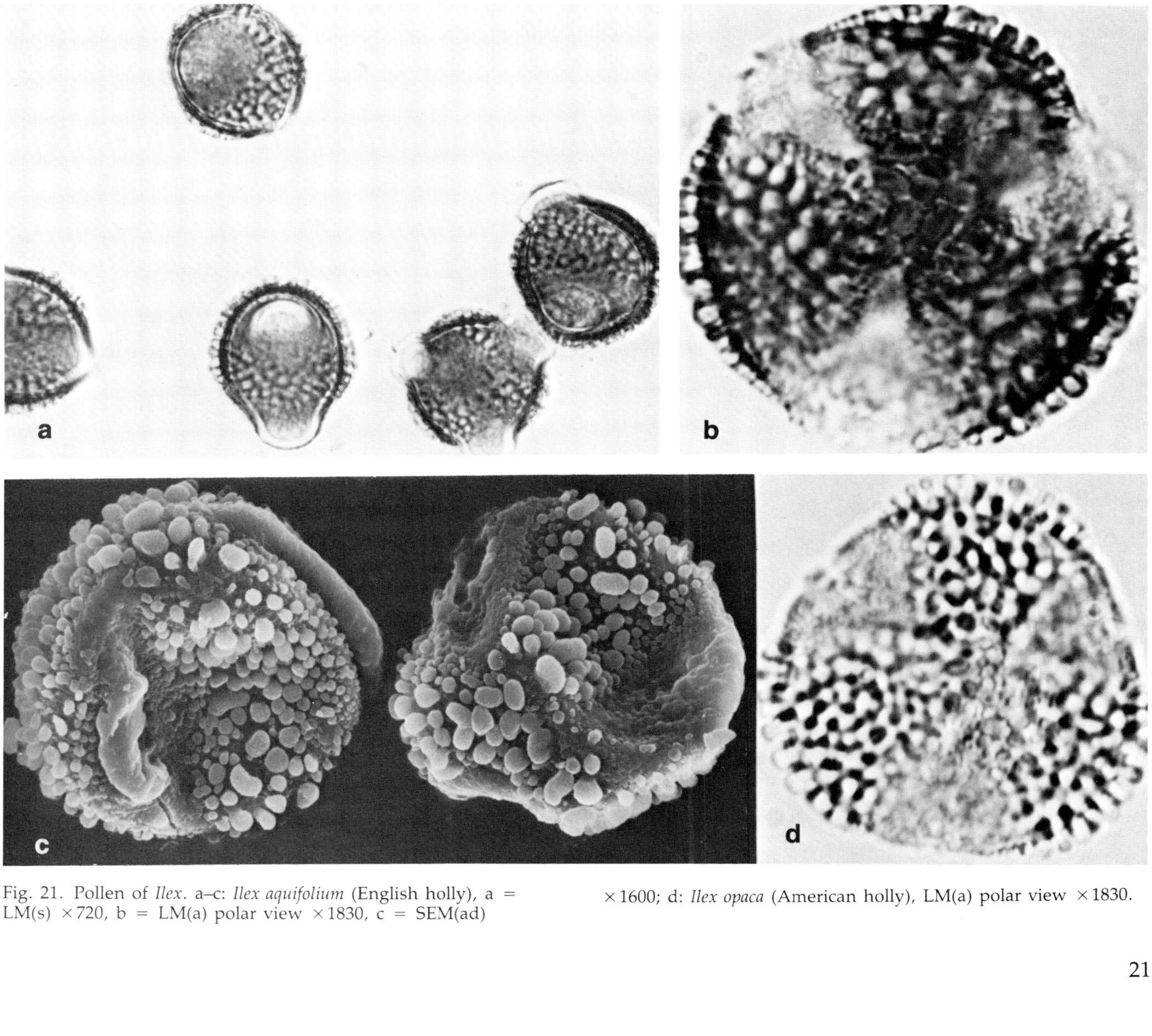

Fig. 21. Pollen of *Ilex*. a–c: *Ilex aquifolium* (English holly), a = LM(s) ×720, b = LM(a) polar view ×1830, c = SEM(ad) ×1600; d: *Ilex opaca* (American holly), LM(a) polar view ×1830.

Fig. 22. *Hedera helix* (ivy) branch with leaves and flowers, ×⅔.

Dioecious plants are obligate outbreeders and many are adapted to pollination by wind. Shedding of abundant powdery pollen and synchronous and short-term flowering of numerous and relatively well exposed flowers are typical. *Elaeis* (oil palm), for example, is exclusively wind-pollinated; the pollen is dry and is transported at least 30 m, even though the staminate flowers are anise-scented and are visited by insects, particularly bees. They do not, however, visit the pistillate flowers (Purseglove 1972). *Cocos* (coconut) and *Phoenix* (date palm) are also anemophilous, although insect-pollination may occur. Exposure to the pollen of palms is therefore fairly restricted to the vicinity of the taller, dioecious plants.

The palms, to which limited exposure in the United States is possible, include *Acoelorraphe* (Everglades

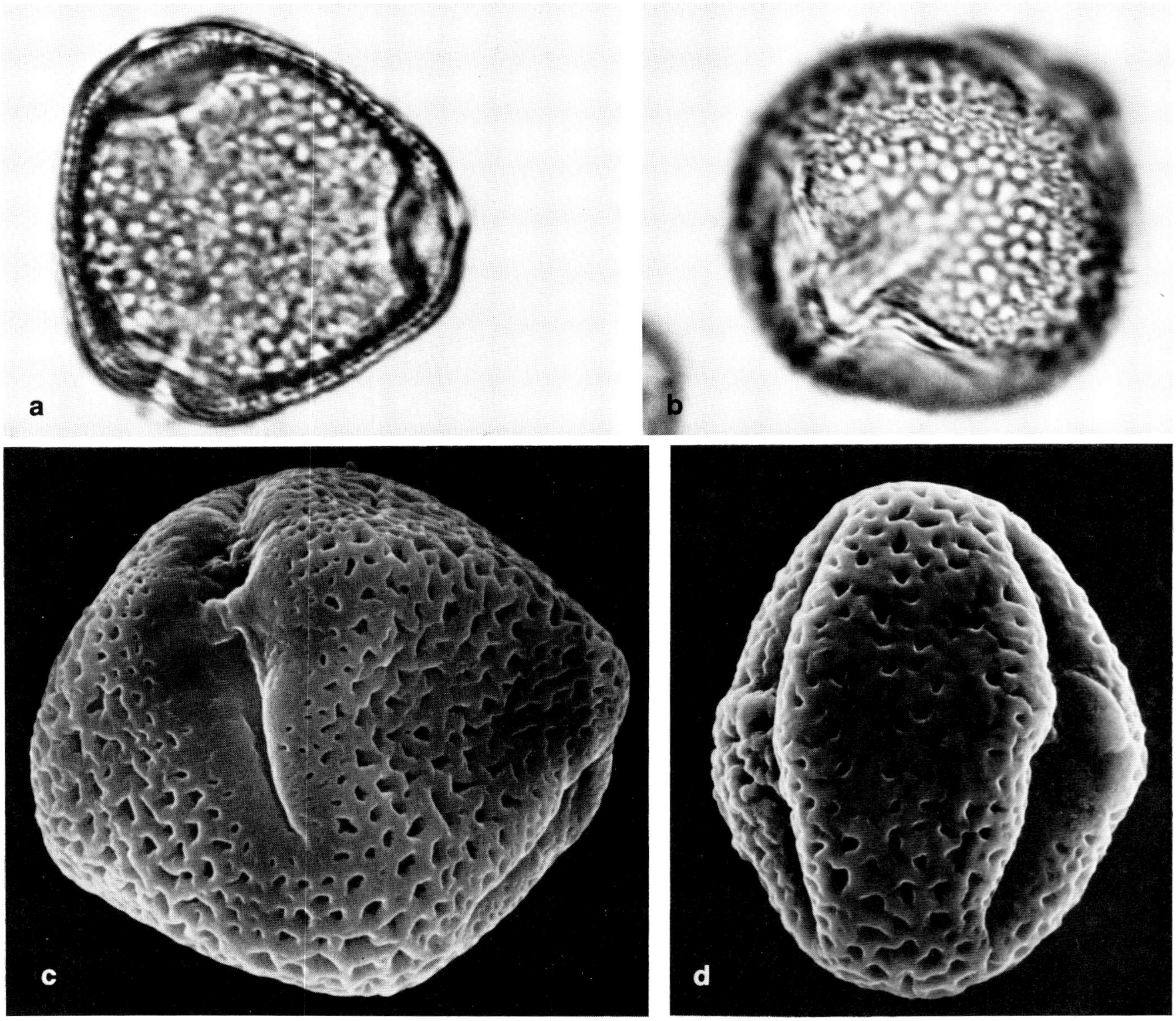

Fig. 23. Pollen of *Hedera helix*. a = LM(a) polar view ×1830, b = LM(a) oblique-equatorial view ×1830, c = SEM(a) ×2900, d = SEM(a) ×2200.

Fig. 24. *Chamaedorea erumpens* (bamboo palm) showing leaf and inflorescence, × ⅓.

palm, Florida), *Chamaedorea* (bamboo and parlor palms, cultivated in Zone 10) (Fig. 24), *Cocos* (Florida, Hawaii), *Phoenix* (Florida, Gulf Coast, Arizona, California, Hawaii), *Pseudophoenix* (Florida cherry palm), *Rhapidophyllum* (blue palmetto, Southeast, cultivated in Zone 8), *Roystonea* (royal palms, Florida), *Sabal* (Southeast, cultivated in Zone 9), *Serenoa* (saw palmetto, Southeast, cultivated in Zone 9), *Trachycarpus* (fan palm, cultivated to southern Zone 8), *Thrinax* (thatch palm, Florida), and *Washingtonia* (Washington palm, California and Arizona, cultivated from Gulf Coast to Florida). Some of these palms are short, monoecious, and insect-pollinated where only incidental exposure to pollen could occur.

Flowering

Flowering of the two common indigenous palm genera occurs from May to July in the southeastern region and about a month earlier in southern Florida. Many of the introduced palms flower from summer to autumn, sometimes extending to the winter months.

Pollen Morphology (Fig. 25)

Grains are heteropolar, bilateral, ovoidal to cymbiform, and 1-sulcate; the sulci long and narrow, smooth or unevenly margined; and ends tapering or rounded. The sexine is either tegillate with a smooth or undulating tegillum and provided with irregularly spaced pits, granules, or small excrescences, or reticulate, thin (usually 1 μm thick or less, although that of *Cocos* is thicker, ca. 1.8 μm); the nexine thinner or as thick as the sexine; and the intine often thickened below the aperture (to 4 μm in *Phoenix* and *Roystonea*), but otherwise 0.7–2.4 μm thick.

Pollen of *Roystonea* is distinguished by its large ovoidal size (22–26 × 48–58 × 30–33 μm) with a sharply undulating pitted sexine. Pollen of *Cocos* is larger (30 × 56–64 × 40–48 μm), having a granular sexinous surface and undulating tegillum. By comparison, pollen of *Phoenix* is smaller (9–11 × 24–25 × 14–16 μm), with a finely reticulate sexine.

Allergenicity

The most common palms grown in the United States are *Phoenix dactylifera*, cultivated commercially for the fruit, and *P. canariensis* (Canary Island date palm), grown for street planting from Florida to southern Texas and the warmer regions of the Southwest. Reports of allergenicity are sharply localized to those persons living near areas where street plantings are heavy and among those workers who pollinate date flowers or who are otherwise employed with the crop. Positive skin test reactions (2+–4+) were elicited among exposed individuals in California and Hawaii who had symptoms of pollinosis, confirming an earlier report from Hawaii of rhinitis due to date pollen (Roth & Shia 1966). *Cocos* also elicited positive skin test reactions from those suffering from pollinosis in Hawaii (3+, approximately 50 patients, A.W. Neilson, Jr., personal communication), even though coconut is sometimes regarded as nonallergenic (Roth & Shia 1966). *Roystonea regis* (Cuban royal palm), though entomophilous, regularly sheds some pollen that becomes airborne and in Cuba may elicit symptoms of hay fever (Quintero 1955).

BERBERIDACEAE

Barberry Family

Barberries (*Berberis*, including *Mahonia*) (Fig. 26) are a large genus of over 500 species of shrubs and herbs widely distributed in the Northern Hemisphere. Many are valued as garden shrubs and incidental exposure to the largely insect-dispersed pollen near the home is a possibility during flowering from March to May in the Pacific Coast regions and somewhat later east of the Rocky Mountains. Allergenicity has not been reported.

Pollen Morphology (Fig. 27)

Grains are generally spheroidal, ca. 35–43 μm in diameter; the 6–12 colpi are narrow (0.5–0.8 μm wide); the ends connecting at the poles with a gradual transition toward a spiraperturate condition. The sexine is tegillate with puncti or perforations about 0.5 μm; the nexine 0.7–1.0 μm thick; and the intine about 2.4 μm thick.

BETULACEAE*

Birch Family

A small family of six genera of deciduous trees and shrubs found predominantly in north temperate regions, the birches (*Betula*), alders (*Alnus*), hornbeams (*Carpinus*), hazelnuts or filberts (*Corylus*), and hop hornbeams (*Ostrya*) are all wind-pollinated and shed

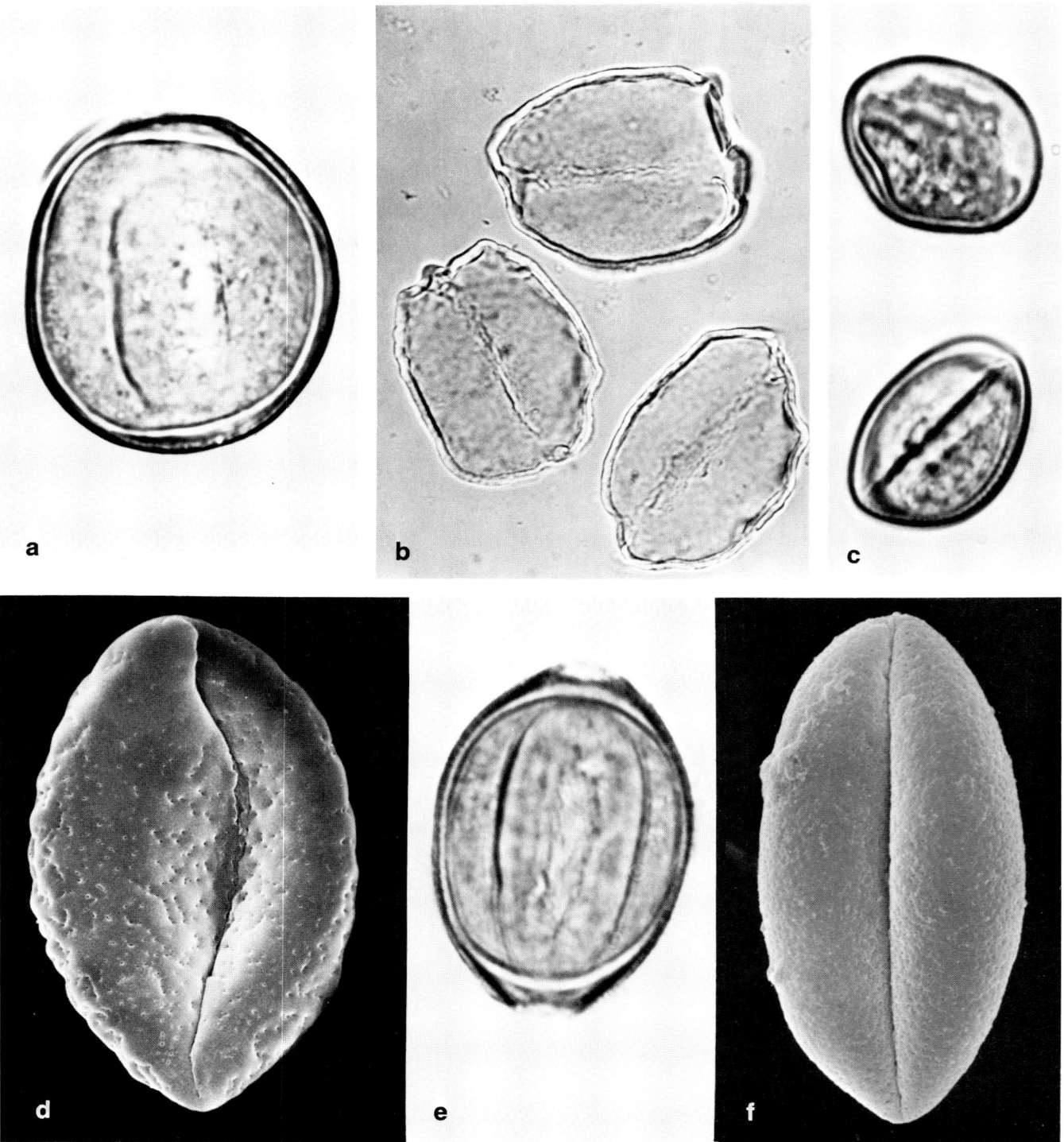

Fig. 25. Pollen of representative Arecaceae. a: *Chamaedorea erumpens* (bamboo palm), LM(s) ×1830; b: *Cocos nucifera* (coconut palm), LM(a) ×720; c: *Phoenix reclinata* (Senegal date palm), LM(s) ×720; d: *Roystonea borinquena* (Puerto Rican royal palm), SEM(a) showing the 1 sulcus ×2100; e–f: *Serenoa repens* (saw palmetto), e = LM(s) ×1830, f = SEM(a) showing the 1 sulcus ×2800.

abundant pollen in the spring. Estimates per catkin range from 3.9 million pollen grains for *Corylus avellana*, 4.5 million for *Alnus glutinosa*, to 6 million per catkin for *Betula pubescens* (Erdtman 1969).

Leaves are simple, alternate, pinnately veined, and stipulate. All species have separate staminate and pistillate flowers borne on the same plant (monoecious) in catkins; the staminate catkins are long and pendulous; and the pistillate ones are on a stiff axis, often held erect. The flowers are highly reduced, with two to twenty stamens per cymule, and the pistil has an inferior ovary of two united carpels, both lacking

Fig. 26. *Berberis aquifolium* (Oregon grape) flowers, ×1½.

a perianth or, if present, highly reduced. The fruit is a 1-seeded nut, often winged for wind-dispersal (Fig. 28).

Distribution

Of the two most frequent genera, *Alnus* is particularly common in New England and adjacent Canada and in the Pacific Northwest to northern California (Fig. 29). Species of *Betula* are also common in the northern part of the continent with a concentration in the New England area (Fig. 30).

Classification

The family is divided into two subfamilies: *Betula* and *Alnus* in the Betuloideae, and *Carpinus, Corylus,* and *Ostrya* in the Coryloideae.

Flowering and Pollen Aerobiology

Members of the birch family flower predominantly in the spring, but *Alnus rhombifolia, Betula occidentalis, Corylus cornuta,* and some introduced species initiate flowering in the winter months, depending on the region. Other species, such as *A. viridis* and *B. glandulosa,* flower to early in the summer and *A. maritima* is a late-summer- to early-autumn-flowering species (Table 2).

Aeropollen of *Alnus* (Table 3) and *Betula* (Table 4)

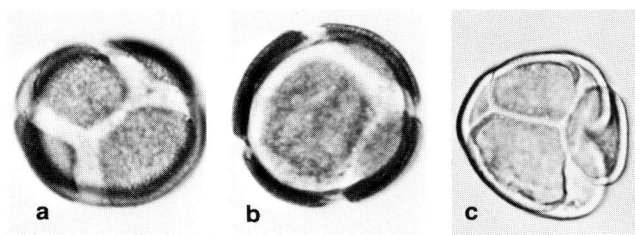

Fig. 27. Pollen of *Berberis aquifolium*. a–b = LM(s) two views ×570, c = LM(a) ×570.

is widely reported in the early months of the year in the northern parts of the continent, becoming less frequent in the southeastern and southcentral regions. Although captured with regularity, *Betula* pollen is not a dominant part of the aeroflora for prolonged periods east of the Rocky Mountains, although it may be locally common, particularly in New England. To the west, *Alnus* pollen is abundant, reaching a frequency of 54.9% of all pollen in the atmosphere of the Northwest during March (see Table 3). In California, alder pollen reaches highest frequencies from December to February.

Little is known about emission of tree pollen, but pollen shed from the catkins of *Betula papyrifera* (white birch) apparently is completed within a few days for each tree without a marked diurnal pattern (Ogden et al. 1974).

Pollen Morphology (Figs. 31–32)

Grains are generally isopolar, occasionally paraisopolar in *Alnus*, suboblate to oblate or oblate-spheroidal; the amb triangular with convex sides, or tetragonal, hexagonal or polygonal, and 3–5(–7)-porate, commonly 2–4 μm in diameter, annulate, distinctly or faintly aspidate; and the pores ±evenly distributed around the equatorial region. The sexine is tectate, smooth or usually appearing granular, but actually (from SEM) with minute verrucae or spinules, 0.7–1.0 μm thick; the nexine thinner than the sexine; and the intine up to 1 μm thick, except below the pores where it is markedly thickened, forming onci 2.5–4.0 × 6–10 μm thick.

Pollen of *Alnus* is readily distinguished by its exinous (sexine and nexine) thickened bands (arci) that extend in pairs from aperture to aperture. Pores of *Betula* are strongly aspidate. Large grains are typical of *Carpinus* (26–31 × 28–35 μm), whereas those of *Alnus* (19–21 × 23–30 μm), *Betula* (18–23 × 21–30 μm), *Corylus* (20–25 × 26–28 μm), and *Ostrya* (16–20 × 21–24 μm) are smaller. *Corylus* pollen is faintly aspidate and the annulus is slight.

ALNUS   alder

Following Furlow (1979), there are seven species native to North America with one, *Alnus glutinosa* (black alder), introduced and naturalized from Eurasia. Alder pollen was regarded a common cause of early pollinosis in the northwestern region (Solomon & Durham 1967), as confirmed by our survey in Oregon, but it was also of major importance in California, where hundreds of patients skin tested moderately to strongly positive (2+–4+) to alder pollen extracts and suffer pollinosis. *A. rhombifolia* (white alder), widely cultivated as well as native to California, was considered the serious offender. Cross-reactions between most species and with those of the allied *Betula* occur (Vaughan & Black 1948).

Fig. 28. Morphology of representative Betulaceae. a: *Alnus rubra* (red alder) branch with staminate catkins, ×⅔; b: *Betula papyrifera* (paper birch) branch with 2 staminate and several pistillate (small) catkins, ×1½; c: *Ostrya virginiana* (American hop hornbeam) branch with fruit, ×⅔; d–e: *Corylus avellana* (European hazelnut or filbert) with staminate catkins and small female flowers (d, ×⅔) and with leaves and fruit (e, ×½).

TREES AND SHRUBS

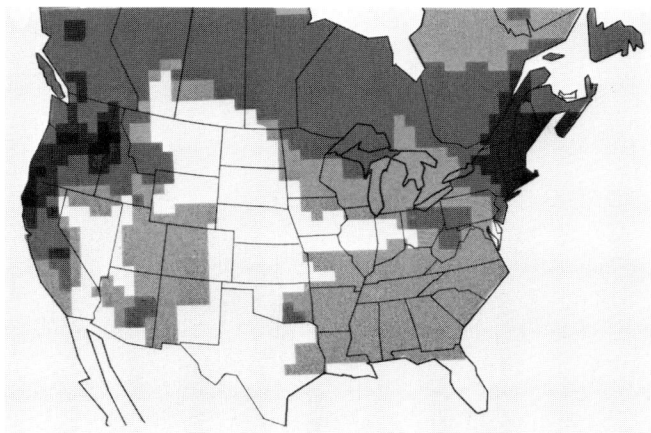

Fig. 29. Generalized composite distribution of seven indigenous species of *Alnus* (☐ 1 sp., ▨ 2 spp., ▩ 3 spp., ■ 4 spp.).

BETULA *birch*

Whereas alder pollen is the greater offender west of the Rocky Mountains, we found in our survey that birch is of greater importance east of the mountains, where hundreds of atopic patients in Connecticut and New Jersey averaged 3+ to skin tests using birch pollen extracts. Milder and fewer reactions were reported to birch than to alder from California and Oregon, but nevertheless of concern there (2+–3+ common, 4+ much less frequent). *Betula nigra*, *B. papyrifera*, and *B. populifolia* were found to elicit moderate reactions on skin testing in St. Louis (Lewis & Imber 1975b).

Allergenicity to birch pollen and other members of the family is well known in Europe, where research on allergenic fractions is advanced. *Betula verrucosa* pollen, for example, has been partially purified and the isolated fraction has a molecular weight of 20,000 daltons (Belin 1972a,b). In Japan *B. platyphylla* was

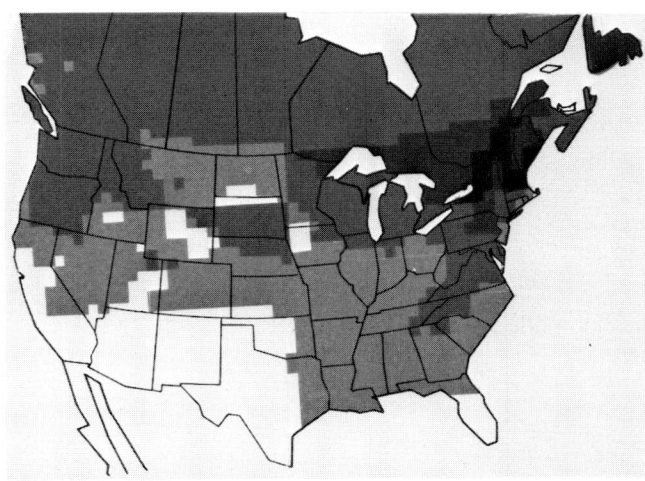

Fig. 30. Generalized composite distribution of eight indigenous species of *Betula* (☐ 1–2 spp., ▨ 3–4 spp., ▩ 5 spp., ■ 6–7 spp.).

TABLE 2. Flowering by month of the Betulaceae.

SPECIES	REGION						
	nNe	sNE	SE	NC	SC	NW	SW
Alnus incana	3–5	3–4	2–3	3–4	2–3	3–5	
A. maritima	8–9	9–10			9–10		
A. rhombifolia						1–4	1–4
A. rubra						3–4	
A. serrulata	2–5	3–4	2–3		3–4	3–5	
A. viridis	5–8		5–6			5–7	5–7
Betula glandulosa	6–8			5		4–7	4–6
B. lenta	4–5		3–4				
B. nigra	4–5	4–5	3–4	5	2–5		
B. occidentalis	4–6			5		2–6	3–5
B. papyrifera	4–6		5–8	4–5		3–5	
B. populifolia	4–5		5				
B. pumila	5–8			4	4–5		
Carpinus caroliniana	4	4–5	3–4			3–5	
Corylus americana	4–5	2–4	1–3	4–5			
C. cornuta	4–5		2–3	4–5		1–3	1–4
Ostrya virginiana	4–5	4–5	4–5	4–5	3–4		

TABLE 3. *Alnus* (alder) aeropollen frequency in regions of North America based on percentage of pollen captured, by month (1974–78).

REGION	MONTH								
	1	2	3	4	5	6	7	8	12
nNE		2.9	2.4	1.1	0.3	0.1			
sNE	0.4	4.2	2.3	3.2	3.0	0.9	0.7		
SE	5.1	6.8	0.3						
NC		4.0	1.8	0.9	0.2				
SC								3.2	
NW	5.5	19.5	54.9	21.8	0.2				
SW	9.5	4.5	0.7	0.1	0.1				4.6
sCA	8.0	0.7	0.1						4.3

TABLE 4. *Betula* (birch) aeropollen frequency in regions of North America based on percentage of pollen captured, by month (1974–78).

REGION	MONTH										
	1	2	3	4	5	6	7	8	9	10	11
nNE	1.9	0.1	1.8	9.2	7.8	1.3	0.1				
sNE	1.9	1.7	5.0	3.9	2.5	1.5					
SE		7.8	1.0	0.2	1.0						
NC				0.6	0.6	0.1					
NW	2.8	8.7	8.4	3.7	0.7						
SW	2.4	1.8	3.0	2.9	1.1	0.2		0.1	0.6	1.3	0.4
sCA		0.1									

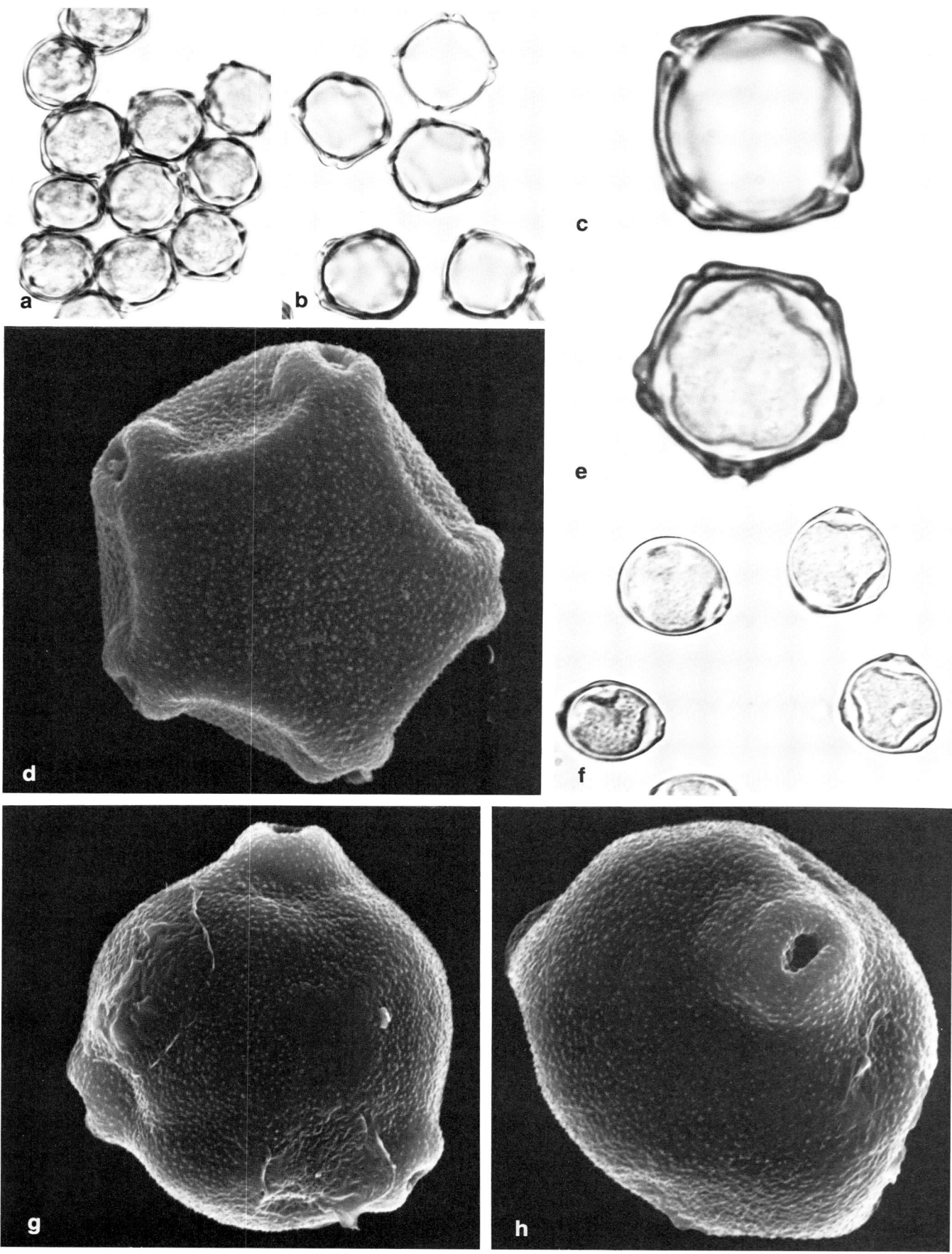

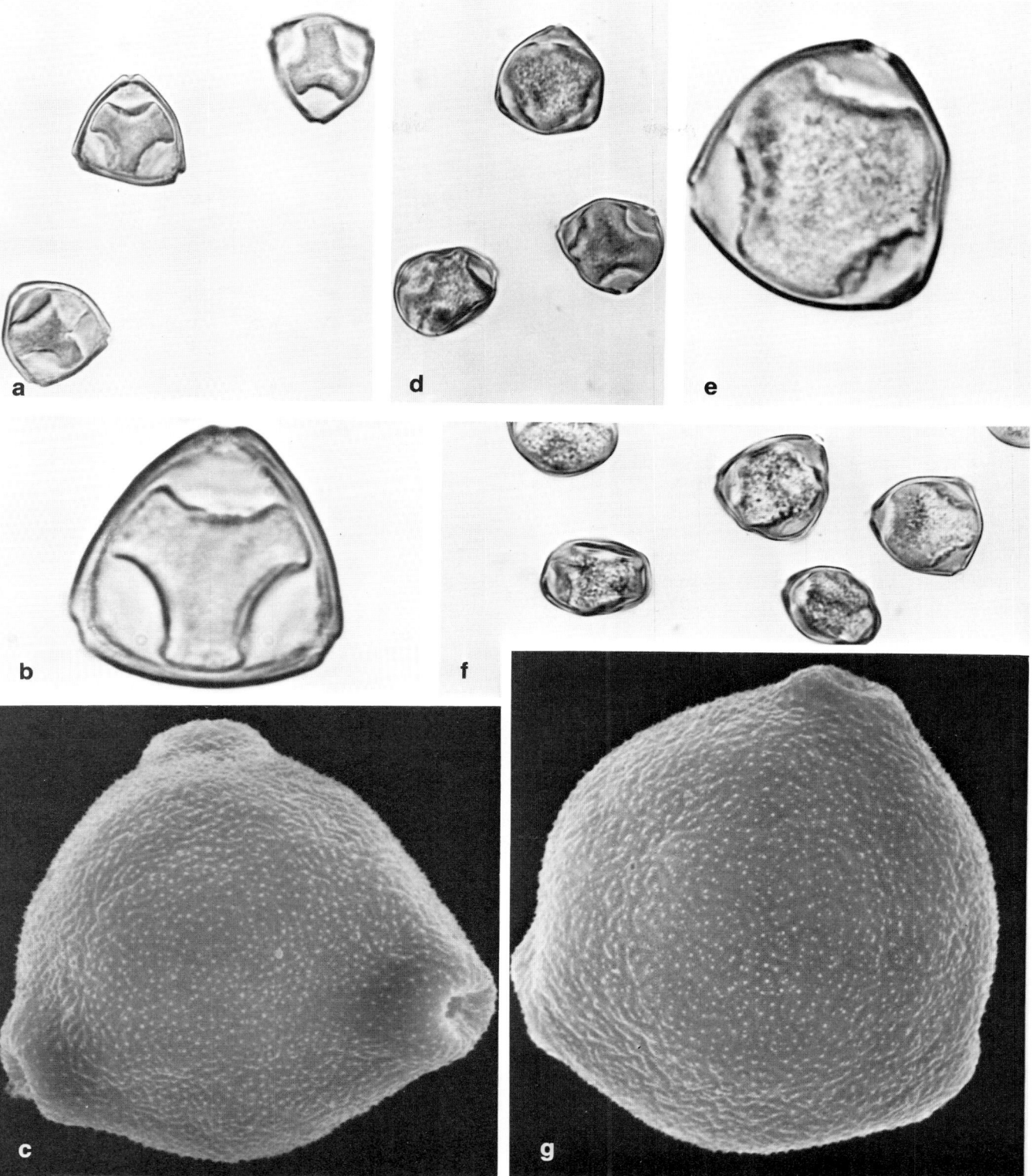

Fig. 31. Pollen of representative Betulaceae (*Alnus* and *Betula*). a–c: *Alnus serrulata* (hazel elder), a = LM(s) ×720, b = LM(a) ×720, c = LM(a) ×1830; d–e: *Alnus rubra* (red alder), d = SEM(a) showing 5 pores, e = LM(s) showing markedly thickened intine below each pore (oncus) ×1830; f: *Betula papyrifera* (paper birch), LM(s) ×720; g–h: *Betula nigra* (red or river birch), g = SEM(ad) polar view showing 3 strongly aspidate pores ×3350, h = SEM(ad) oblique-equatorial view showing 1 of 3 pores ×4000.

Fig. 32. Pollen of representative Betulaceae (*Carpinus-Ostrya*). a–c: *Corylus avellana* (European hazelnut), a = LM(s) polar views x720, b = LM(s) polar view showing 3 large onci (thickened intine below pores) ×1830, c = SEM(a) polar view showing 3 strongly aspidate pores ×3800; d–e: *Carpinus caroliniana* (American hornbeam), d = LM(s) x720, e = LM(s) polar view showing 3 large onci ×1830; f–g: *Ostrya virginiana* (American hop hornbeam), f = LM(s) ×720, g = SEM(a) polar view ×3600.

CORYLUS
hazelnut, filbert

Exposure to *Corylus* and other betulaceous genera is more limited than to the alders and birches and less is known of their allergenic pollen. Nevertheless, two allergists from Oregon report treating approximately 50 patients annually for hazelnut pollinosis who scored either 3+ or 4+ on skin testing, with the majority at 4+ (N.M. Kudelko and A.D. Wert, personal communications). Similar instances can be expected wherever sufficient exposure occurs among individuals sensitized to hazelnuts and the allied hornbeams (*Carpinus, Ostrya*) as reported in Minnesota (Ellis & Rosendahl 1933).

High correlations of cross-reactivity on skin testing were reported by Dalen & Voorhorst (1981) between *Alnus, Betula,* and *Corylus* pollen extracts, as well as between extracts of these genera and those of *Fagus* and *Quercus* (Fagaceae). In the currently accepted Angiosperm phylogeny, the Betulaceae and Fagaceae are allied families with an immediate common ancestor within Phylogenetic Group 2 (see Fig. 1), a classification confirmed by the presumed presence of common pollen allergens in the two families.

BIGNONIACEAE
Bignonia Family

Centered in the tropics, particularly South America, this family of trees, shrubs, and woody vines is represented in North America foremost by *Catalpa* (cigar tree) and *Campsis* (trumpet vine). Plants are primarily, if not exclusively, zoophilous, but the numerous showy flowers of *Catalpa* (Fig. 33) may shed pollen incidentally and be caught airborne in the vicinity of the trees to 50 m distant (Swineford 1940). Such an exposure was sufficient, however, to elicit strong allergic reactions among nine patients suffering from spring pollinosis in a Virginia community where *Catalpa* trees were common. This well-documented report stresses the importance of unsuspected flowering plants in the immediate environment of home and work causing acute allergic symptoms.

Two *Catalpa* species native to eastern North America flower principally during May and June and from March in the southcentral region.

Pollen Morphology (Fig. 34)

Grains are united in tetrads, the tetrad diameter being 48–56 μm and inaperturate. The sexine is about 1.5 μm thick, areolate; the areole ±polygonal, reticulate, and separated by narrow granular grooves; the lumina polygonal to irregular, 0.5–1.2 μm in diameter; the nexine about 1 μm thick; and the intine ±uniformly 1.6 μm thick.

Fig. 33. *Catalpa speciosa* (northern catalpa or cigar tree) showing leaves and large, showy flowers, × ⅓.

BUXACEAE
Boxwood Family

A small family of shrubs typified in North America by the introduced *Buxus* (boxwood) (Fig. 35), these plants are widely used for evergreen plantings around homes and buildings. The monoecious flowers produce abundant nectar and are pollinated by insects, but they are also facultatively or incidentally anemophilous (Knuth 1909). In Switzerland, for example, *Buxus* pollen has been captured in aerosamplers well beyond the region where plants were flowering (Markgraf 1980). Occasional exposure to pollen during the spring can thus be expected, but allergenicity has not been reported.

Perhaps more significant is the dioecious wind-pollinated *Simmondsia chinensis* (jojoba) of the southwestern region that is being planted commercially for its oil. Exposure to its pollen from March to May, and accompanying sensitivities, should increase as plantations expand in Arizona and California.

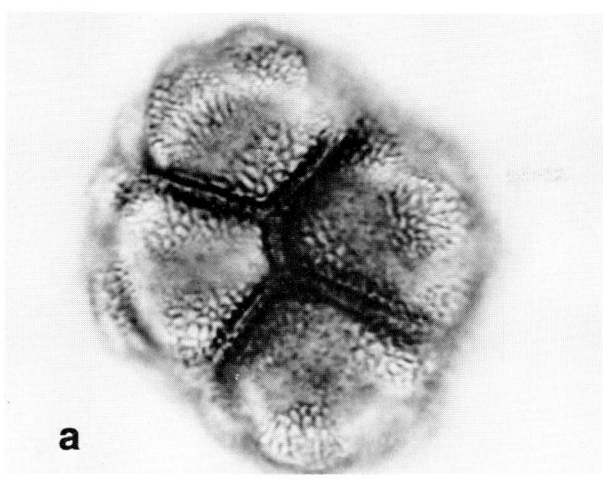

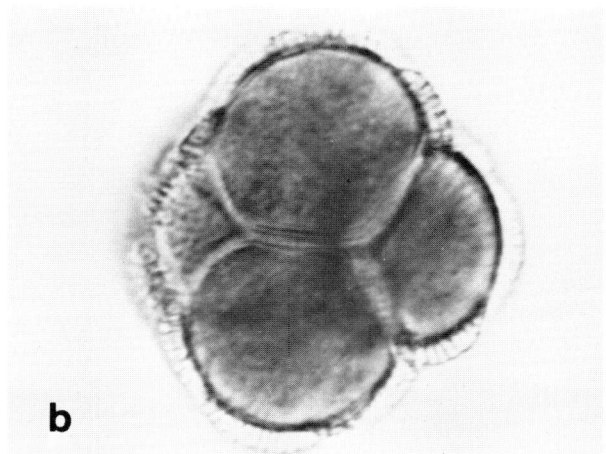

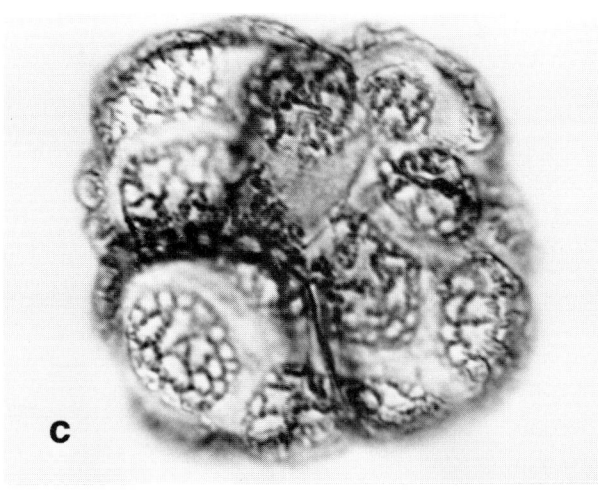

Fig. 34. Pollen tetrads of representative Bignoniaceae. a–b: *Catalpa speciosa* (northern cigar tree), LM(s) ×900 showing surface (a) and optical (b) views; c: *Chilopsis linearis* (desert-willow), LM(s) ×900.

Pollen Morphology (Fig. 36)

Grains are spheroidal to prolate-spheroidal, 29–30 × 28–34 μm; the amb triangular to rhomboidal with convex sides; and the apertures either polyporate (*Buxus*) or 3(–4)-colpate (*Simmondsia*); the pores irregularly outlined, 1.5–2.0 μm in diameter; and the colpi short and broad, ca. 11 × 6 μm, irregularly margined. The sexine is reticulate (*Buxus*) with small, irregular to polygonal lumina 0.5–0.8 μm in size, or punctitegillate (*Simmondsia*) with the tegillum slightly undulating; the nexine slightly thinner than the sexine; and the intine about 0.7 μm (*Simmondsia*) or 1.8 μm (*Buxus*) thick, but thickened to 1.7 and 2.0 μm, respectively, below the apertures.

Fig. 35. *Buxus sempervirens* (boxwood) branch with axillary flowers, ×2.

CAPRIFOLIACEAE
Honeysuckle Family

The honeysuckles are well represented in eastern North America, where many species of small trees, shrubs, and climbers have become familiar ornamentals (Fig. 37). The most troublesome is the weedy vine *Lonicera japonica* (Japanese honeysuckle), but better enjoyed around the home are the hardy ornamental shrubs *Abelia*, other *Lonicera*, *Sambucus* (elderberry), *Symphoricarpos* (snowberry), *Viburnum* (snowball), and *Weigela*.

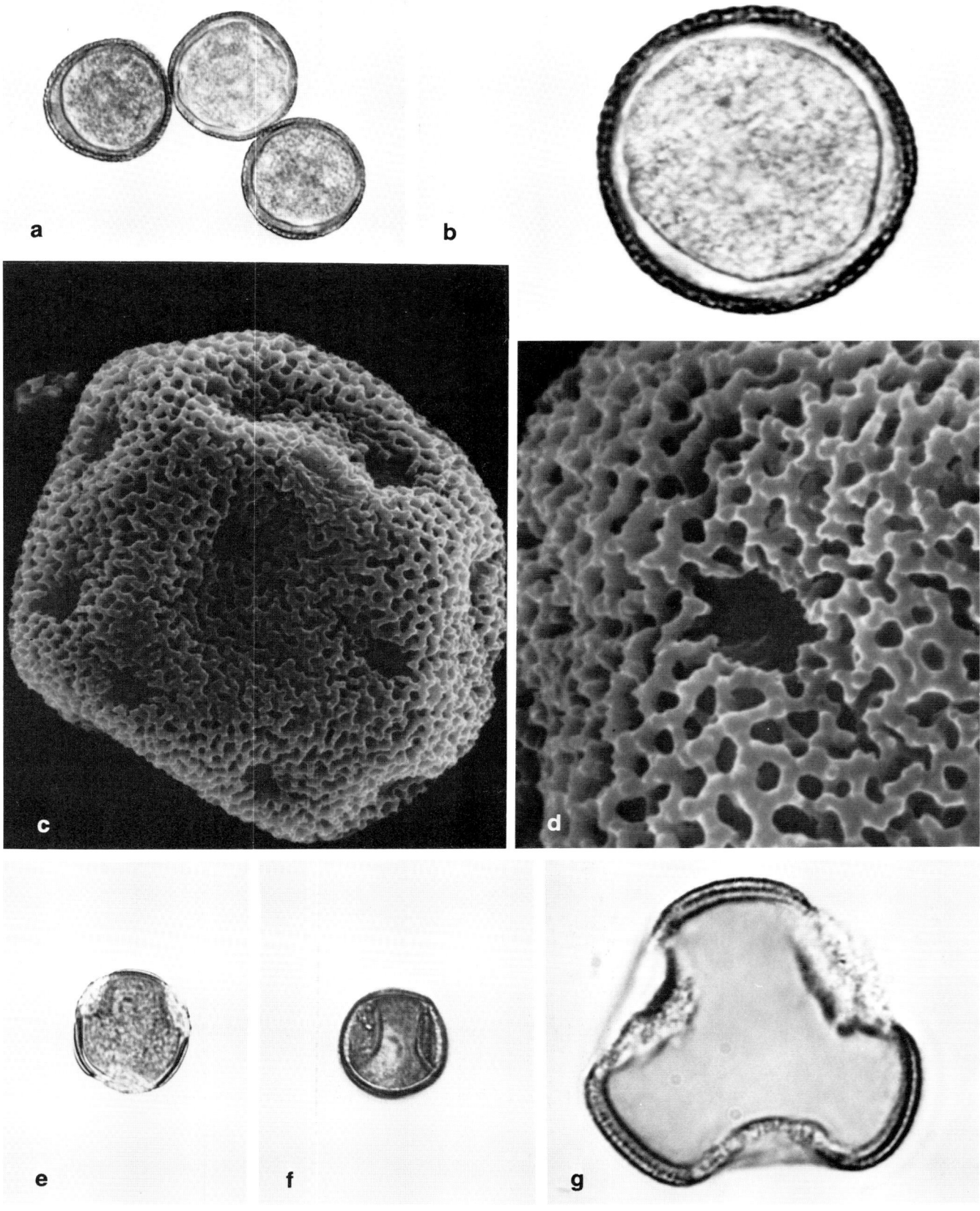

Fig. 36. Pollen of *Buxus* and *Simmondsia*. a–d: *B. sempervirens* (common boxwood), a = LM(s) ×720, b = LM(s) with thick (light) ring of intine ×1830, c = SEM(ad) ×3300, d = SEM(ad) showing pore and sexine structure ×8000; e–g: *S. chinensis* (jojoba), e = LM(s) polar view ×720, f = LM(s) near equatorial view ×720, g = LM(a) polar view ×1830.

All are insect-pollinated, but the usually exposed anthers release small amounts of airborne pollen primarily from April through July (Lewis & Vinay 1979). As exposures are limited so are reports of allergenicity, although allergic patients with 3+ skin tests to pollen extracts of *Lonicera* were reported from Oregon and Texas (Lewis & Vinay 1979).

Pollen Morphology (Fig. 38)

Grains vary in shape from spheroidal to prolate and suboblate; the amb triangular to rounded-triangular, parasyncolpate or angulaperturate, and 3-colporate or 3-porate; and the pores circular, 7–8 μm in diameter and annulate (ca. 4 μm wide) (*Weigela*). The sexine is tegillate with spines (*Lonicera*), reticulate (*Sambucus*) to retipilate (*Viburnum*), or with fingerlike excrescences 2.4–7 μm long and a few spinules ca. 2 μm long (*Weigela*), 1 μm or less in thickness, except *Lonicera* where ca. 3.2 μm thick. The nexine is about as thick as the sexine, much thinner (ca. 0.8 μm) in *Lonicera*, or somewhat thicker in *Viburnum* and *Weigela*, often slightly thickened at the apertures. The intine is only about 0.8 μm thick, but, as in *Lonicera*, with large, lens-shaped onci below the apertures (ca. 6.4 × 11.2 μm).

The spheroidal, spinulate grains of *Lonicera* are among the largest (ca. 45–70 μm in diameter), whereas the prolate to subprolate, reticulate grains of *Sambucus* and *Viburnum* are much smaller. *Viburnum* grains are somewhat larger (28–32 × 17–22 μm) than those of *Sambucus* (20–22 × 16–17 μm). Most distinctive of all are the porate grains of *Weigela*, which are intermediate in size (38–48 × 42–56 μm) and possess the unusual sculpturing described earlier.

Fig. 37. Morphology of representative Caprifoliaceae. a: *Lonicera fragrantissima* branch showing axillary flowers, ×1¼; b–c: *Sambucus racemosa* (red-berried or stinking elder), b = compound leaves and inflorescence, ×½, c = details of inflorescence showing exposed anthers, ×¾; d: *Weigela florida* branch with flowers, ×½.

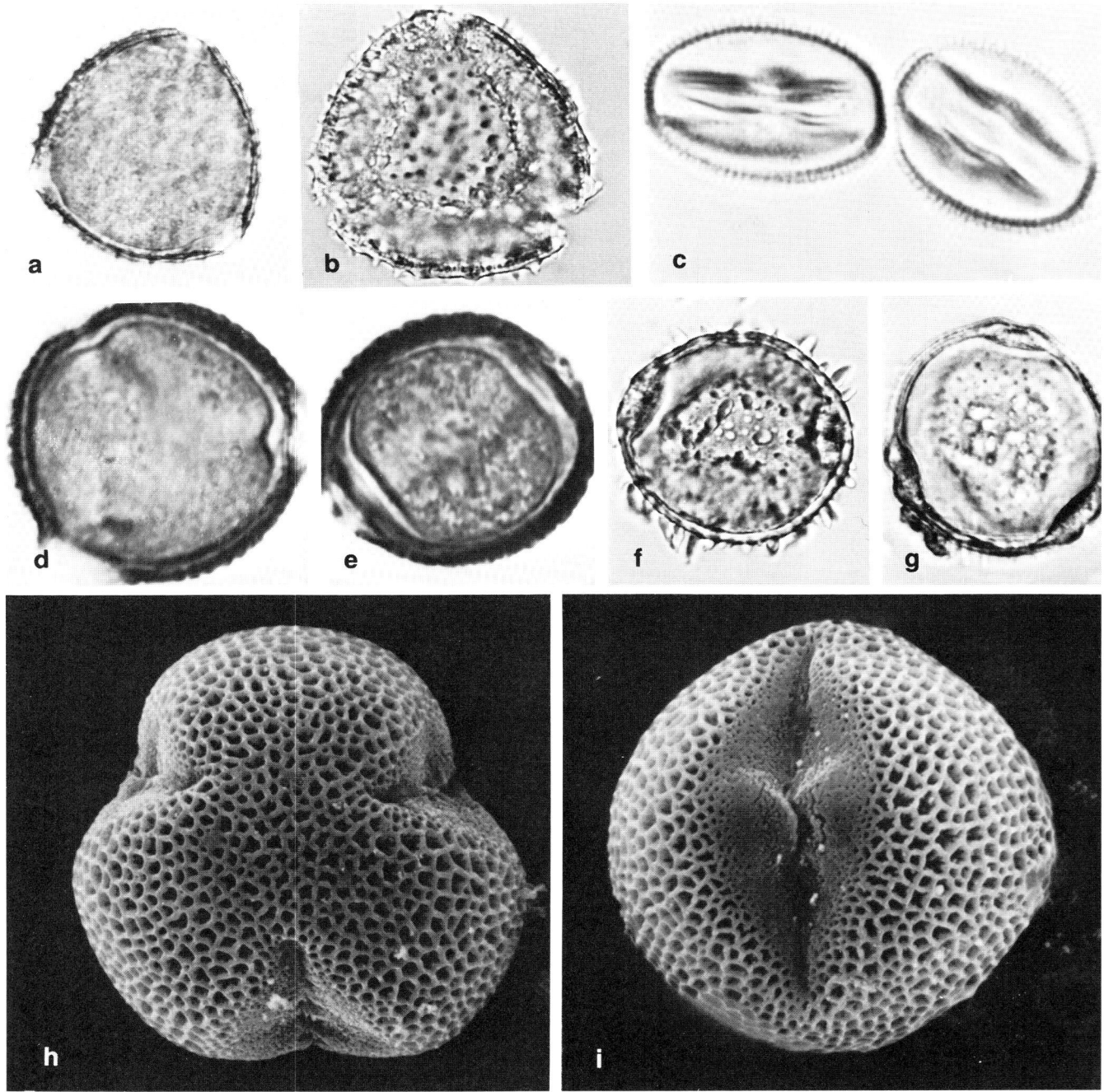

Fig. 38. Pollen of representative Caprifoliaceae. a–b: *Lonicera fragrantissima*, a = LM(s) polar view ×720, b = LM(a) polar view ×720; c: *Sambucus canadensis* (American elder), LM(a) ×1830; d–e: *Viburnum lanatum* (wayfaring tree), d = LM(s) polar view ×1830, e = LM(s) equatorial view ×1830; f–g: *Weigela florida*, f = LM(a) equatorial view ×720, g = LM(a) polar view ×720; h–i: *Sambucus canadensis*, SEM(a) showing polar (h) and equatorial views (i) ×4500.

CASUARINACEAE*
Australian Pine Family

The Casuarinaceae are mostly tall, rapidly growing trees with a characteristic weeping habit caused by their jointed branches with short internodes. They look like wispy pines, but the resemblance is superficial. *Casuarina* (Australian pine, beefwood, she-oak) bears very reduced flowers, the staminate ones at the tip of branches toward the top of the tree (Fig. 39), and the pistillate ones tending to grow on side branches lower down. The leaves are also reduced to scales closely appressed in whorls around the nodes of the branchlets. Pollination is by wind.

The single genus of about 65 species is native to

TREES AND SHRUBS

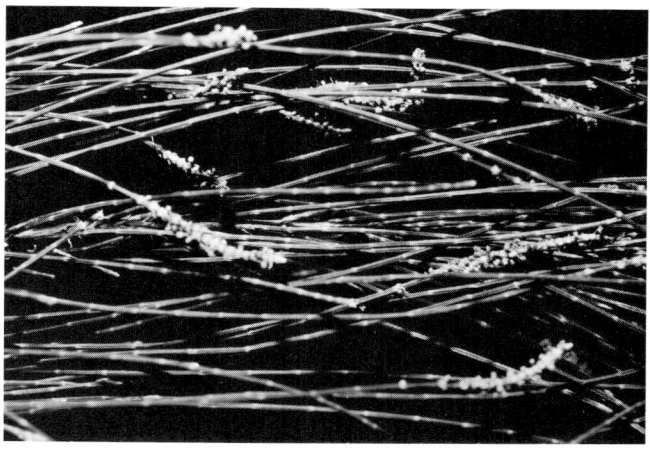

Fig. 39. *Casuarina equisetifolia* (Australian pine) showing reduced staminate flowers at the top of branches, ×1½.

Australia and southeastern Asia, but it is widely adapted in the tropics to dry habitats, often along seacoasts. Several species are much planted in Florida and California as street trees and as hedges, windbreaks, and for seaside plantings. In Florida *Casuarina equisetifolia* flowers from January to April and *C. cunninghamiana* in October and November, but there is considerable annual variation (Zivitz 1942). In Cuba *Casuarina* contributes more pollen to the air than any other tree (Quintero 1955).

Pollen Morphology (Fig. 40)

Grains are usually suboblate, 19–21 × 22–30 μm; the amb triangular with convex sides, and 3(–4)-porate, aspidate, ca. 8–9 μm in diameter including the thickened annulus, and slightly lolongate. The sexine is

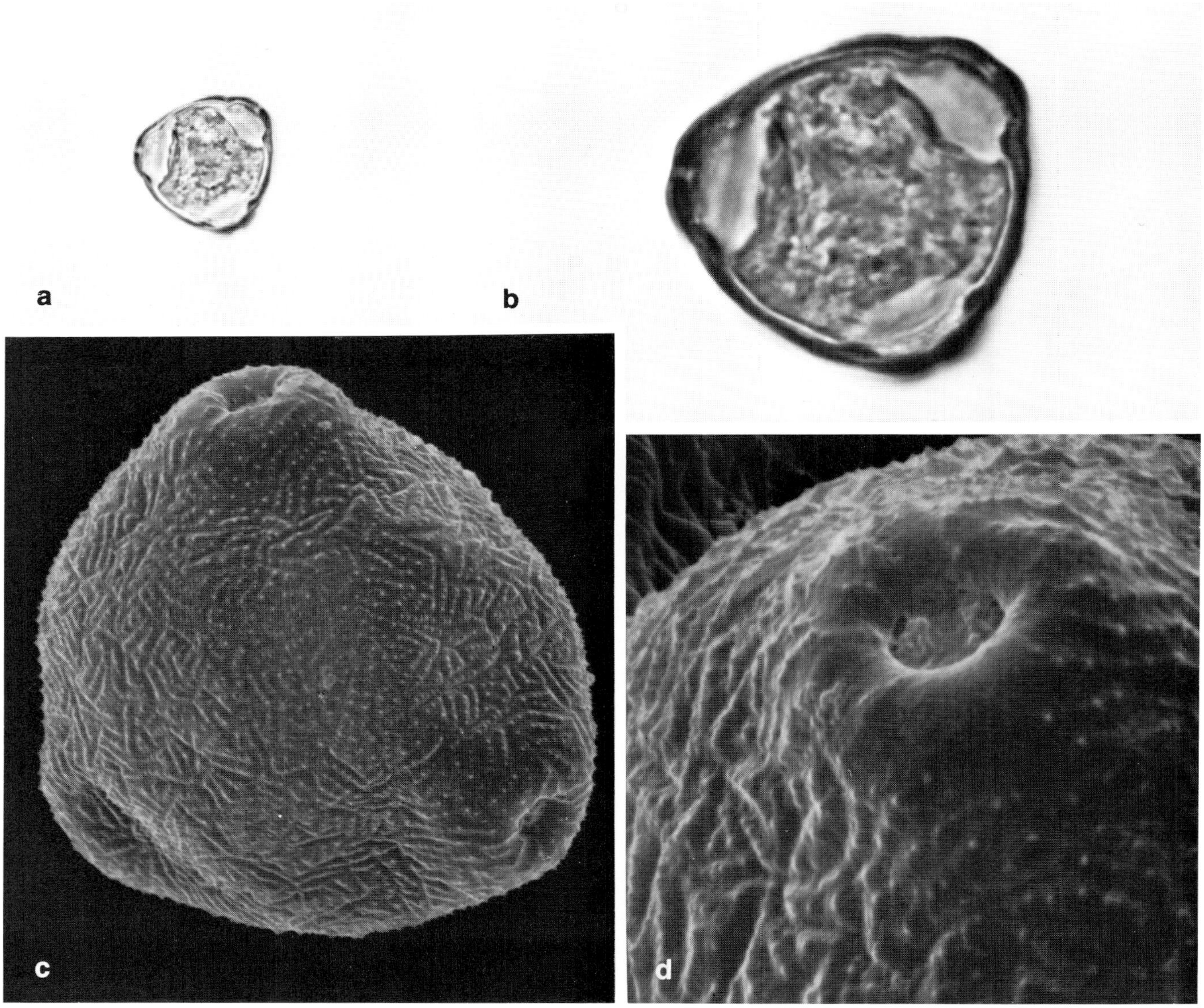

Fig. 40. Pollen of *Casuarina equisetifolia*. a = LM(s) polar view ×720, b = LM(s) polar view showing 3 large onci (thickened intine below pores) ×1830, c = SEM(a) polar view showing 3 aspidate pores ×3200, d = SEM(a) showing pore morphology and sexine surface ×8600.

tectate with the surface transgressed by narrow, irregular ridges having minute superimposed spinules (from SEM), ca. 1.1 μm thick; the nexine thinner than the sexine, ca. 0.5 μm thick and thicker at the apertures; and the intine much thickened to about 7.2 μm below the apertures.

Allergenicity

Exposure to the pollen of *Casuarina* found in Florida produced rhinitis and bronchial asthma among three sensitized patients (Zivitz 1942). In our survey a few allergists reported instances of pollinosis associated with *Casuarina* flowering in Florida and California. For example, in central Florida only 10 to 20 patients allergic to *Casuarina* are seen annually, but they are acutely sensitive (E.J. Luippold, personal communication). Several species cultivated in Israel were found to be allergenic (Gutmann 1950).

CELASTRACEAE

Spindle-tree Family

The spindle-trees are valued as ornamentals and are often found around homes. Although insect-pollinated, their open flowers with exposed anthers (Fig. 41) may shed some pollen incidentally and consequently be of local concern in pollinosis. In California and Texas, for example, pollen extracts of *Euonymus* elicited 3+–4+ skin test reactions from atopic patients suffering from pollinosis (Lewis & Vinay 1979). This report and the close proximity of many of these plants to homes suggest potentially important localized sources of allergens.

The principal genera, *Celastrus* (bittersweet) and *Euonymus* (spindle-tree, burning bush), flower in May and June and a month earlier in southern latitudes.

Pollen Morphology (Fig. 42)

Grains of *Euonymus* are prolate-spheroidal, 21–30 × 23–29 μm; the amb ±circular and 3-colporate; the colpi long; and the ora lalongate. The sexine is 2–3 μm thick, reticulate with small but distinct lumina (<1 μm); and the nexine thinner than the sexine.

CISTACEAE

Rockrose Family

A medium-sized family of shrubs and subshrubs with large showy flowers and numerous stamens, the widely cultivated rockroses may in part be secondarily anemophilous following staminal stimulation (as found for *Helianthemum chamaecistus*) (Knuth 1909). Pollination data for the American species of *Helianthemum* (frostweed, rockrose), *Hudsonia* (golden heather, beach heath), and *Lechea* (pinweed) are unknown, but some may shed airborne pollen that could be allergenic.

Pollen Morphology (Fig. 43)

Grains are prolate, 34–42 × 23–24 μm; the amb triangular with convex sides and 3-colporate; the colpi

Fig. 41. *Euonymus fortunei*. a = branch, ×⅘; and b = flowers, ×2¾, showing open flowers with exposed anthers.

TREES AND SHRUBS

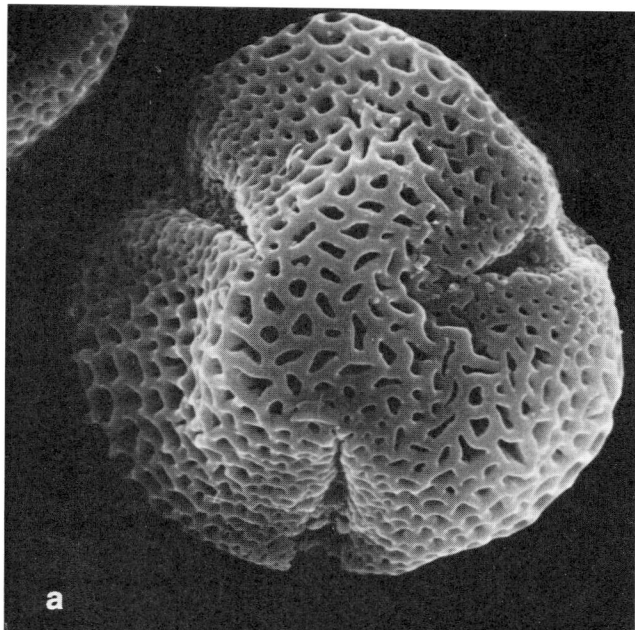

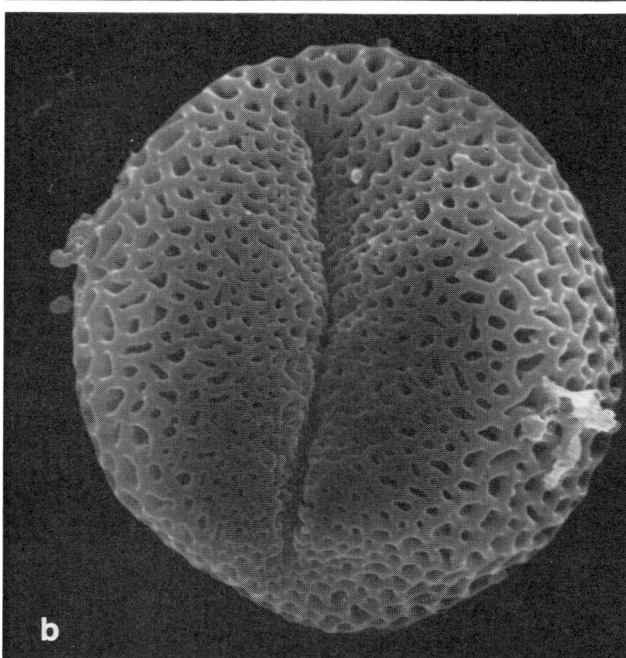

Fig. 42. Pollen of *Euonymus fortunei*. a = SEM(a) polar view ×4200, b = SEM(a) equatorial view ×4200.

siaceae are represented in temperate areas by the widespread genus *Hypericum* (St. John's wort). *H. calycinus*, for example, is commonly planted in North America as a ground cover, particularly in the West (Fig. 44). The species has numerous stamens and,

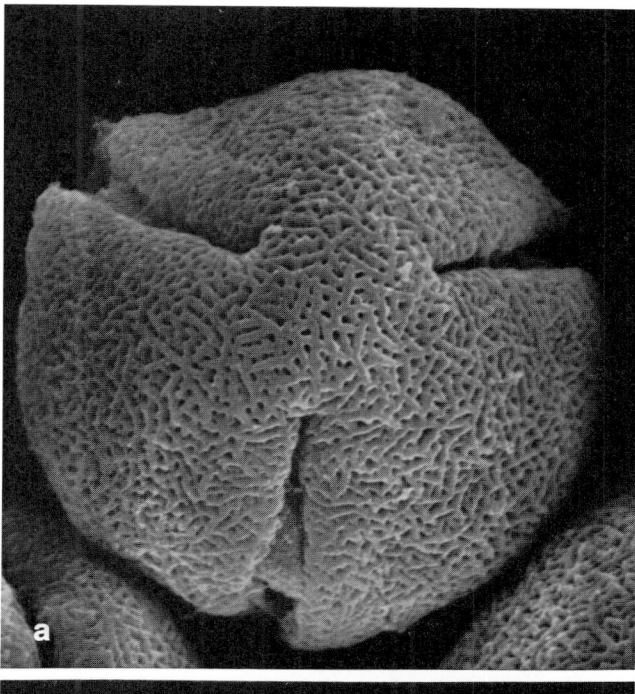

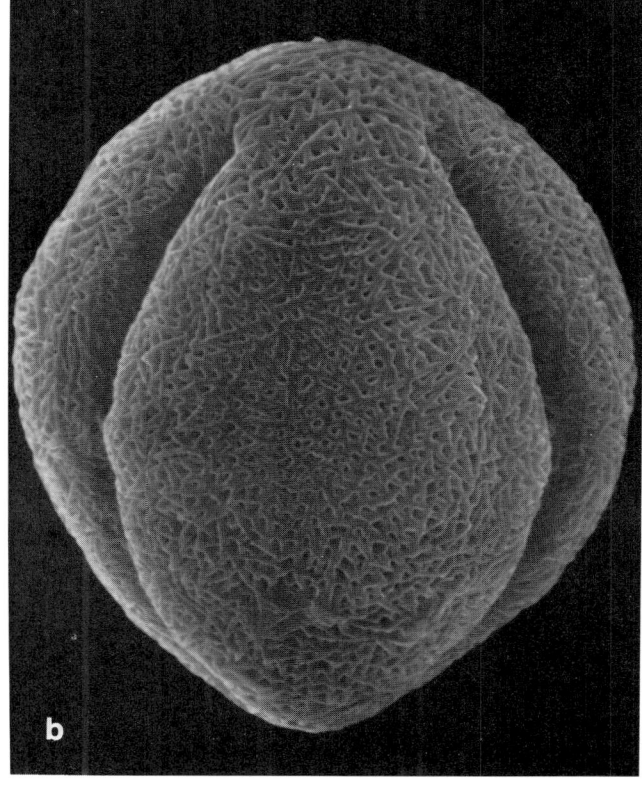

Fig. 43. Pollen of *Hudsonia ericoides* (golden-heather). a = SEM(a) polar view ×3200, b = SEM(a) equatorial view showing 2 apertures ×3600.

long and narrow, 26 × 0.5 μm, with thickened margins; and the ora faintly demarked, ca. 3.2 × 4.8 μm. The sexine is finely reticulate with small lumina, ca. 1.6 μm thick; and the nexine only ca. 0.8 μm thick, but thicker at the apertures.

CLUSIACEAE (*Guttiferae*)

Mangosteen Family

A largely tropical family of woody plants, the Clu-

although entomophilous, incidental release of pollen and exposure during the summer months can be expected in most urban communities. There are many native species as well.

Pollen Morphology (Fig. 45)

Grains of *Hypericum* are subprolate, 21–24 × 18–19 μm; the amb triangular and 3-colporate; the colpi long and narrow (15 × 1.5–2.0 μm), and slightly constricted equatorially; the ora lalongate, ca. 1.5–1.8 μm in diameter; and the opercula granular. The sexine is reticulate, ca. 0.8 μm thick; the lumina small and polygonal, <0.5 μm in diameter; the nexine as thick as the sexine; and the intine thin, ca. 0.5 μm, somewhat thickened below the apertures.

Allergenicity is not known.

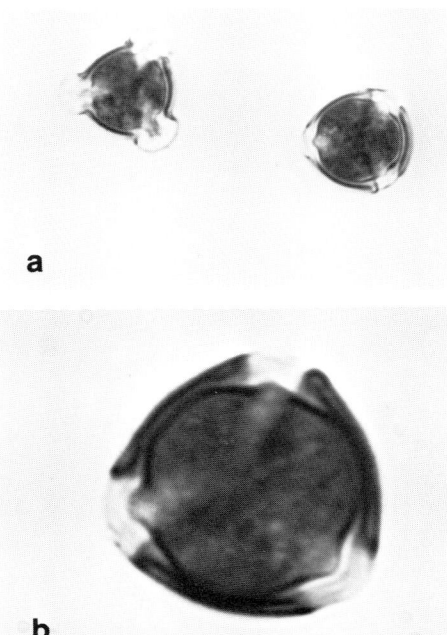

Fig. 45. Pollen of *Hypericum perforatum* (common St. John's wort). a = LM(s) polar views ×720, b = LM(s) polar view ×1830.

Fig. 44. *Hypericum calycinum* (St. John's wort). a = as a creeping ground cover, ×1/10; and b = as a flower showing numerous stamens, ×1.

Fig. 46. *Cornus alba* (Tartarian dogwood) branch with leaves and flowers (a, ×2/3) and *Cornus florida* (flowering dogwood) inflorescence with large white bracts (b, ×1).

CORNACEAE

Dogwood Family

The dogwoods (*Cornus*) are mostly trees and shrubs of north temperate regions that are popular ornamentals (Fig. 46). Pollen is occasionally found airborne (Massachusetts, Missouri, Oklahoma), but its frequency is low. Flowering begins in March (southeast and southcentral), April (midlatitudes and Pacific regions), and May (northern northeast and northcentral regions).

Atopic patients in New Jersey skin tested strongly positive using dogwood pollen extracts (Lewis & Vinay 1979). Further research is needed to determine airborne pollen frequency in this entomophilous family and the level and frequency of allergenicity correlated with pollinosis.

Pollen Morphology (Fig. 47)

Grains are subprolate to prolate, 24–62 × 20–50 μm; the amb triangular with ±straight sides and the three long apertures nearly fusing and 3-colporate; the colpi long and narrow, 19–48 × 0.6–2.0 μm, with smooth and thickened margins; the ora ±faint, lalongate, ca. 2 × 5 μm; and the opercula somewhat granular. The sexine is tectate, densely covered irregularly with spinules (from SEM), appearing ±granular, 0.6–1.0 μm thick; the nexine 0.8–1.0 μm thick, but thicker at the colpi margins; and the intine 0.8–1.0 μm thick and thickened to 4 μm below the apertures.

ELAEAGNACEAE

Oleaster Family

The oleasters are a small family of branching shrubs covered with silvery or golden scales. *Elaeagnus angustifolia* (Russian or wild olive, oleaster, or silverberry, Fig. 48) is a common ornamental that flowers in May or June; it is entomophilous but may shed some airborne pollen (Lewis & Vinay 1979). Limited

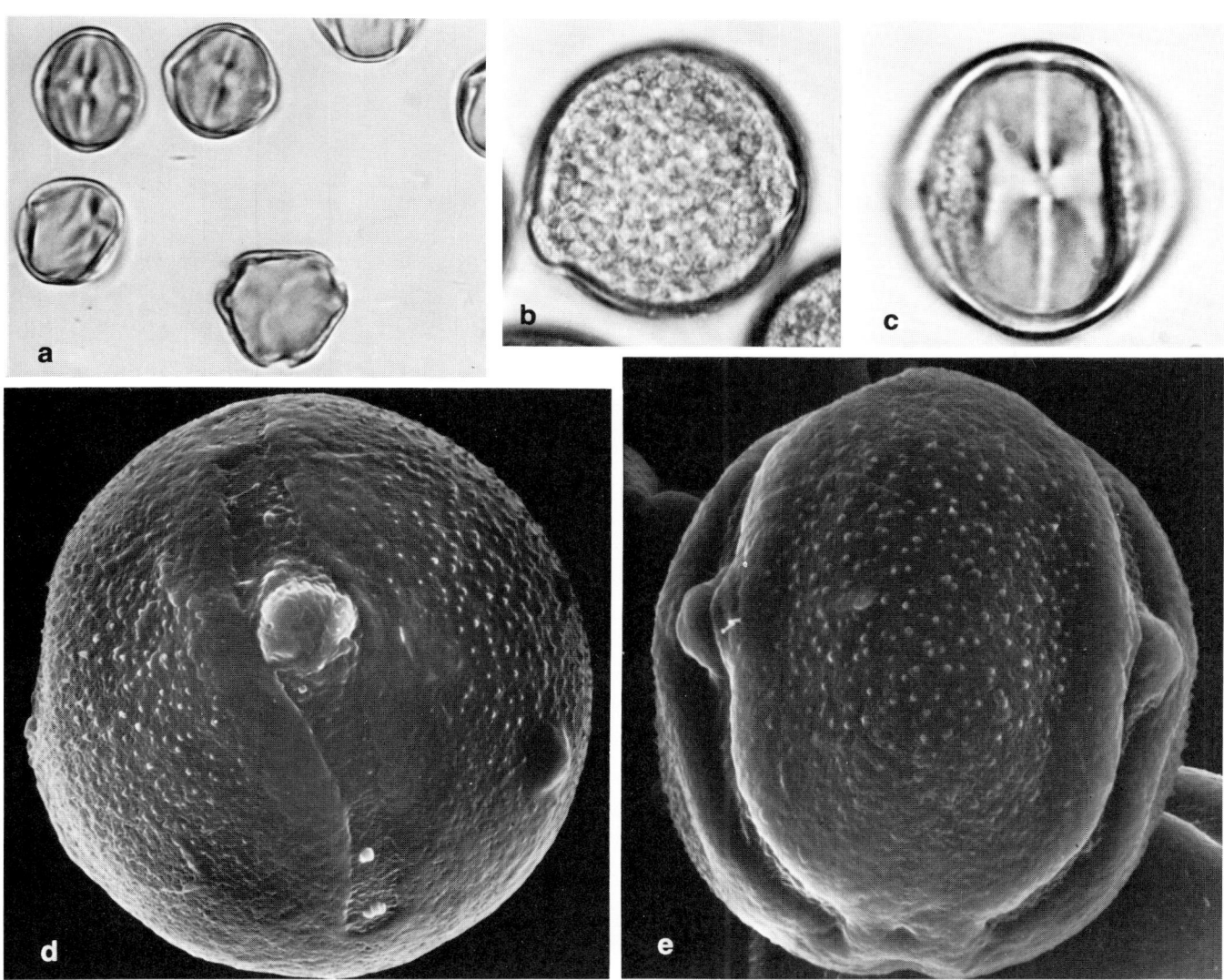

Fig. 47. Pollen of *Cornus*. a–b: *C. mas* (Cornelian cherry), a = LM(a) ×720, b = LM(s) ×1830, c–e: *C. stolonifera* (red-osier dogwood), c = LM(a) equatorial view ×720, d–e = SEM(a) equatorial views showing 1 (d) and 2 (e) apertures ×2100.

wind-dispersed pollen was also found for *Shepherdia* (Bassett et al. 1978). The European *Hippophae* is, however, anemophilous. Low positive (2+) skin test responses to *Elaeagnus* pollen among hay fever patients in New Mexico and Wisconsin suggest only moderate clinical significance of its pollen.

Pollen Morphology (Fig. 49)

Grains are oblate (*Elaeagnus*, 24–30 × 40–44 μm) or prolate (*Shepherdia*, 31–37 × 20–21 μm); the amb triangular with convex sides and 3-colporate; the colpi short in *Elaeagnus*, 8.8 × 3.5 μm, or long in *Shepherdia*, 26–30 × 1.6–2.0 μm; and the ora lalongate and about 8.8 × 10.4 μm in *Elaeagnus*, or circular, small (ca. 1.2 μm) with thickened margins and slightly protruding granular opercula in *Shepherdia*. The sexine is only 0.6–0.8 μm thick, and either tectate and conspicuously thickened at the apertures (*Elaeagnus*) or rugulate (*Shepherdia*); the nexine about 1 μm thick, somewhat thickened adjacent the apertures; and the intine about 0.8 μm thick, but to 2.4 μm below the apertures.

EMPETRACEAE

Crowberry Family

The crowberries are a small family of heathlike evergreen shrubs; all three genera of the family are found in eastern North America. *Empetrum* (crowberry) is anemophilous, but records of airborne pollen and allergenicity are wanting; *Ceratiola* and *Corema* are even less well known.

Pollen Morphology (Fig. 50)

Grains of *Ceratiola* are united into tetrahedral tetrads 30–34 μm in diameter and 3-colporate; the colpi long and narrow; and the ora lalongate. The sexine is tectate with a granular surface and about 0.6 μm thick; the nexine about 0.8 μm thick, but conspicuously thickened adjacent the apertures; and the intine about 1 μm thick, slightly thickened below the apertures.

The tetrads show a close resemblance to those of the Ericaceae.

ERICACEAE

Heath Family

This large family of shrubs with many well-known plants (Fig. 51), such as *Arbutus* (madrone), *Calluna* (heather), *Erica* (heath), *Kalmia* (mountain laurel), *Rhododendron* (azalea), and *Vaccinium* (blueberry), is primarily insect-pollinated and has not been impli-

Fig. 48. *Elaeagnus angustifolia* (Russian or wild olive) shrub (a) and branch showing leaves and flowers (b, ×1½).

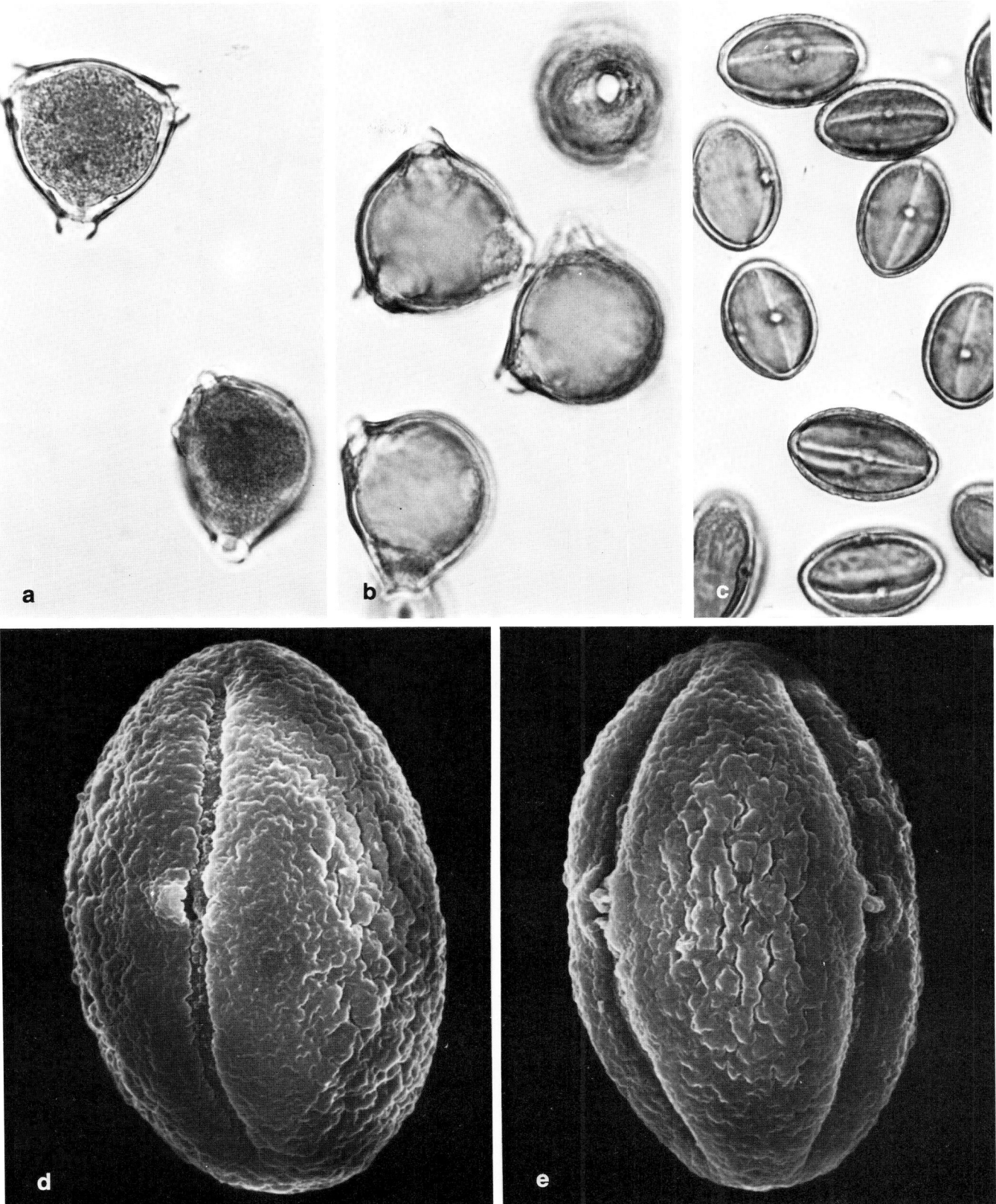

Fig. 49. Pollen of representative Elaeagnaceae. a–b: *Elaeagnus angustifolia* polar and equatorial views, a = LM(s) ×720, b = LM(a) ×720; c–e: *Shepherdia canadensis* (buffalo or soapberry), c = LM(a) ×720, d–e = SEM(ad) equatorial views showing 1 (d, ×3800) and 2 (e, ×3200) apertures.

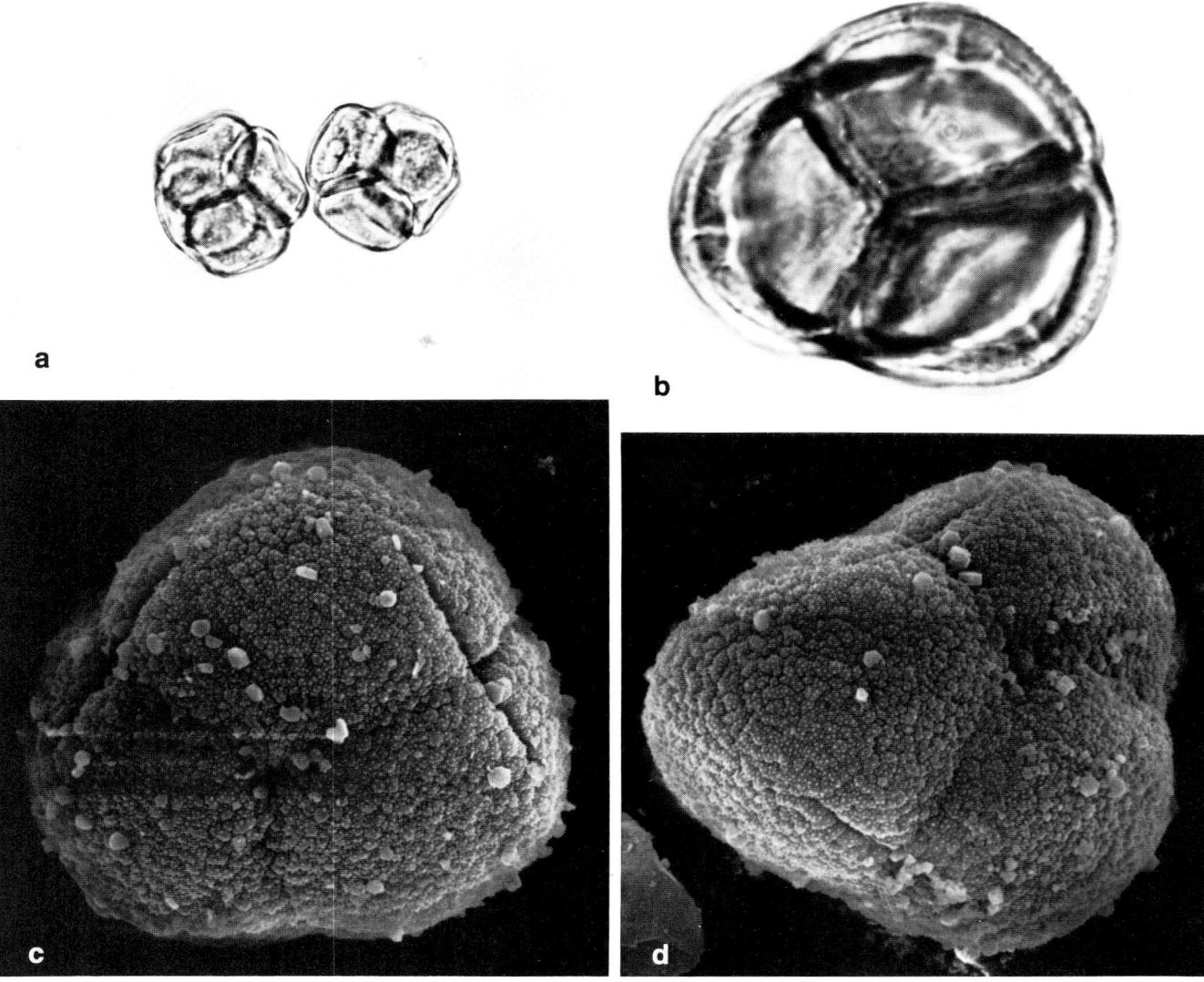

Fig. 50. Pollen tetrads of *Ceratiola ericoides* (rosemary). a = LM(s) ×720, b = LM(a) ×1830, c = SEM(a) ×2900, d = SEM(a) ×2600.

cated in any significant way in pollinosis. Yet, following an initial phase of entomophily, secretion of nectar ceases in *Calluna*, the filaments elongate so that the anthers become exserted, and the pollen is then carried away by the wind (Knuth 1909). In warm sunshine, the elasticity of filaments of *Kalmia* stamens increases, so that the slightest touch causes them to be released and to scatter a cloud of pollen from the dehisced anthers (Knuth 1909). Insects are not needed to trigger this mechanism. How general these examples are is unknown, but both insect- and wind-pollination have been reported also for *Bruckenthalia, Erica,* and *Ledum* (Knuth 1909). Ericaceous pollen has been identified from airborne samples chiefly in Britain (Hyde 1950; Tinsley & Smith 1974) and Scandinavia (Käpylä 1981) and more occasionally in North America (Lewis & Vinay 1979), but the extent of concomitant exposure and sensitization is only conjectural. A low level of allergenicity of questionable clinical importance was reported for azalea pollen extracts from Texas (Lewis & Vinay 1979).

Flowering in the family begins in May (March for *Arbutus* in the southwest) for one or two months, but the heaths and heathers begin later in midsummer to late summer and extend into the autumn.

Pollen Morphology (Fig. 52)

Grains are united into tetrahedral or rhomboidal tetrads 26–56 μm in diameter and usually 3-colporate or indistinct; the colpi long (*Vaccinium*) or short (*Calluna, Kalmia*); and the ora primarily lalongate, 1.0–2.4 × 3–12 μm. The sexine is tectate, scabrate, rugulate, or with scattered gemmae or minute verrucae (*Ledum*), 0.5–0.9 μm thick; the nexine often thicker than the sexine (to 1.5 μm) and usually thickened near the apertures; and the intine about 0.8 μm thick being particularly thickened (to 3.2 μm) at the apertures.

Tetrads of *Arbutus* are comparatively large (48–56 μm).

Fig. 51. Morphology of representative Ericaceae. a–b: *Arbutus menziesii* (Pacific madrone) tree (a) and branch with leaves and inflorescence (b, ×⅔); c: *Calluna vulgaris* (common heather) plant from herbarium specimen, ×1½; d: *Kalmia latifolia* (mountain laurel) flower, ×1½.

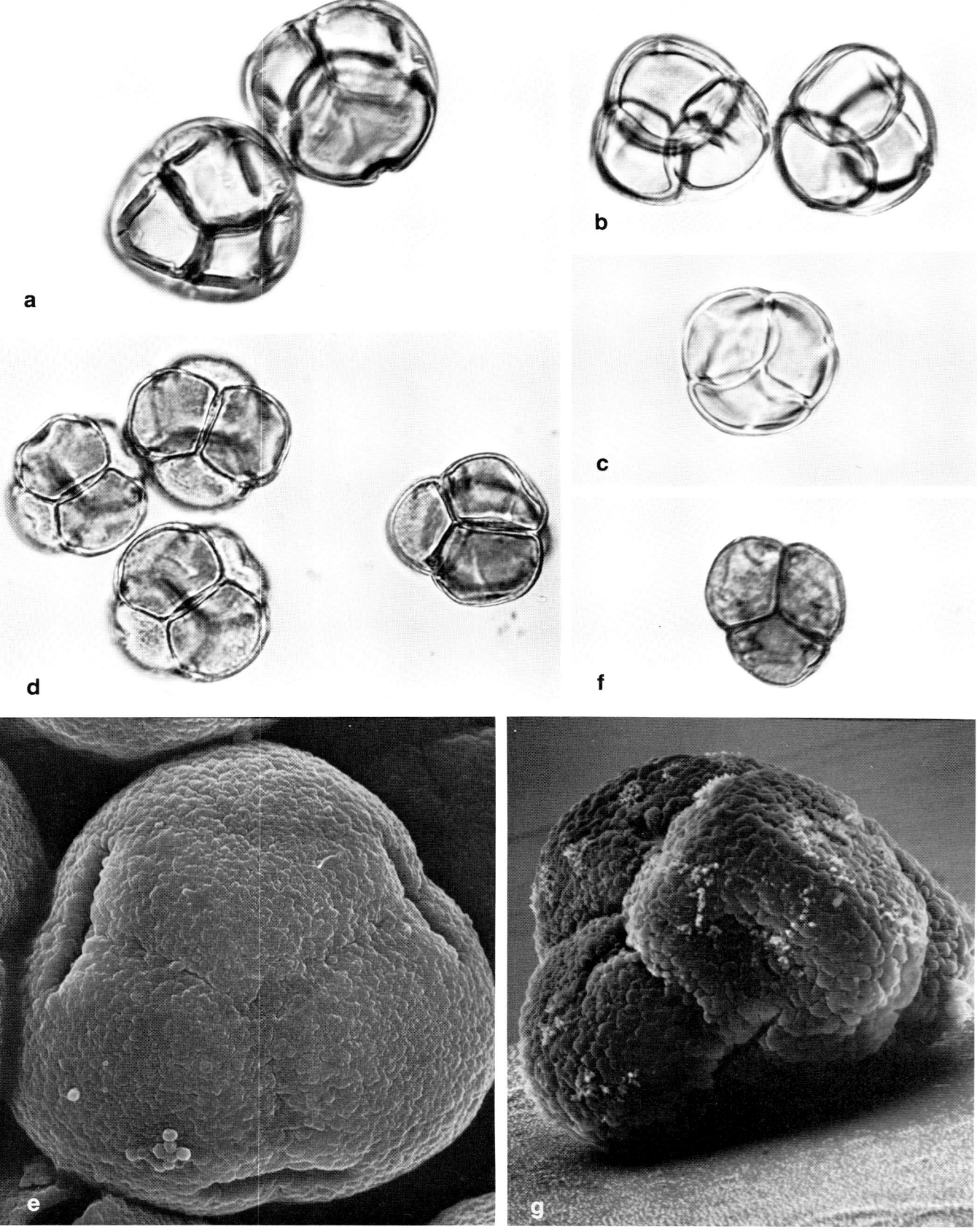

Fig. 52. Pollen tetrads of representative Ericaceae. a: *Arbutus × andrachnoides*, LM(a) ×720; b: *Rhododendron* hybrid, LM(a) ×720; c: *Vaccinium oreophilum*, LM(a) ×720; d–e: *Kalmia latifolia* (mountain laurel), d = LM(s) ×720, e = SEM(a), ×3200; f–g: *Calluna vulgaris* (common heather), f = LM(s) ×720, g = SEM(a) ×2800.

EUPHORBIACEAE

Spurge Family

A large, predominantly tropical family of trees, shrubs, and herbs, the spurges of North America are primarily insect-pollinated and have not been widely observed in aerosamples or implicated in pollinosis. However, *Ricinus* (castor bean) (Fig. 53) is strongly anemophilous by the explosive dehiscence of its anthers and subsequent scattering of pollen. *Ateramnus* (crabwood) is also probably wind-pollinated (Tomlinson 1980), but this plant is found only south of Miami.

Ricinus may persist from year to year in southern latitudes where it reaches treelike proportions (to 12 m), but elsewhere it is annually planted. Pollen in Missouri, for example, is shed from June throughout the summer, but plants do not persist. There is no report of pollen allergenicity in North America, but in Israel there is a well-documented case of pollinosis involving rhinitis, sinusitis, and asthma that became asymptomatic following hyposensitization (Lindenbaum 1966a). Allergic, particularly asthmatic, reactions to castor bean and its pomace can be common (Small 1952) and toxicity to pollen extracts has been reported. Exposure to pollen of other ±woody Euphorbiaceae may also trigger pollinosis, but this is limited to close association with the offending entomophilous plants shedding occasional pollen (Chen & Huang 1980), as recently reported following exposure to *Sapium* in Texas (D.S. Seigler, personal communication).

For a discussion of herbaceous Euphorbiaceae, see Chapter 3.

Pollen Morphology (Fig. 54)

Grains are prolate-spheroidal to subspheroidal (*Ricinus*) or prolate (*Sapium*); the amb triangular with convex (*Ricinus*) or slightly convex (*Sapium*) sides and 3-colporate; the colpi long and narrow (26–37 × 0.5–1.5 µm); the ora lalongate, 3 × 4–6 µm in *Sapium*, 5–7 × 16 µm in *Ricinus*; and the opercula granular. The sexine is finely reticulate (*Ricinus*) or striato-reticulate (*Sapium*); the lumina small, about 0.5 µm, polygonal, ca. 0.7 µm thick in *Ricinus* and ca. 2.2 µm thick in *Sapium*; the nexine also thin in *Ricinus* (0.9 µm) and thicker in *Sapium* (1.8 µm); and the intine thin (0.8 µm), but conspicuously thickened to ca. 3 µm below the apertures.

The grains of *Ricinus* are also decidedly smaller (26–32 × 27–28 µm) than those of *Sapium* (42–43 × 32 µm).

FABACEAE (*Leguminosae*)

Legume Family

A large, widely dispersed and economically important family, the legumes are primarily zoophilous and only of minor significance in pollinosis. However, pollination in the Mimosoideae often is facultatively anemophilous, and thus members of the subfamily do shed sufficient airborne pollen to elicit sensitivities among some atopic individuals. Reports of allergenicity of legumes from the other subfamilies is a consequence of incidental exposure to pollen, such as contact in hay containing alfalfa or clover in which dried, easily broken flowers of these papilionaceous plants release dry pollen when disturbed.

SUBFAMILY MIMOSOIDEAE

Mimosas (Fig. 55) include those trees and shrubs mainly of tropical regions that have bipinnately compound leaves, capitate inflorescences, and numerous stamens (10 or often many more) that are long and conspicuous. The important mimosas in the warmer regions of the United States include: *Acacia* (acacia, wattle), *Albizia* (mimosa or silk tree), *Calliandra* (powderpuff), *Leucaena* (lead tree), *Mimosa* (cat's claw, mimosa), *Pithecellobium* (blackbead), and *Prosopis* (mesquite). Many are cultivated trees or shrubs and contact with them is greater than would be among those from desolate regions of the Southwest or those, like *Desmanthus*, that are short, trailing subshrubs.

Pollen Morphology (Figs. 56–57)

Grains are monads or polyads. The polyads consist of 12 to 24 grains having a bilateral diameter 38–140 µm, and the grains square, rectangular, or angular, with small circular 1(–4)-porate or indistinct apertures. The sexine is either tectate, scabrate, punctitegillate, or tegillate, supported by slender bacula, 0.8–1.6 µm thick; the nexine as thick as or thinner than the sexine; and the intine thin (1 µm or less), except *Calliandra*, where 1.8–2.2 µm thick.

The monads (typical pollen) are prolate, subprolate, or prolate-spheroidal, 30–51 × 21–37 µm; the amb triangular or rounded-triangular and 3-colporate (3-colpate in some genera not included); the colpi long and narrow (syncolpi in *Prosopis*); and the ora lolongate, ca. 8 × 2–3 µm. The sexine is faintly striate, punctitegillate, reticulate, or scrobiculate; the nexine and intine are like those of the polyads.

Acacia, *Albizia*, *Calliandra*, and *Pithecellobium* form polyads, the polyads of *Calliandra* being distinctly cone-shaped and large (130–140 µm long), and those of *Acacia* small (38–52 µm in diameter). Pollen of *Desmanthus*, *Leucaena*, and *Prosopis* is 3-colporate and all are monads.

Allergenicity

Allergenicity is associated with at least three genera. *Acacia*, a predominantly ornamental genus grown in the Southwest, flowers there from January through October and sheds sufficient airborne pollen to produce positive skin test reactions in patients suffering

Fig. 53. *Ricinus communis* (castor bean) plant (a), inflorescence showing staminate (lower) and pistillate (upper) flowers (b, ×⅔) with the staminate and one pistillate flower in greater detail (c, ×2½), and maturing fruit (d, ×⅙).

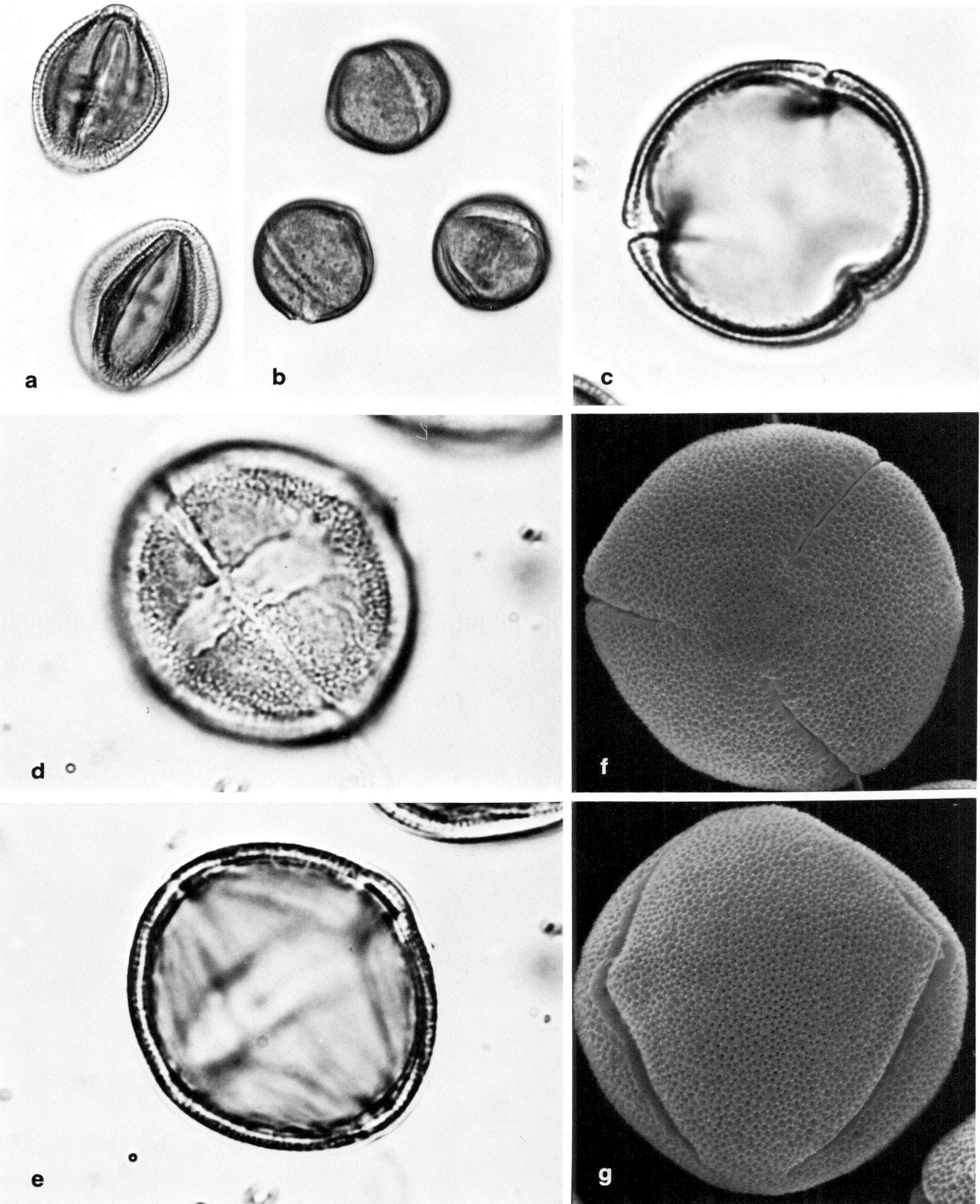

Fig. 54. Pollen of *Sapium* and *Ricinus*. a: *Sapium sebiferum* (Chinese tallow tree), LM(a) ×720; b–g: *Ricinus communis* (castor bean), b = LM(s) ×720, c = LM(a) polar view ×1830, d–e = LM(a) equatorial views ×1830, f = SEM(a) polar view ×3200, g = SEM(a) equatorial view showing 2 apertures ×3200.

Fig. 55. Morphology of representative Fabaceae. a: *Acacia cinemia* branch, ×½; b: *Albizia julibrissin* (mimosa or silk tree) showing flowers with numerous stamens, ×½; c: *Desmanthus illinoensis* (prairie mimosa) with flowers and fruit, ×1¼; d: *Prosopis glandulosa* (glandular mesquite) with inflorescence (photo from dried herbarium sheet), ×1; e: *Cassia fasciculata* (partridge pea) showing leaves and flower, ×1; f: *Cercis canadensis* (redbud) branch with flowers and fruit, ×⅔.

from pollinosis (Lewis & Vinay 1979). In Israel *Acacia* pollen is considered strongly allergenic. Airborne pollen of *Albizia* in the Miami area is held responsible for pollinosis (Fly 1952). *Prosopis*, from which considerable pollen may be shaken when dry (Simpson 1977), and carried by wind for several miles (Bieberdorf & Swinny 1952), can be a serious offender in the southcentral and southwestern regions from March through July (Wodehouse 1971). Specifically from Texas (Sellers 1929; Bieberdorf & Swinny 1952; Fein & Kamin 1962), California (Novey et al., 1977), Hawaii (Wodehouse 1971), South Africa (Ordman 1959), and India (Menon et al. 1977), case histories involving *Prosopis* pollen allergenicity have been reported. Indeed, sufficient exposure of atopic individuals to pollen of any Mimosoideae listed, and to others, may elicit allergic reactions (e.g., following exposure to *Ceratonia* [carob] in Spain) (Wodehouse 1971).

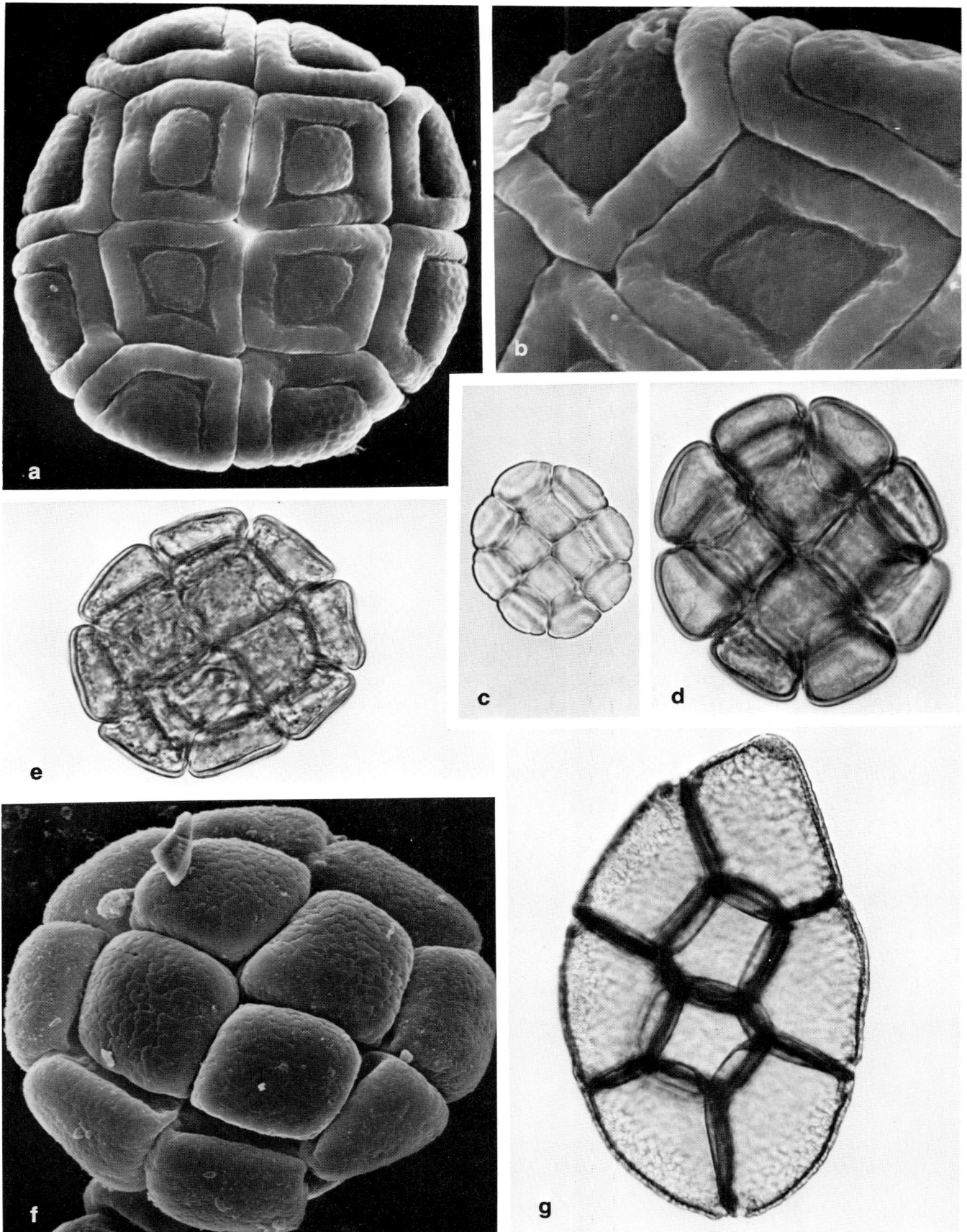

Fig. 56. Pollen polyads of representative Fabaceae, subfamily Mimosoideae. a–b: *Acacia baileyana* (golden mimosa), a = SEM(ad) ×1000, b = SEM(ad) showing sculpturing of polyad surface ×3800; c: *Acacia podalyriifolia* (Queensland silver wattle), LM(s) ×720; d: *Albizia julibrissin* (mimosa or silk tree), LM(s) ×720; e–f: *Pithecellobium guadalupense* (blackbead), e = LM(s) ×720, f = SEM(ad) ×1600; g: *Calliandra haematocephala* (red powderpuff tree), LM(s) ×450.

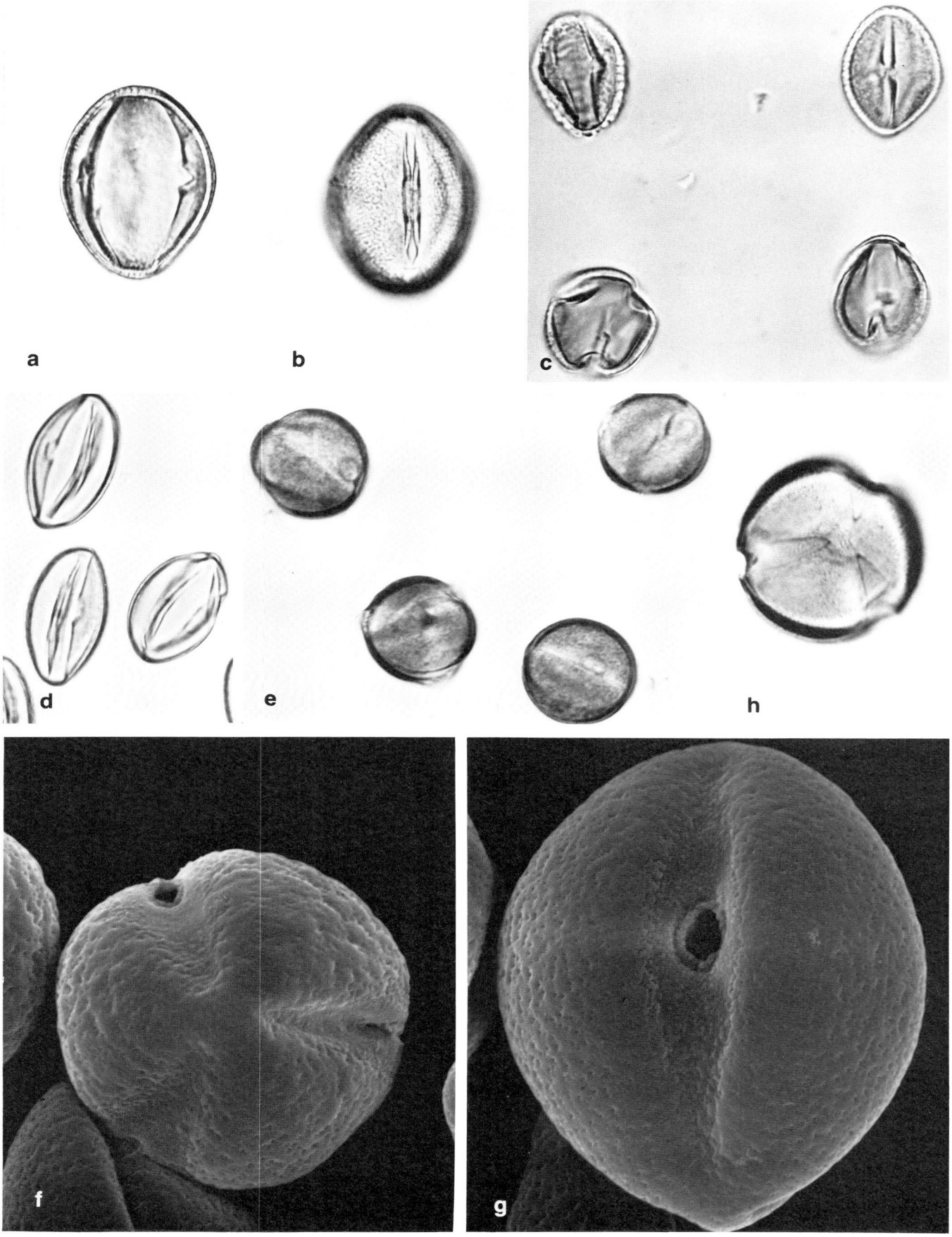

SUBFAMILY CAESALPINIOIDEAE

Confirmation of allergenicity of *Cassia* (senna), *Cercis* (redbud), *Cercidium* (palo-verde), *Gleditsia* (honey locust), and *Gymnocladus* (Kentucky coffee tree), for example, is wanting. Exposure and eventual sensitization would be a local event that could be expected on an infrequent basis (Lewis & Vinay 1979).

Pollen Morphology

Based on *Gymnocladus* (see Fig. 57), grains are prolate (32–44 × 24 μm) to subprolate (30 × 26 μm) and 3-colporate; the colpi long and narrow; the ora lolongate; the sexine ca. 0.9 μm thick, scrobiculate; and the nexine and intine slightly thinner.

SUBFAMILY FABOIDEAE (syn. *Papilionoideae*)

Members of this subfamily with enclosed stamens having strict vector transfer of pollen are not prime candidates for consideration in this volume. Yet allergic reactions are known from pollen of herbaceous plants, particularly among agricultural workers, as discussed in Chapter 3.

FAGACEAE*

Beech Family

The beeches, oaks, and chestnuts are an important, predominantly temperate group of hardwood trees and shrubs (Fig. 58). Their leaves are deciduous or less frequently evergreen, alternate, simple, and entire, serrate, or pinnately lobed. The plants are monoecious with both staminate and pistillate flowers in catkins; the bractlike perianth having four to seven lobes; the staminate flowers having as many as 40 stamens (but usually 6 in oaks, 12 in chestnuts); and the pistillate flowers often in groups of two or three subtended by a 3-lobed bract and consisting of two united carpels, inferior ovary, and three to six styles. The fruit is a 1-seeded nut surrounded or enclosed by an often hardened cupule.

The family includes 8 genera and about 1,000 species. Many are used for hardwood timber, edible fruit, and ornamentals. The beeches, temperate oaks, and the South American *Nothofagus* (southern beeches) are wind-pollinated, but other members of the family, including the tropical oaks, chestnuts, chinquapins, and tanbark oaks, are primarily insect-pollinated, and pollen is only secondarily dispersed by wind if at all.

Flowering and Pollen Aerobiology

Chestnuts flower primarily in the summer, but *Castanea alnifolia* may flower as early as March to June in the southeastern and southcentral regions, and *C. pumila* in September and October in the northeastern and southeastern regions. Flowering of *Castanopsis* is also typical during the summer, although *C. chrysophylla* may extend to September in the West. The eastern native beech flowers during March and April in the South and April and May in the North. The western tanbark oak (*Lithocarpus*) flowers from June to October.

The native oaks are a large and diverse genus divisible into five groups that flower primarily in the late winter, spring, and early summer (Table 5). The most common flowering months overall are April and May. They shed enormous amounts of pollen over prolonged periods and even though oak pollen has been credited with inciting more hay fever cases than most trees, the number is in no way commensurate with the amount of pollen shed, at least in northern regions of the continent (Wodehouse 1971). This probably relates to more moderate levels of allergenicity for oak pollen compared to many other trees, as found in St. Louis (Lewis & Imber 1975b).

The predominant frequency of pollen in the ambient air by region of the continent correlates well with flowering times (Table 6). High frequencies of pollen are caught by aerosamplers during April and May but commonly earlier in the southeastern and southcentral regions and later particularly in the Southwest. Oak pollen may be found perennially in California, although frequency is reduced in July and August and is very small from September through February.

Pollen Morphology (Figs. 59–60)

Grains are prolate subspheroidal or suboblate; the amb rounded, triangular, or trilobed and 3-colporate, 3-colporoidate, or 3-colpate; the colpi long and narrow; the ora, when present, circular, lolongate, or lalongate, sometimes with thickened margins (*Castanea, Fagus*); and the opercula granular. The sexine is tectate, with coarsely irregularly shaped verrucae, or scabrate (*Quercus*), verrucate-pilate (*Fagus*), or foveolate (*Castanopsis*), 0.6–1.2 μm thick; the nexine usually as thick as the sexine; an the intine thin (0.5–1.2 μm), usually somewhat thickened below the apertures (*Quercus*, 1.5–2.5 μm).

Fig. 57. Pollen of representative Fabaceae, subfamily Mimosoideae (monads) and Caesalpinioideae. a–b: *Desmanthus illinoensis* (prairie mimosa), LM(s) equatorial views showing 2 (a) and 1 (b) apertures; c: *Gymnocladus dioica* (Kentucky coffee tree), LM(a) ×720; d–g: *Prosopis glandulosa* (glandular mesquite), d = LM(a) ×720, e = LM(s) ×720, f = SEM(a) polar view ×3200, g = SEM(a) equatorial view showing 1 aperture ×3600; h = *Leucaena leucocephala* (lead tree), LM(a) near polar view ×720.

TABLE 5. Flowering by month of *Quercus* (oaks), where native, by major groups.

GROUP	REGION							
	nNE	sNE	SE	sFL	NC	SC	NW	SW
True white oaks	5–6	4–5	2–4 (–5)	2–3	5–6	2–4	4–6	4–6
Chestnut oaks	5–6	4–5	4					
True red oaks	5–6	4–5 (–6)	3–5	2–3		3–4	4–6	4–5
Willow oaks		4–5	3–5	2–3		3–5		
Live oaks			2–4	2–4		2–4		3–5

CASTANEA — chestnut

A number of native and cultivated species of *Castanea* are found in North America, particularly the Southeast, but, as entomophily is characteristic, limited pollen is wind-dispersed and its role in pollinosis is minor (e.g., moderate allergenicity in New Jersey where five atopic patients skin tested 2+) (C. Dubovy, personal communication). Nevertheless, pollen of the European chestnut (*C. sativa*) may become windborne and transported long distances in Switzerland (Leuschner & Boehm 1981). The once common American chestnut (*C. dentata*) is now rare because of chestnut blight, but small flowering trees can still be found in its native range, and the species has been widely planted west of the Rocky Mountains where it is still free of blight.

Grains of *Castanea* are small (15–18 × 13–16 μm), prolate, and tectate.

CASTANOPSIS — chinquapin

Like *Castanea*, the two western species of chinquapins are insect-pollinated. Limited pollen may be wind-dispersed as found in Taiwan (Chen & Huang, 1980), which may result in localized cases of pollinosis.

Pollen of *Castanopsis* is similar to that of *Castanea*, except that it is subprolate and its sculpturing foveolate.

FAGUS — beech

Two species are found frequently in eastern North America, one American, the other European in origin. Even though wind-pollinated, reports of pollen shed are infrequent in this continent, in part because its pollen is easily mistaken for the enormously more common oak pollen. Probably for the same reason, few cases of pollinosis are attributed to beech and it is considered of secondary allergenic importance, at least in relation to oak. Even so, it is a source of allergens (average 3+ on skin testing) in New Jersey, where some patients are treated for beech pollinosis (C. Dubovy, personal communication).

Pollen of *Fagus* is distinct because of its large size (31–52 × 32–54 μm) and densely verrucate-pilate surface.

LITHOCARPUS — tanbark oak

A large genus native to southeastern Asia, only *Lithocarpus densiflorus* is indigenous to the western part of the continent. It is found in wooded, dry slopes in redwood and mixed evergreen forests at elevations between 700 and 2,500 m. Pollination is primarily entomophilous and allergenicity is unknown.

TABLE 6. *Quercus* (oak) aeropollen frequency in regions of North America based on percentage of pollen captured, by month (1974–78).

REGION	MONTH											
	1	2	3	4	5	6	7	8	9	10	11	12
nNE			1.0	7.4	34.7	3.4						
sNE			6.7	26.6	27.0	4.0						
SE	0.5	3.3	21.6	35.6	4.3							
NC			0.4	23.9	20.0	0.5						
SC		14.7	24.1	10.0	8.1	1.8						
NW				6.1	7.6	0.9						
SW	0.1	0.7	8.4	30.0	19.6	16.8	2.8	1.4	0.3	0.9	0.1	0.7
sCA	4.4	23.8	29.6	38.4	22.5	0.4						

Fig. 58. Morphology of representative Fagaceae showing branches with staminate catkins and leaves. a: *Castanea dentata* (American chestnut), × ⅓; b: *Fagus sylvatica* (European beech), × ½; c: *Quercus alba* (white oak), × ½; d: *Quercus prinoides* (chinquapin or dwarf-chestnut oak), × ½; e: *Quercus marilandica* (blackjack oak), × ½; f: *Quercus palustris* (pin oak), × ½; g: *Quercus phellos* (willow oak), × ½; h: *Quercus virginiana* (live oak), × ½.

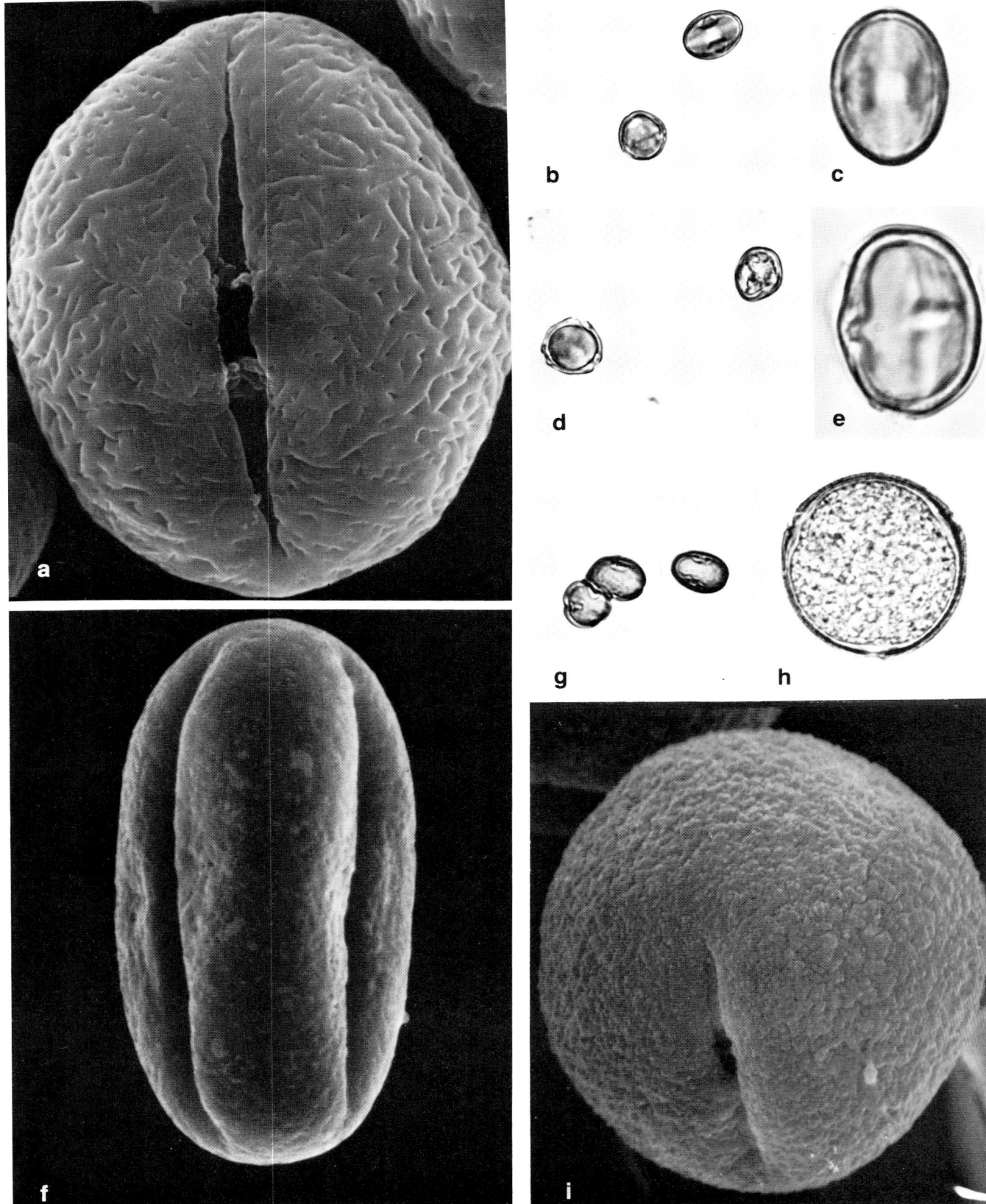

Fig. 59. Pollen of representative Fagaceae (*Castanea-Fagus*). a–c: *Castanopsis chrysophylla* (giant or golden chinquapin), a = SEM(a) equatorial view showing 1 aperture ×7600, b = LM(s) ×720, c = LM(s) equatorial view ×1830; d–f: *Castanea sativa* (European or Spanish chestnut), d = LM(s) ×720, e = LM(a) equatorial view ×1830, f = SEM(a) equatorial view showing 2 apertures ×7600; g: *Lithocarpus densiflorus* (tanbark oak), LM(s) ×720; h: *Fagus sylvatica* (European beech), LM(s) polar view ×720; i: *Fagus grandifolia* (American beech), SEM(a) oblique-equatorial view showing 1 aperture ×3600.

TREES AND SHRUBS

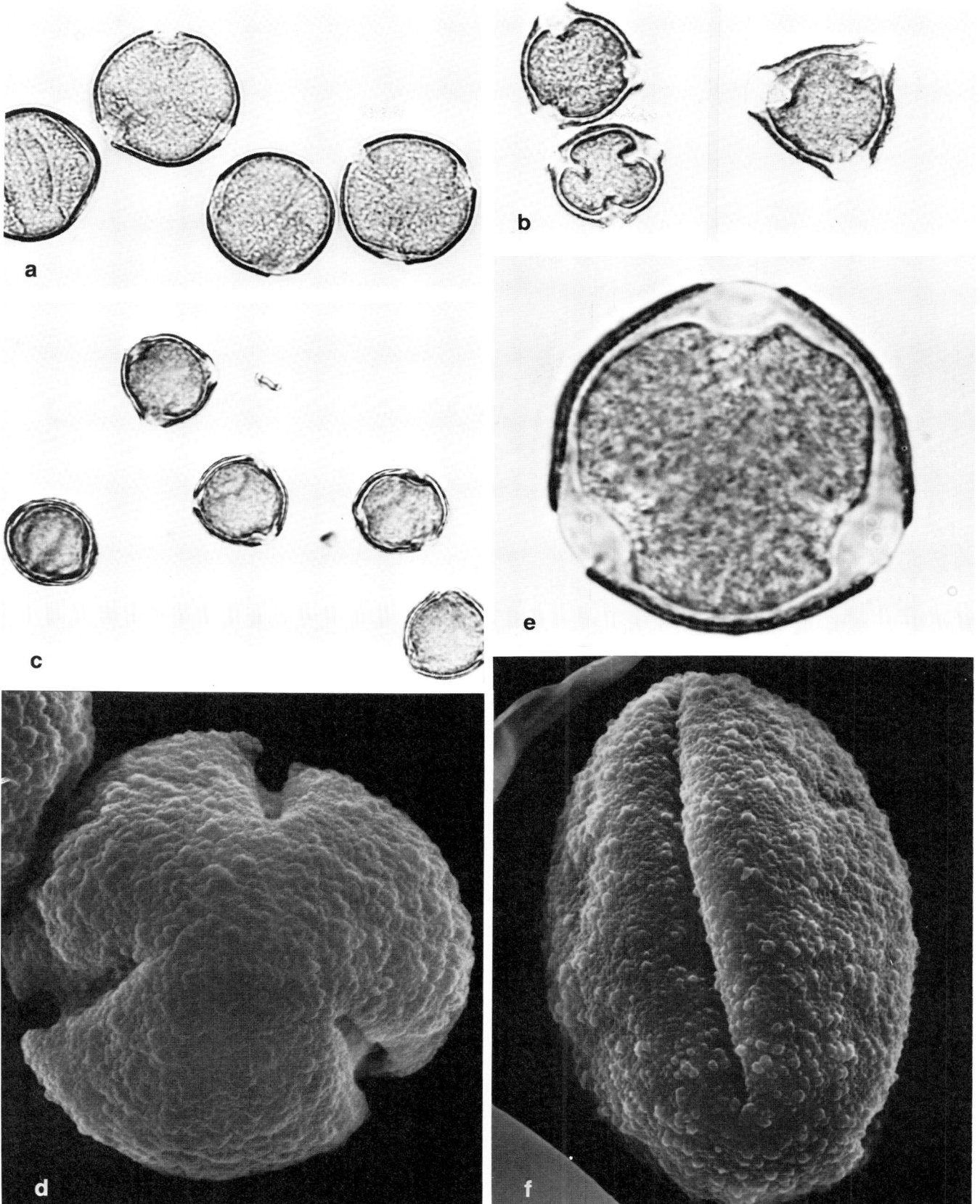

Fig. 60. Pollen of *Quercus*. a: *Q. alba* (white oak), LM(s) ×720; b: *Q. marilandica* (blackjack oak), LM(s) equatorial views ×720; c–d: *Q. virginiana* (live oak), c = LM(s) ×720, d = SEM(a) polar view ×4200; e: *Q. laurifolia* (laurel oak), LM(s) polar view showing thickened intine below apertures ×1830; f: *Q. phellos* (willow oak), SEM(ad) equatorial view showing 1 aperture ×3600.

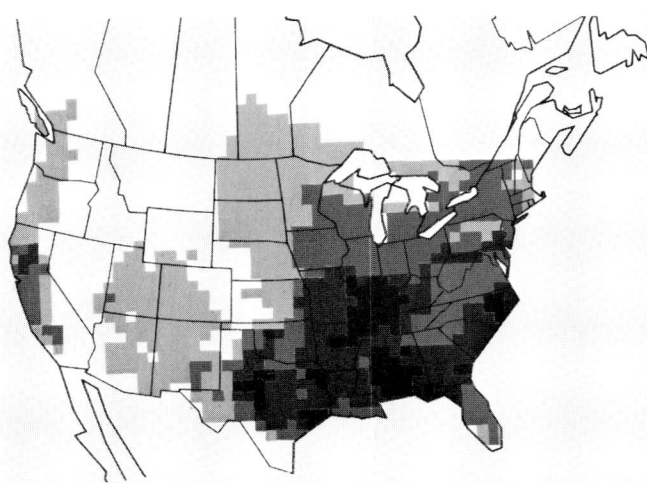

Fig. 61. Generalized composite distribution of 12 indigenous species of true white oaks (☐ 1 sp., ▨ 2 spp., ▧ 3 spp., ■ 4–5 spp.).

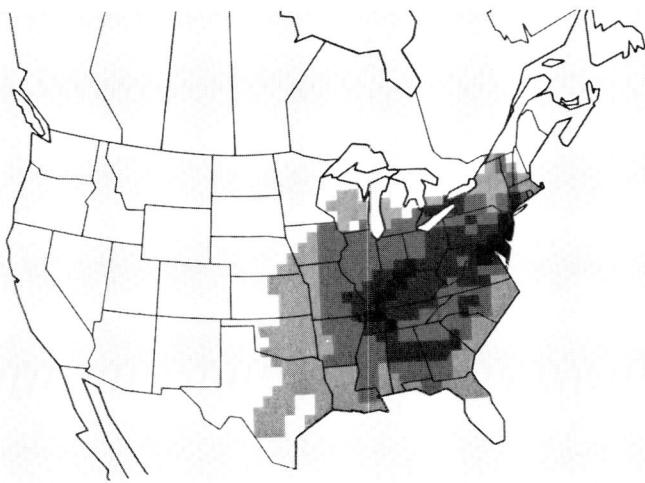

Fig. 62. Generalized composite distribution of 4 indigenous species of chestnut oaks (☐ 1 sp., ▨ 2 spp., ▧ 3 spp., ■ 4 spp.).

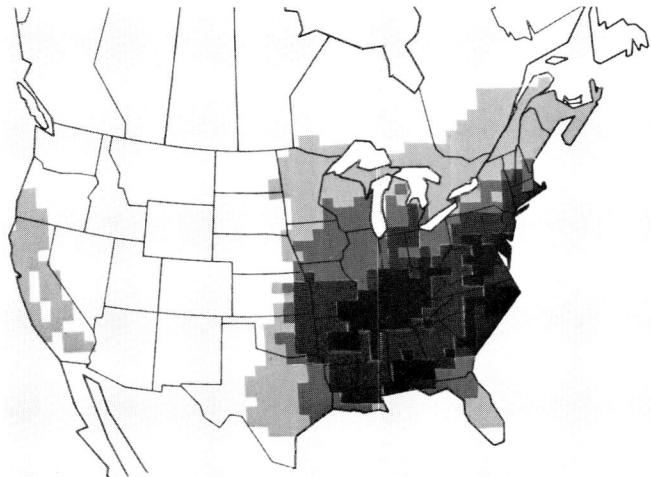

Fig. 63. Generalized composite distribution of 12 indigenous species of true red oaks (☐ 1–2 spp., ▨ 3–4 spp., ▧ 5–6 spp., ☐ 7–8 spp.).

Pollen of *Lithocarpus* resembles that of *Castanea* and *Castanopsis* (prolate and 16–18 × 10–12 μm).

QUERCUS *oak*

A large genus in North America, the oaks are readily subdivided into five natural groups that may correspond to allergenic classes. Oak pollen is responsible throughout the continent for many cases of moderate to moderately severe pollinosis, and it is invariably considered a potential incitant of tree inhalant allergy. The groups and their species are:

(1) True white oak (Fig. 61)—*Quercus alba, Q. chapmanii, Q. douglasii, Q. durandii, Q. gambelii, Q. garryana, Q. havardii, Q. lobata, Q. lyrata, Q. macrocarpa, Q. mohriana,* and *Q. stellata.* The group is most prominent in the southeastern half of the continent.

(2) Chestnut oaks (Fig. 62)—*Quercus bicolor, Q. michauxii, Q. prinus,* and *Q. prinoides.* The chestnut oaks are found throughout the eastern U.S., most frequently in the mid-Atlantic and Ohio River valley regions.

(3) True red oaks (Fig. 63)—*Quercus arkansana, Q. coccinea, Q. ellipsoidalis, Q. falcata, Q. ilicifolia, Q. kelloggii, Q. laevis, Q. marilandica, Q. nuttallii, Q. palustris, Q. rubra, Q. shumardii,* and *Q. velutina.* The group is common in eastern North America, with many species found from the mid-Atlantic region to the southeastern United States.

(4) Willow oaks (Fig. 64)—*Quercus imbricaria, Q. incana, Q. laurifolia, Q. nigra,* and *Q. phellos.* The willow oaks are very prominent from the southeastern United States to central Florida.

(5) Live oaks (Fig. 65)—*Quercus agrifolia, Q. arizonica, Q. chrysolepis, Q. dumosa, Q. grisea, Q. hypoleucoides, Q. myrtifolia, Q. turbinella, Q. virginiana,* and *Q. wislizenii.* The live oaks are particularly well developed in California and Arizona, but they also extend to the Pacific Northwest and to the southeastern coastal states.

Unfortunately, there are no studies relating these major oak groups to frequency or level of pollen allergenicity. A hint that differences may exist comes from California and the Gulf Coast states, where species of live oaks are common and where many reports of allergenicity involving strongly positive 4+ skin tests correlate with serious pollinosis.

Pollen of *Quercus* is intermediate in size (22–36 × 19–39 μm) and is either 3-colpate or 3-colporoidate. True white oak pollen is 3-colpate (*Q. garryana,* 3–4-colpate) with a verrucate sexine; chestnut oak is 3-colporoidate and verrucate; true red oak is either 3-colpate or 3-colporoidate, depending on species, and verrucate; willow oak is 3-colpate and scabrate; and live oak is also scabrate but 3-colporoidate.

TREES AND SHRUBS

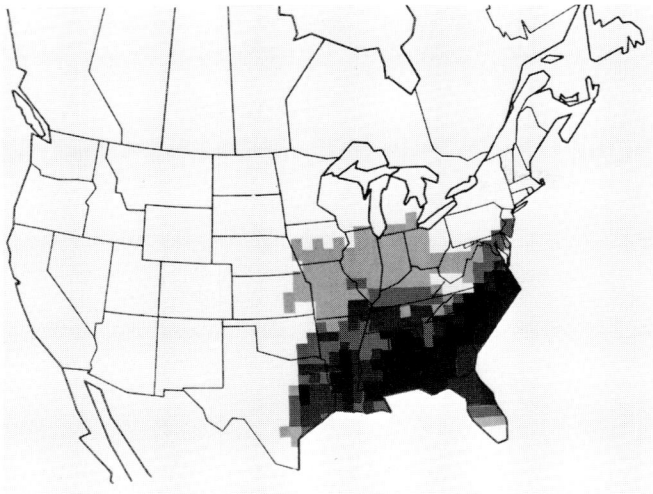

Fig. 64. Generalized composite distribution of 5 indigenous species of willow oaks (☐ 1 sp., ▨ 2 spp., ▥ 3 spp., ■ 4–5 spp.).

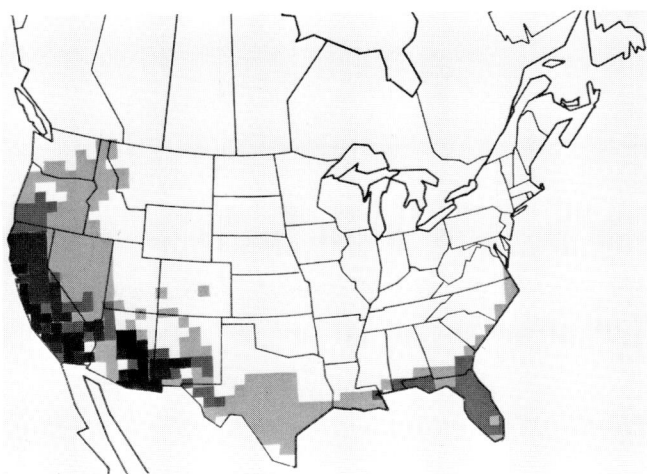

Fig. 65. Generalized composite distribution of 9 indigenous species of live oaks (☐ 1 sp., ▨ 2 spp., ▥ 3 spp., ■ 4–5 spp.).

GARRYACEAE
Silk-tassel Family

A small family of one genus, *Garrya* (Fig. 66) species pollinate by wind in coastal regions of California and Oregon from January to April, and May to August from western Texas to Arizona. This endemic family is related to the dogwoods. Allergenicity is not known, but as abundant pollen is dispersed, it may be an offender in areas where plants are found commonly.

The genus predominates in the southwestern region but extends to southern Washington and western Texas (Fig. 67).

Fig. 66. *Garrya elliptica* (wavy-leaved silk-tassel) shrub with staminate catkinlike inflorescences (a, ×⅔) and inflorescence with staminate flowers (b, 2½).

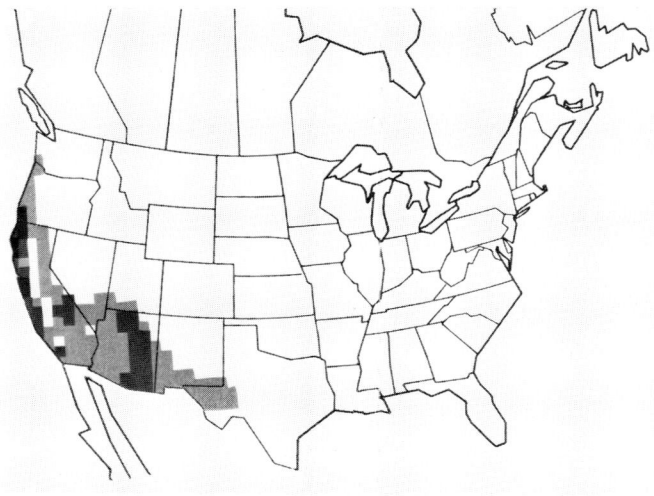

Fig. 67. Generalized composite distribution of 5 species of *Garrya* (☐ 1 sp., ▨ 2 spp., ■ 3 spp.).

Pollen Morphology (Fig. 68)

Grains are oblate-spheroidal (31–33 × 32–37 µm) or suboblate (ca. 30 × 37 µm); the amb triangular with convex sides and 3–4-colporate; the colpi short (ca. 15 × 1.6 µm) with thickened margins and pointed ends; the ora lalongate but faintly demarked, ca. 3.2 × 1.6 µm, with narrow thickened ends due to incrassate nexine; and the opercula granular. The sexine is thin, ca. 0.8 µm, reticulate, with small, polygonal lumina, 0.5–0.8 µm in diameter; the nexine thick, ca. 2.4 µm, particularly thickened at the apertures; and the intine ca. 2.4 µm thick, and also thickened to 3.5 µm below the apertures.

HAMAMELIDACEAE
Witch Hazel Family

Pollen allergenicity among the witch hazels (*Hamamelis*) and certainly the sweet gum (*Liquidambar*) (Fig. 69) is anticipated, for the latter sheds abundant airborne pollen in the spring, beginning in March in Florida and Texas and as late as May in New York. Heretofore, reports of involvement have been very rare, but we now record 325 patients who were suffering from spring pollinosis from Alabama, California, and Florida responding an average of 2.7+ to skin tests using sweet gum extract. American sweet gum (*L. styraciflua*) pollen is, therefore, moderately allergenic and should be considered a minor offending spring-flowering tree in inhalant allergy.

Pollen Morphology (Fig. 70)

Pollen of *Liquidambar* is distinguished by its spheroidal shape, about 34–43 µm in diameter; pantoporate apertures with 18–20 globally distributed, circular pores, 4.8–8.0 µm in diameter, with an inner uneven margin and outer thickened margin; and a densely granular or microverrucate operculum. The sexine is ca. 0.9 µm thick, slightly thickened at the pores, and punctitegillate; the nexine as thick as the sexine and thickened at the pores; and the intine 0.8 µm thick, thickened to ca. 2.2 µm below the apertures. Pollen of *Hamamelis* is smaller (19–22 × 14–19 µm), ±prolate, 3-colpate, and reticulate.

HIPPOCASTANACEAE
Horse Chestnut Family

Horse chestnuts and buckeyes (*Aesculus*) (Fig. 71) are primarily, if not exclusively, zoophilous, but because of the large number of flowers with exposed anthers, occasional airborne pollen has been captured in both Europe (Leuschner & Boehm 1981; Nilsson & Persson 1981) and North America (Lewis & Vinay 1979). Limited allergic reactions in the spring can be expected, but at present allergenicity is not known.

Pollen Morphology (Fig. 72)

Grains are prolate, ca. 26–28 × 21 µm; the amb triangular with convex sides and 3-colporate; the colpi long, ca. 24 × 2.5–5.0 µm, with thickened margins and occasionally syncolpate at one pole; the ora ±circular, 4.5 × 5 µm, with thickened margins; and the opercula densely granular. The sexine is striato-reticulate, ca. 0.8 µm thick; the lumina small (<0.5 µm); the nexine as thick as the sexine; and the intine ca. 0.8 µm, thickened to ca. 2.2 µm below the apertures.

JUGLANDACEAE*
Walnut Family

Hickories (*Carya*) and walnuts (*Juglans*) are easily recognized: they are deciduous trees with pinnately compound, alternate, aromatic leaves, having unisexual flowers, the staminate ones in pendulous catkins of the previous year, the pistillate ones in small erect spikes formed on the new branches. Stamens number from only 3 to as many as 100, pollination is by wind, and pistils consist of two to three united carpels with an inferior ovary and two often plumose stigmas. The perianth is typically 4-lobed but is often reduced or wanting by abortion. The fruit is usually a nut (Fig. 73).

This is a small family mainly of north temperate and subtropical distributions, the plants being important for edible nuts and oil (walnut, hickory, pecan), timber, and ornamentals. In North America the native hickories are limited in distribution to the East, with a preponderance of species in the Mississippi

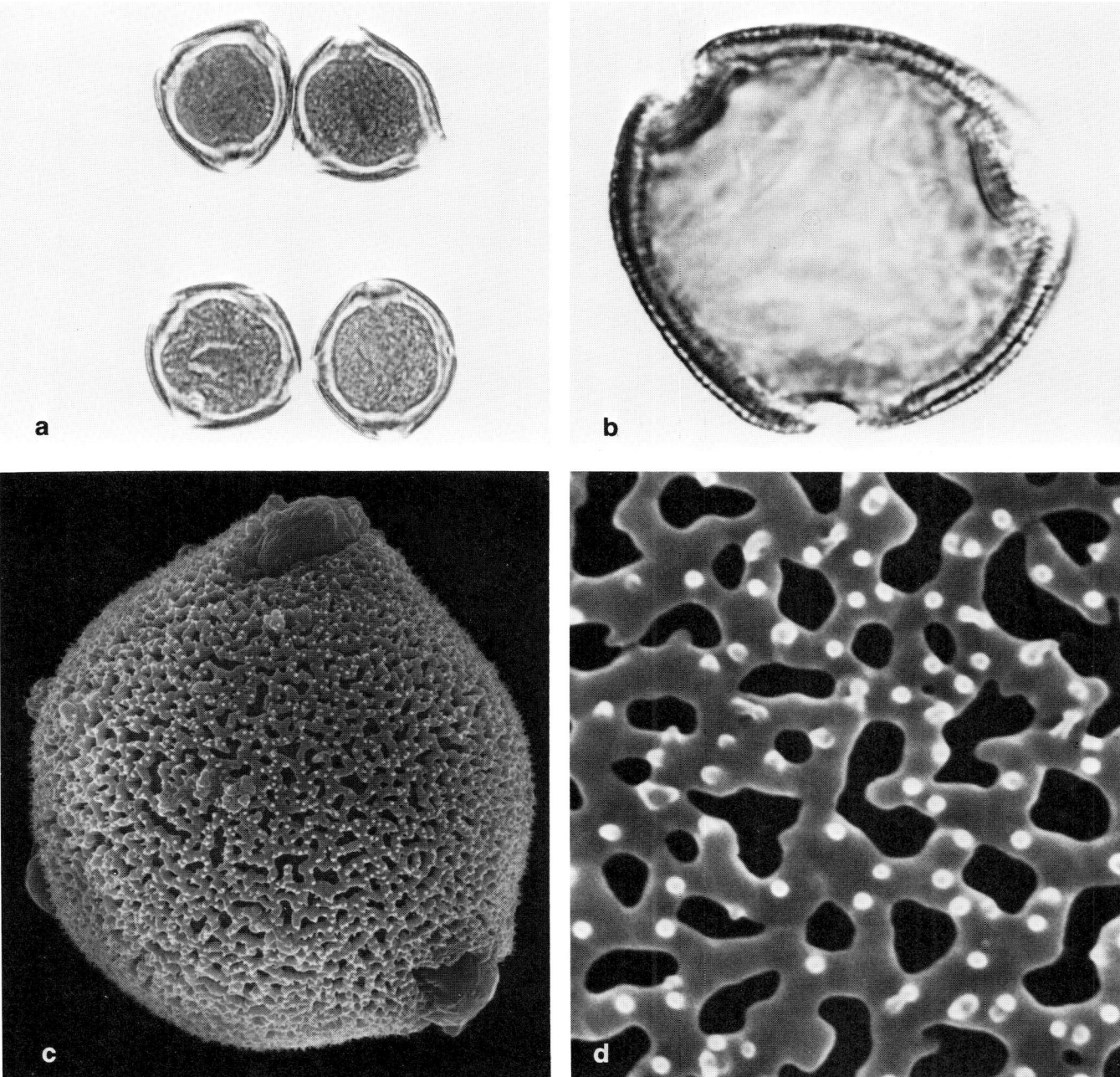

Fig. 68. Pollen of *Garrya elliptica*. a = LM(s) ×720, b = LM(a) ×1830, c = SEM(ad) oblique view ×3000, d = SEM(ad) showing sexine reticulation ×16000.

valley from Louisiana to Illinois and Missouri (Fig. 74). *Juglans* is also indigenous to eastern North America as well as to the southwestern region, but few species are known from a single locality (Fig. 75).

Flowering and Pollen Aerobiology

The hickories and walnuts of eastern North America flower primarily in May and June in their northern ranges, April and May in midlatitude areas, and as early as January and February along the Gulf Coast and in Florida. In the West walnuts flower principally in April and May.

Members of the family shed enormous amounts of pollen, but because of its large size, the pollen is not well adapted to long-distance dispersal (Wodehouse 1971). Nevertheless, significant quantities of hickory pollen have been captured on atmospheric slides throughout the northeastern and southern regions of the continent, particularly in May and June in the Northeast and in February and May in the Southeast (Table 7).

Fig. 69. Morphology of representative Hamamelidaceae. A: *Hamamelis mollis* (Chinese witch-hazel) flower clusters, ×1½; b: *Liquidambar styraciflua* (sweet gum) branch showing globose heads of staminate flowers and one of pistillate flowers (pendant), ×¾.

Pollen Morphology (Fig. 76)

Grains are isopolar, heteropolar, or paraisopolar, 10–31 × 11–68 μm; the amb triangular, rounded-triangular, polygonal, or circular and 3(–4)-por(or)ate (*Carya*) or polypor(or)ate (*Juglans, Pterocarya*); and the pores 0.5–4.0 μm in diameter, circular to meridionally elongate, aspidate or flattened-conical, located on the equator and distal (but not proximal) face. The exine is less than 3 μm thick; the sexine thicker than the nexine, tectate, and spinulose; the spinules shorter than 0.5 μm and evenly spaced; and the intine 0.5–3.0 μm thick, much thickened in *Carya* and *Juglans* beneath the apertures, forming a lens-shaped oncus to 2.4–3.5 × 9–11 μm. For amplification of pollen morphology, see Stone and Brome (1975).

The 3-porate grains of *Carya* are easily distinguished from the polyporate grains of *Juglans*.

Allergenicity

Based on skin test reactions of many atopic patients in St. Louis, *Carya laciniosa* (shellbark hickory) pollen extracts proved among the most allergenic of tree pollen (±equivalent in level and frequency of allergenicity with box-elder and willow). Those of *Juglans nigra* (black walnut), however, fell among a second group of trees that were only moderately allergenic. Where cultivated in orchards and as street trees, *C. illinoensis* (pecan) is a cause of widespread inhalant allergy, often ranking next in importance to ragweed in the Southeast (Wodehouse 1971). Other species of hickory are scattered naturally in woods away from population centers and, as exposure to their pollen is much reduced, they are often considered of secondary importance (Sellers 1935).

In the eastern part of the continent exposure to pollen of *Juglans* is in general less frequent than to *Carya* pollen, but in the West this level of exposure is reversed, and correlated with this reversal is a heightened significance of *Juglans* allergens. In California and Oregon, for example, ten allergists reported a moderately to strongly positive reaction (2+–4+) to *Juglans* pollen extracts among 766 patients; so acute were the symptoms of pollinosis for many individuals that walnut allergens were considered among the most serious offenders of inhalant allergy. Apparently, both cultivated (*J. regia*) and indigenous species were involved. One allergist reported strong cross-reactions with *Carya* pollen extracts.

All other genera in the family release airborne pollen in sufficient quantities to be offenders in inhalant allergy wherever planted. Only *Platycarya* and *Pterocarya* are occasionally found in North America.

LAURACEAE

Laurel Family

Primarily a tropical family of aromatic trees and shrubs, the Lauraceae have a few representatives in North America (Fig. 77). Vector pollination is the rule so that pollen dispersal by wind is limited to incidental occurrences (Chen & Huang 1980). *Cinnamomum camphora* (camphor), for example, has been considered an offender in southern California, where the species is planted in parks and as avenue trees (Small & Small 1946). Other plants in the family that might be similarly implicated where close association is possible are *Lindera* (spicebush), *Nectandra* (Jamaica nectandro), *Persea* (red or swamp bay), *Sassafras,* and *Umbellularia* (California bay or laurel).

Pollen Morphology (Fig. 78)

Grains are spheroidal, 24–50 μm in diameter, and inaperturate. The sexine is 0.4–0.9 μm thick and provided with short spinules or spinuloid processes that are dense or scattered over the surface; the nexine thinner than the sexine; and the intine 2–4 μm thick.

Grains within the family are basically similar. Those of *Cinnamomum* and *Lindera* are small (24–28 μm) and

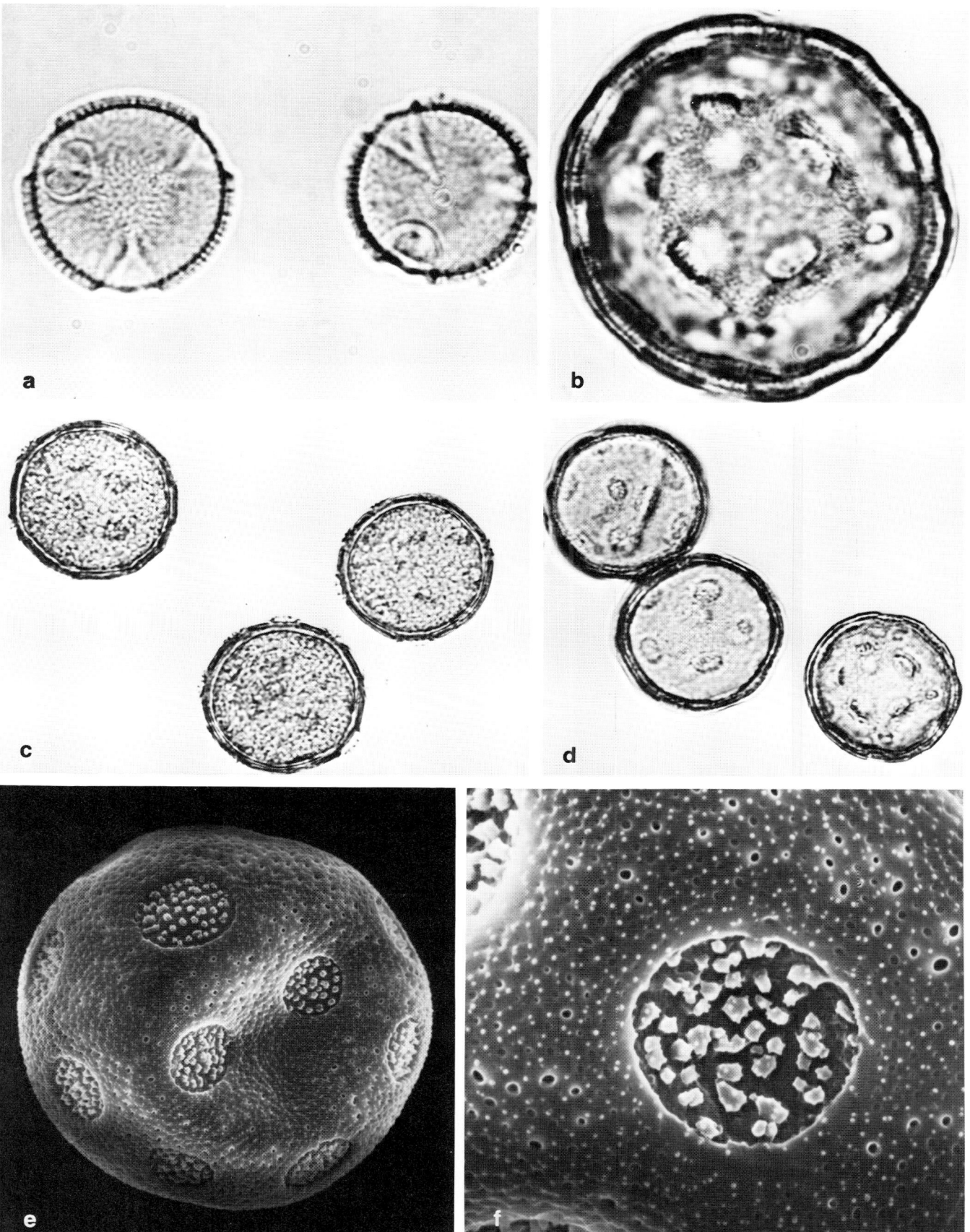

Fig. 70. Pollen of *Hamamelis* and *Liquidambar*. a: *H. mollis*, LM(a) polar views ×1830; b–f: *L. styraciflua*, b = LM(a) ×1830, c = LM(s) ×720, d = LM(a) ×720, e = SEM(ad) ×2100, f = SEM(ad) showing pore morphology and sexine surface ×5600.

Fig. 71. *Aesculus glabra* (fetid, or Ohio, buckeye) branch showing leaves and inflorescence, ×½.

those of *Persea, Sassafras,* and *Umbellularia* are larger (36–50 μm).

LEITNERIACEAE
Corkwood Family

A family of only one species, *Leitneria floridana* (Florida corkwood) is found in swampy areas of the South. It is dioecious with staminate catkins. Pollination is by wind beginning in March; allergenicity is not known. Its closest affinity is with the Myricaceae, plants known to be allergenic.

Pollen Morphology (Fig. 79)

Grains are suboblate to oblate-spheroidal, 26–28 × 29–32 μm; the amb rounded-triangular to circular and 3(–4–6)-colporate; the colpi long and narrow, ca. 13 × 0.5–1 μm; the ora large, ca. 3.2 × 9 μm, lalongate; and the opercula granular. The sexine is 0.7 μm thick, reticulate; the lumina small (<0.5 μm), polygonal, scabrate; the nexine as thick as the sexine, slightly thickened at the pores; and the intine 0.8 μm thick, thickened to 2.2 μm below the apertures.

MAGNOLIACEAE
Magnolia Family

Magnolia (Fig. 80) and *Liriodendron* (tulip tree) are important ornamentals, particularly of eastern North America. They flower predominantly in April and May, some extending until August. Although insect-pollinated, airborne pollen of *Magnolia* has been iden-

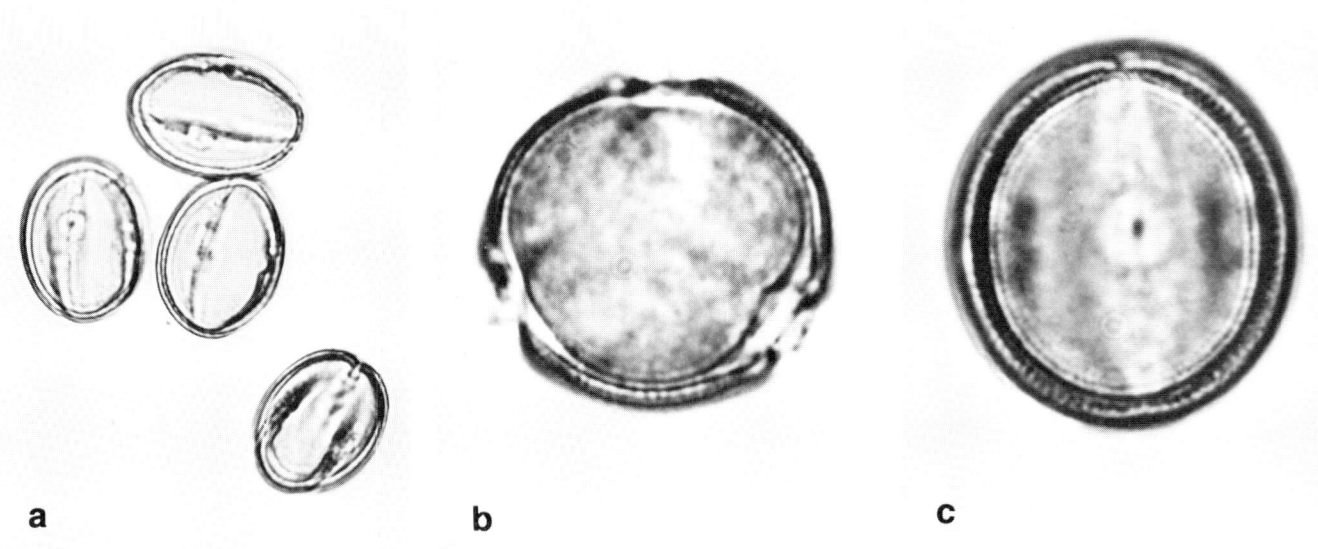

Fig. 72. Pollen of *Aesculus glabra*. a = LM(a) ×720, b = LM(s) polar view ×1830, c = LM(s) equatorial view ×1830.

Fig. 73. Morphology of representative Juglandaceae. a–b: *Carya* x *laneyi* branch with leaves and staminate catkins (a, ×⅖) and developing fruit (b, ×½); c: *Pterocarya* × *redheriana* branch with leaves and staminate catkins, (×⅗); d: *Juglans regia* (English walnut) branch with leaves and staminate catkins, ×¾.

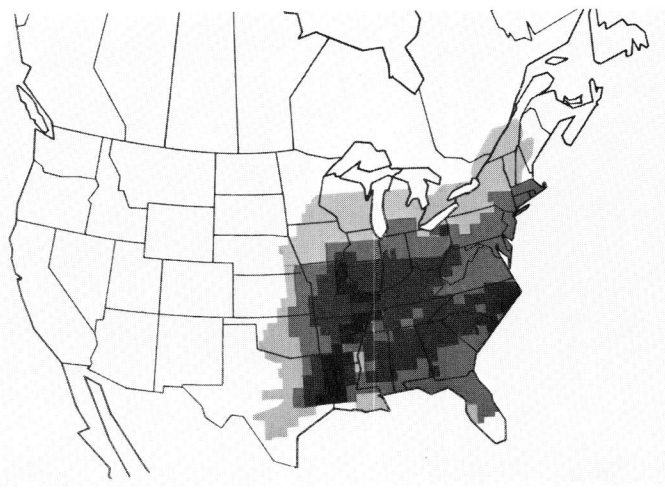

Fig. 74. Generalized composite distribution of 11 indigenous species of *Carya* (☐ 1–2 spp., ▨ 3–4 spp., ▨ 5–6 spp., ■ 7–8 spp.).

TABLE 7. *Carya* (hickory, pecan) aeropollen frequency in regions of North America based on percentage of pollen captured, by month (1974-78).

REGION	MONTH						
	1	2	3	4	5	6	7
nNE			0.2	0.6	2.7	1.5	0.1
sNE			1.8	2.5	11.0	14.7	4.6
SE	1.4	7.6	1.0	1.2	7.6	1.0	
NC			0.2	0.6	0.8		
SC				1.1	5.2		
SW				0.8	0.5		

tified in small quantities in California and Britain, and of *Liriodendron* in Missouri and Oklahoma (Lewis & Vinay 1979). Using pollen extracts, moderately positive skin test reactions (2+) among some atopic patients were obtained in New Jersey (Lewis & Vinay 1979); allergenic effects of the pollen can be expected only infrequently and locally.

Pollen Morphology (Fig. 81)

Grains are heteropolar, bilateral, cymbiform, and 1-sulcate; the sulcus long and narrow, ±extending the entire length, and with a smooth or wavy margin. The sexine is punctitegillate, supported by slender bacula and provided with scattered tubercles (*Liriodendron*), or tectate with a corrugated sculpturing (*Magnolia*), 1.2–1.3 μm thick; the nexine thinner (ca. 0.5 μm) than the sexine; and the intine thin (ca. 0.5 μm).

Grains of *Liriodendron* are larger (61 × 32 × 34 μm) than those of *Magnolia* (40 × 24 × 17–20 μm).

MORACEAE*

Mulberry Family

Trees and shrubs centered in the tropics, the mulberries are nevertheless sometimes temperate, as *Broussonetia*, *Maclura* (Osage orange), and *Morus* (mulberry) (Fig. 82). They are distinguished by their milky latex, alternate, simple leaves that are deciduous (in North America), unisexual flowers arranged in heads or catkins with four sepals and four stamens, and a compound pistil of two carpels and two styles. The fruit is often an aggregation of druplets.

Flowering and Pollen Aerobiology

Most representatives in North America are wind-pollinated and shed enormous amounts of pollen, often explosively, as in the paper mulberry, whose "rather thick filaments lie in the bud like a bent watch-spring, and suddenly straighten when the perianth expands, thus scattering the pollen into the air" (Knuth 1909). This plant is sometimes called the "smoking mulberry" because of the clouds of light pollen that suddenly emit from the tree during spring pollination. Osage orange is amphiphilous, being pollinated by insects and the wind.

Typical of trees and shrubs, the majority of mulberries flower in the spring (March-May) throughout the continent, although extending to June in northern latitudes. Osage orange usually flowers somewhat later, from April or May to June (Table 8).

Pollen Morphology (Fig. 83)

Grains are ±spheroidal, 11–25 μm in diameter, and 2–3(–4)-porate; the pores usually circular, 1.5–2.0 μm in size but larger in *Morus* (2.5–4.0 μm), sometimes slightly aspidate; the margin in some genera slightly thickened (*Broussonetia*) or distinctly annulate (*Maclura*); the annulus ca. 1.8 μm wide; and the opercula centrally thickened and often protruding. The sexine is tectate and usually scabrate, the surface appearing rough and thin (0.5–0.8 μm); the nexine as thick as or thinner than the sexine, slightly thickened at the

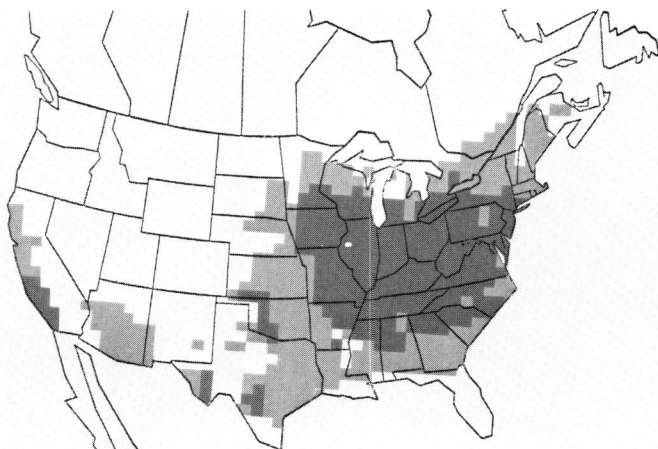

Fig. 75. Generalized composite distribution of 6 indigenous species of *Juglans* (☐ 1 sp., ▨ 2 spp.).

TREES AND SHRUBS

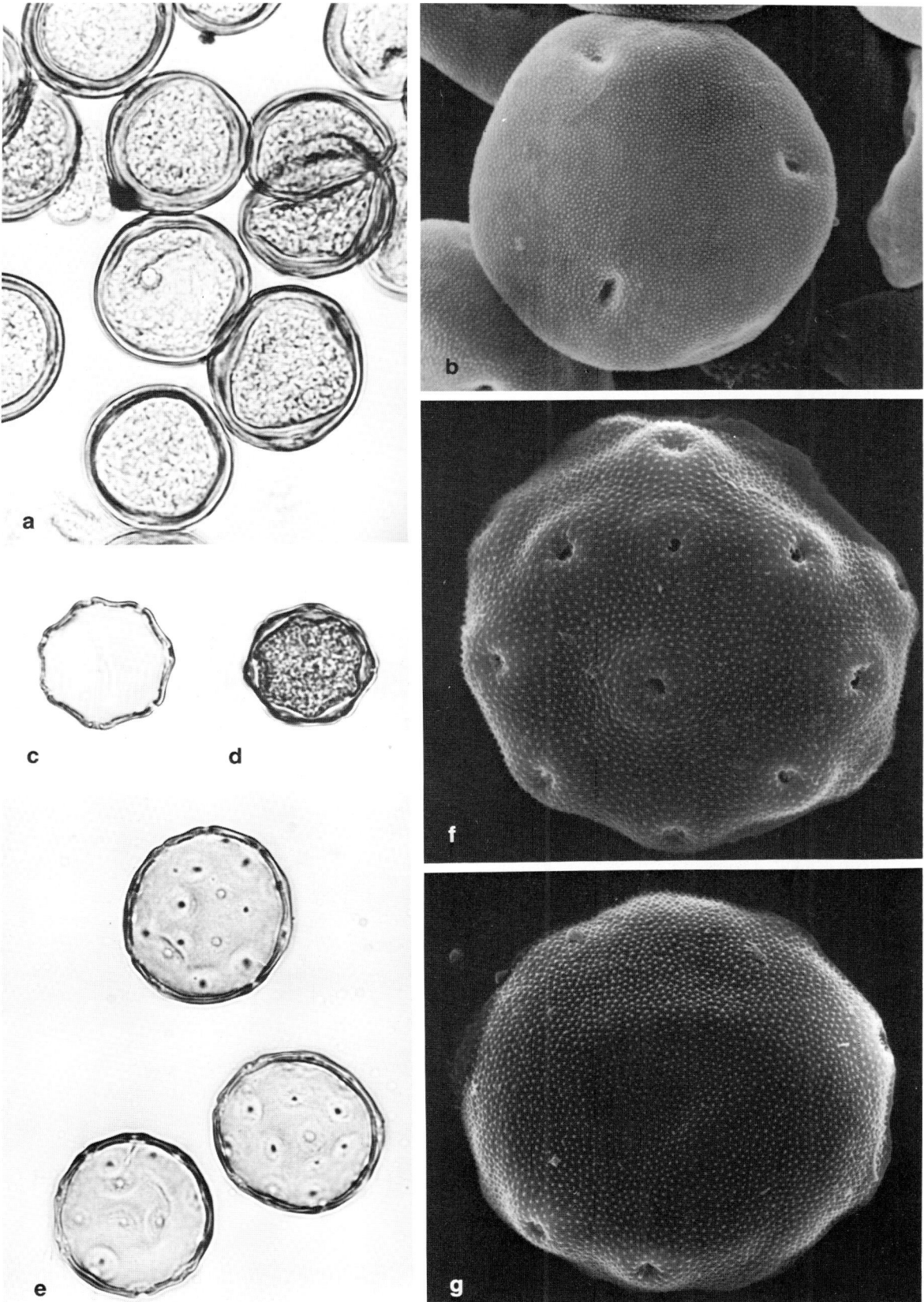

Fig. 76. Pollen of representative Juglandaceae. a: *Carya × laneyi*, LM(s) ×720; b: *Carya ovata* (shagbark hickory), SEM(a) showing the 3 distal pores ×1800; c: *Juglans nigra* (black walnut), LM(a) ×720; d–e: *Pterocarya × redheriana*, d = LM(s) ×720, e = LM(a) ×720; f–g: *Juglans cinerea* (butternut), f = SEM(a) distal face with pores ×2800, g = SEM(a) proximal face without pores although equatorial pores can just be observed ×2800.

65

Fig. 77. Morphology of representative Lauraceae. a: *Lindera benzion* (spicebush) branch with emerging flowers, ×1½; b: *Sassafras albidum* (sassafras) branch with flowers, ×1½; c: *Umbellularia californica* (California bay or laurel) branch with flowers, ×1½.

pores; and the intine thin except below the pores, where characterized by an oncus to 5.6 × 7.0 μm in thickness.

Grains of *Cudrania* (ca. 11 × 13 μm) and *Broussonetia* (ca. 14 × 15 μm) are small, while those of *Maclura* (ca. 21 × 25 μm) and *Morus* (ca. 21 × 17 μm) are somewhat larger. Aperture frequency also varies: *Broussonetia* and *Cudrania* are 2-porate, *Maclura* is (2–)3(–4)-porate, and *Morus* is 2–3(–4)-porate. Similar pollen is found in the Cannabaceae, Ulmaceae, and Urticaceae.

Allergenicity

Where common, the wind-pollinated mulberries are capable of causing severe pollinosis (Wodehouse 1971). Cases documenting inhalant allergy in southern California elicited by pollen of *Broussonetia papyrifera* (paper mulberry) and *Morus alba* (white mulberry) (Targow 1971) and in Oklahoma and Washington, D.C., by *B. papyrifera* (Bernton 1928; Balyeat & Rinkel 1931) illustrate the significance of mulberry allergens. Cross-sensitivities between them, and also with *Maclura pomifera* (Bernton 1928), suggest common allergens among all three genera. Occasionally *M. pomifera* is a serious offender (Wodehouse 1971).

Additional data confirming allergenicity in the family is provided by the following examples. In Kansas approximately 20 atopic patients proved positive to tests using *Maclura pomifera* pollen extracts; this plant is commonly grown along hedgerows (R. Hale, personal communication). In western Texas about 500 atopic patients treated using *Morus* pollen extracts reacted strongly positive (4+); there the annual incidence of pollinosis is increasing (R.L. Don, personal communication). Species of *Morus* tend to be weedy and their frequency sometimes rises rapidly, particularly in urban and suburban areas, with a concomitant increased incidence of pollinosis.

MYRICACEAE*
Wax-myrtle Family

A small, widely distributed family of aromatic trees and shrubs, the wax-myrtles or bayberries (Fig. 84) are readily distinguished by their yellow, glandular, dotted leaves and indehiscent, waxy-coated, 1-seeded fruit. The flowers are unisexual and borne on axillary, catkinlike spikes, monoecious or dioecious, with the stamens four to eight, and the pistil with two carpels and two stigmas.

In North America there are six native and several introduced species of *Myrica* and also the monotypic *Comptonia peregrina* (sweet fern). The native species of *Myrica* occur in northern latitudes and along the Pacific and Atlantic and Gulf coasts (Fig. 85). In the south the genus extends inland to the Appalachian Mountains and west to Arkansas.

Flowering and Pollen Aerobiology

Species of *Myrica* flower during February and March in Florida, during March and April in most other states of the Southeast, and from April to July in the northeastern region. Flowering of western species is parallel: during March and April in the Southwest

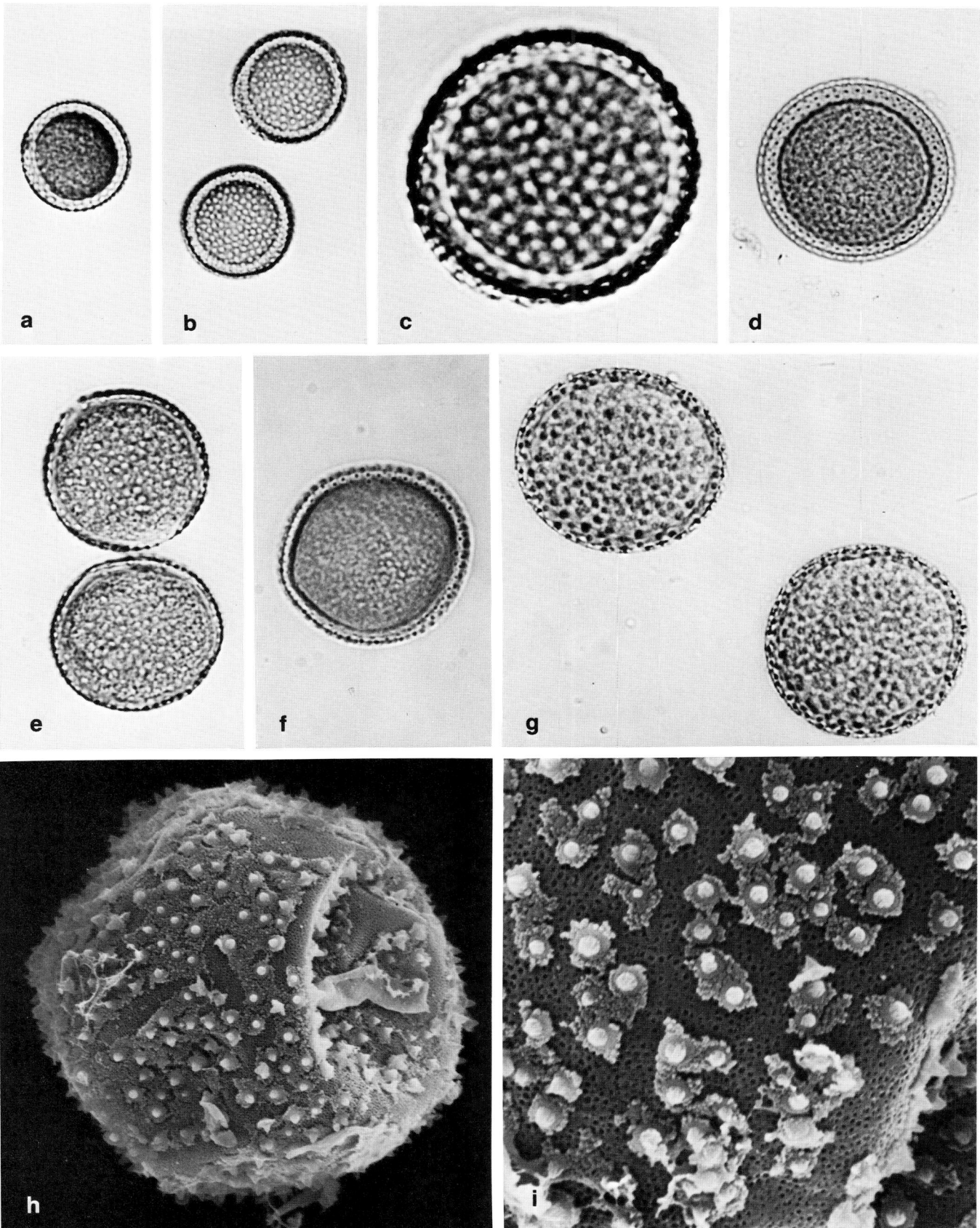

Fig. 78. Pollen of representative Lauraceae. a: *Cinnamomum camphora* (camphor tree), LM(s) ×720; b–c: *Lindera benzoin*, b = LM(s) ×720, c = LM(s) ×1830; d: *Persea americana* (avocado), LM(s) ×720; e–f: *Umbellularia californica*, LM(s) surface (e) and optical (f) views ×720; g–i: *Sassafras albidum*, g = LM(s) ×720, h = SEM(a) ×2400, i = SEM(a) showing sexine surface ×5800.

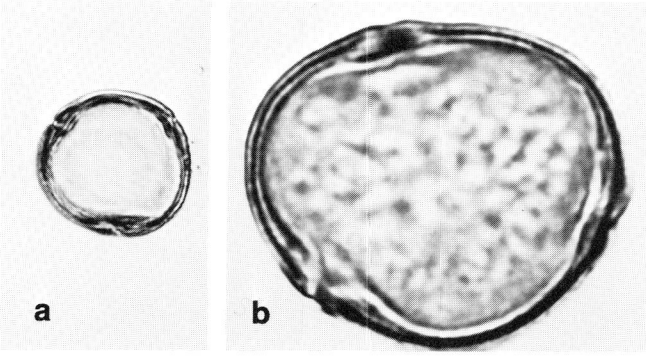

Fig. 79. Pollen of *Leitneria floridana* (Florida corkwood). a = LM(a) polar view ×720, b = LM(s) polar view ×1830.

and from April to June in the Northwest. *Comptonia* flowers during March and April in the Southeast and from April to June in the Northeast.

All species in the family are wind-pollinated and shed abundant pollen.

Pollen Morphology (Fig. 86)

Grains are suboblate, 20–24 × 25–30 μm; the amb rounded-triangular with slightly convex sides and 3–4-porate, usually distinctly aspidate, circular and ca. 3.2 μm wide (*Myrica*) or lolongate and ca. 4.8 × 1.6 μm (*Comptonia*); and the opercula granular. The sexine is tectate, granular or minutely scabrate, 0.8–1.0 μm thick, often thicker adjacent the pores; the nexine thinner than the sexine (0.6–0.8 μm); and the intine conspicuously thickened beneath the pores, forming an oncus ca. 6 × 14 μm. Myricaceous pollen superficially resembles that of the Betulaceae, particularly *Betula* and *Corylus*, and also that of the Casuarinaceae.

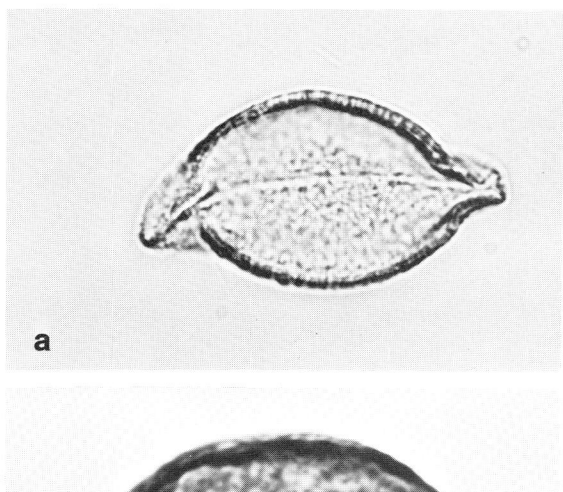

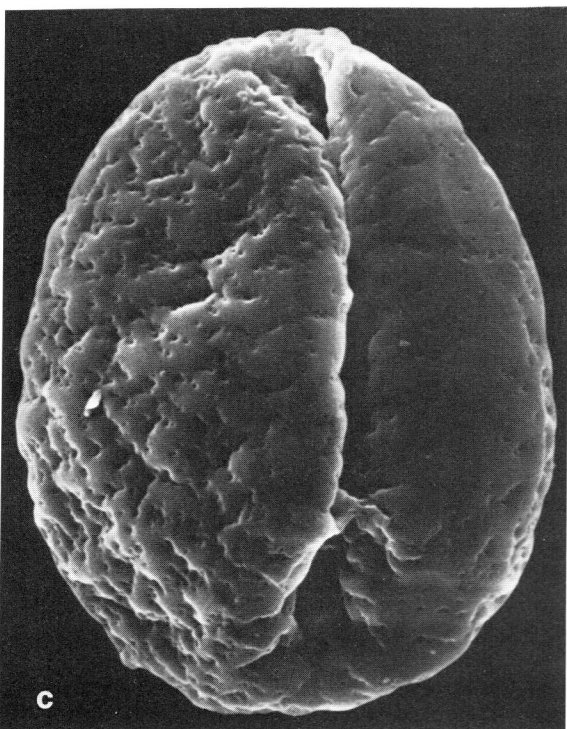

Fig. 81. Pollen of *Liriodendron* and *Magnolia*. a: *L. tulipifera* (tulip tree), LM(a) ×720; b: *M. grandiflora*, LM(s) ×1830; c: *M. kobus*, SEM(a) ×3900.

Fig. 80. *Magnolia grandiflora* (southern magnolia) flower, ×⅓.

TREES AND SHRUBS

Fig. 82. Morphology of representative Moraceae. a: *Maclura pomifera* (Osage orange) branch with heads of staminate flowers, × ½; b: *Morus alba* (white mulberry) branch with catkinlike inflorescence of staminate flowers, × 2½.

TABLE 8. Flowering by month of the Moraceae.

SPECIES	REGION						
	nNE	sNE	SE	sFL	NC	SC	NW	SW
Broussonetia papyrifera	4–5	4–5	4			4–5		3–4
Maclura pomifera	6	5–6	4–5		5–6	4–5	5–6	
Morus alba		4–5	3–5	3–5[a]	5–6	3–4[b]	4–6	3–4
Morus rubra	4–6	4–5	4–5			3–5		

[a]*Morus nigra* (black mulberry).
[b]*Morus microphylla* (mountain or Texas mulberry).

Allergenicity

The best documented case of severe wax-myrtle induced inhalant allergy is that elicited by pollen of *Myrica cerifera* (southern bayberry or wax-myrtle) (Prince & Meyer 1977). The patient from eastern Texas, who had earlier experienced similar episodes in northern Florida and Arkansas, suffered from sneezing, severe ocular irritation, profuse rhinorrhea, and nasal occlusion. He responded favorably to hyposensitization using *M. cerifera* pollen extracts. In central Florida, where *M. cerifera* is very common and initiates pollination by February, over 100 atopic patients annually respond strongly positive (4+) to *Myrica* pollen extracts (R.F. Lockey, personal communication). Clearly the southern wax-myrtle is a significant offender in pollinosis; the remaining species in the continent, as well as *Comptonia*, are undoubtedly equally as important where common. Unfortunately for the patient, this relevant source of allergens is not well recognized, particularly when it is realized that the shrubs invade and readily displace other plants in the eastern Piedmont (e.g., *M. pensylvanica*, Collins & Quinn 1982).

MYRTACEAE
Myrtle Family

The myrtles are a large family concentrated in tropical America and Australia. The few native North American species are of no known allergenic importance, but some plants introduced to warmer regions, notably *Callistemon* (bottlebrush), many species of *Eucalyptus* (eucalypt, gum tree) (Fig. 87), and *Melaleuca* (cajeput), are of secondary relevance in pollinosis.

Flowering and Pollen Aerobiology

Most cultivated myrtles flower in the winter months and some into the spring in both the Southwest and Florida. Airborne *Eucalyptus* pollen has been found in California and Florida in addition to many localities abroad; pollen of *Melaleuca* has been caught airborne in Florida and particularly in Taiwan (Lewis & Vinay 1979; Chen & Huang 1980). Nonetheless, pollination is largely zoophilous and only secondarily or incidently anemophilous, so limited airborne pollen and exposures are expected.

Pollen Morphology (Fig. 88)

Grains are suboblate, 20–24(E) μm; the amb triangular and usually with concave sides, angulaperturate and parasyncolpate with a triangular apocolpium, 3–4-colporate, and somewhat aspidate; the colpi narrow, long and fused; and the ora lalongate and 1.2 × 4.0 μm (in *Callistemon*). The sexine is ±smooth at the apertures, roughened or somewhat scabrate elsewhere, 0.6–0.7 μm thick, much thicker (to 2 μm) adjacent the apertures; the nexine as thick as or thinner than the sexine; and the intine 0.6–0.8 μm thick, much thicker beneath the apertures forming onci 1.7 μm thick in *Melaleuca*, ca. 2.4 × 5.6 μm thick in *Callistemon*, and ca. 4.8 × 5.6 μm thick in *Eucalyptus*.

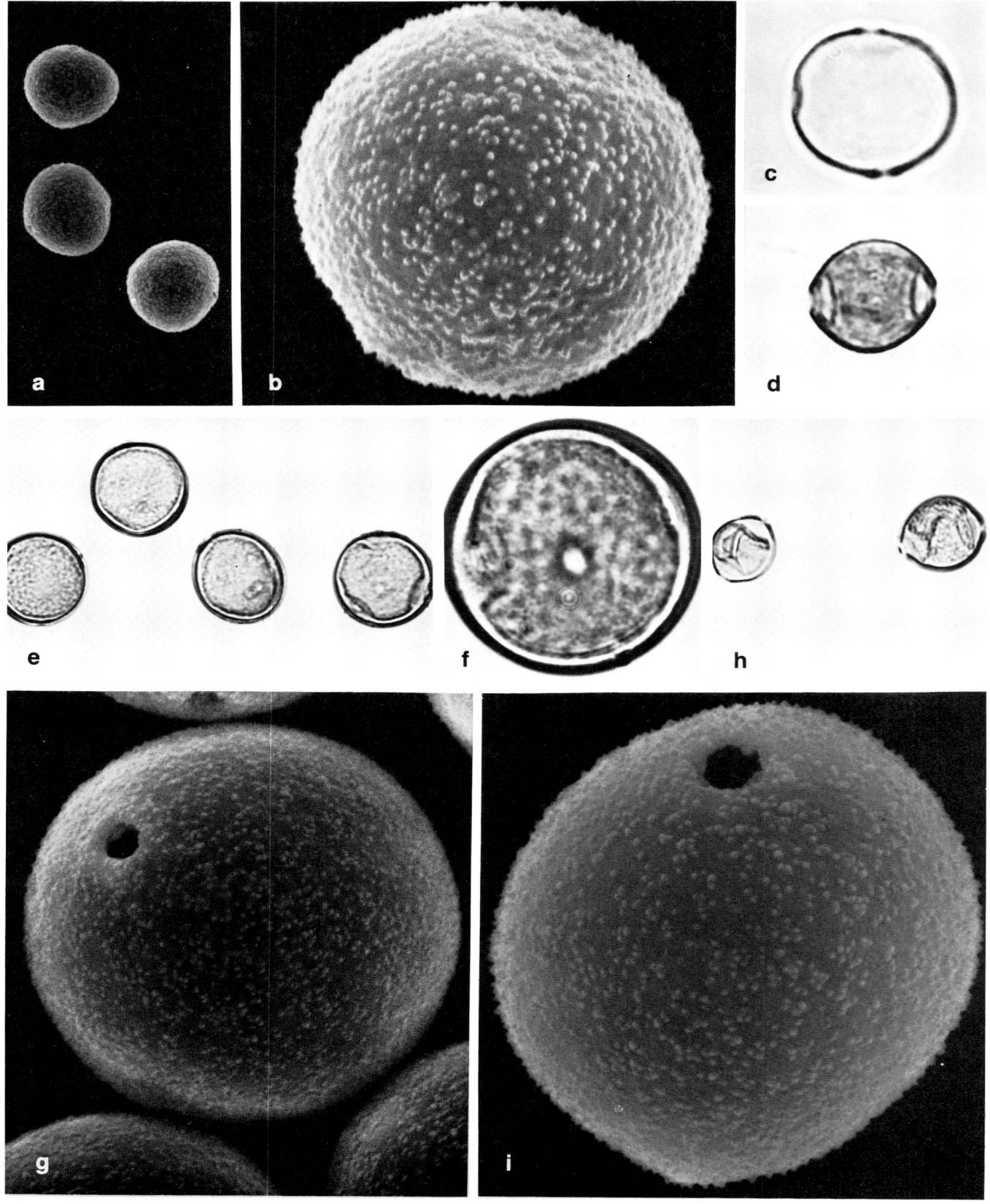

Fig. 83. Pollen of representative Moraceae. a–c: *Broussonetia papyrifera* (paper mulberry), a = SEM(ad) ×1400, b = SEM(ad) ×5600, c = LM(s) ×1830; d: *Cudrania fruticosa*, LM(s) ×1830; e–g: *Maclura pomifera* (Osage orange), LM(s) ×720, f = LM(s) 1830, g = SEM(a) ×4400; h–i: *Morus alba* (white mulberry), h = LM(s) ×720, i = SEM(a) ×5500.

Fig. 84. Morphology of *Myrica*. a: *M. californica* (California wax-myrtle) branch with leaves and catkinlike staminate inflorescences, ×¾; b: *M. cerifera* (southern wax-myrtle) branches with catkinlike staminate inflorescences, ×½; c–d: *M. pensylvanica* (northern bayberry) branches showing terminal leaves and catkinlike staminate inflorescences (c, ×¾) with pollen shedding from them (d, ×2).

Allergenicity

Throughout California allergists report rare to moderate frequencies of allergic reactions to *Eucalyptus* pollen responding 2+–4+ to skin tests. These data parallel those from India, where eucalypt pollen is considered an occasional offender in pollinosis (Shivpuri & Singh 1979). In Hawaii, however, greater numbers are subject to inhalant affects: as many as 150 atopic patients treated annually for pollinosis responded positively (2+–3+) to extracts from eucalypt pollen (A.W. Neilson, Jr., personal communication). Similar, if fewer, cases have been elicited with *Callistemon* pollen in California.

In southern Florida the most common myrtaceous introduction is *Melaleuca*. Pollen and volatile oils of *M. quinquenervia* have been implicated in both immediate and delayed hypersensitivities (Luippold 1974; Morton 1978), but a recent controlled study in Tampa (Lockey et al. 1981) shows that the species is only of minor allergenic importance because of limited pollen shed and negative odor challenges.

NYSSACEAE

Tupelo Family

The Nyssaceae are represented in eastern North America by several species of *Nyssa*, the most frequent being *N. sylvatica* (black gum or tupelo). Flow-

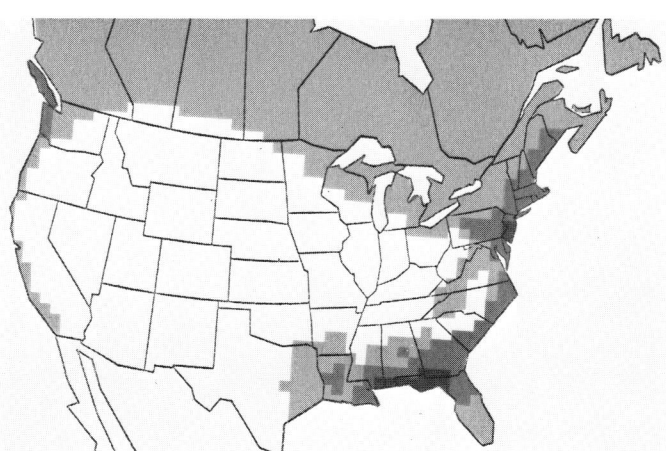

Fig. 85. Generalized composite distribution of 6 indigenous species of *Myrica* (☐ 1 sp., ▨ 2 spp., ▰ 3 spp.).

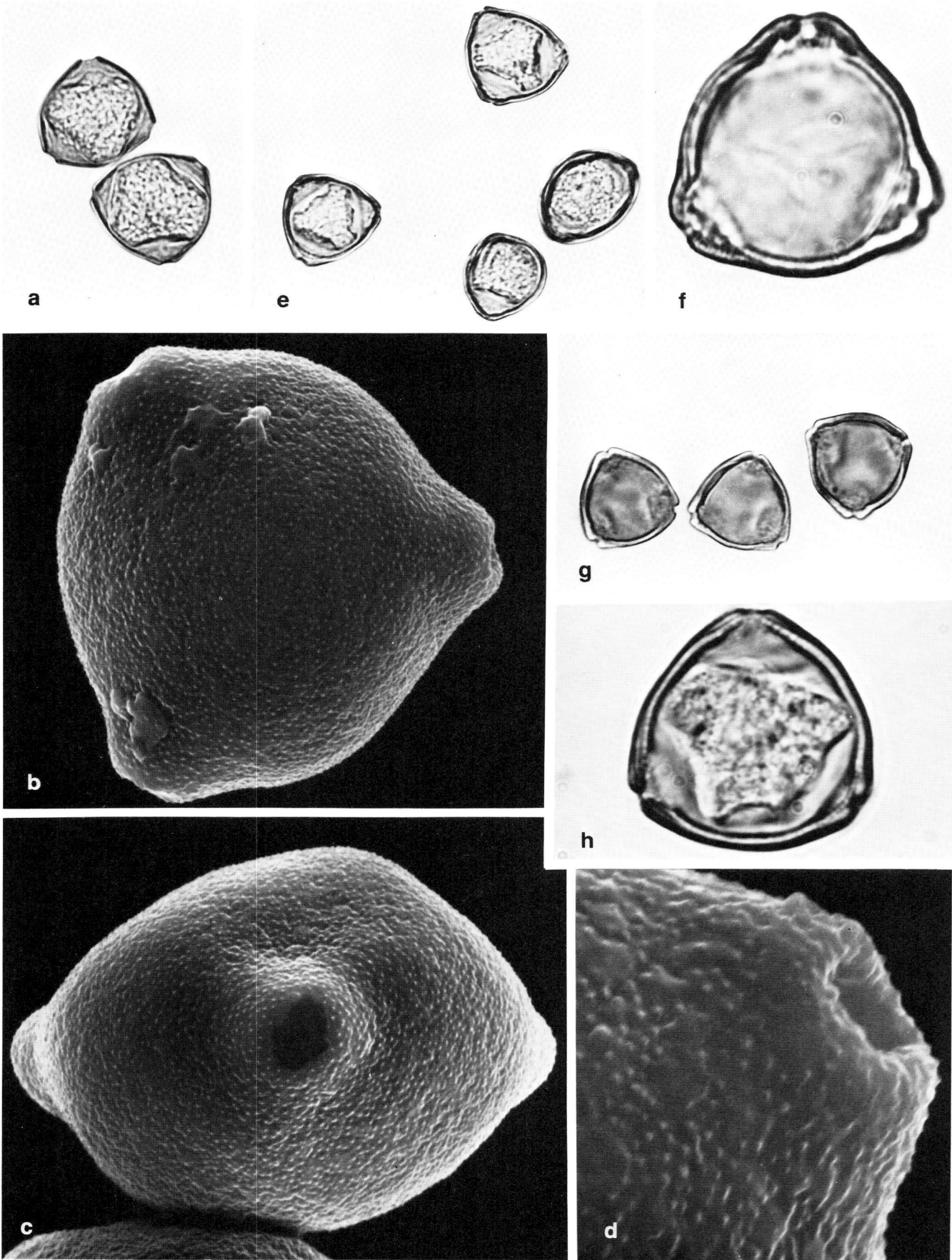

ering occurs from April to June or as early as March in Florida, and although pollination is entomophilous, incidental shedding of pollen may occur with subsequent sensitization of atopic individuals locally and infrequently.

Pollen Morphology (Fig. 89)

Grains are oblate-spheroidal, 29–33 × 30–35 μm; the amb triangular with slightly convex sides and 3-colporate; the colpi long and narrow, ca. 22 × 0.8–1.5 μm; the margins nexinously thickened; the ora lolongate, ca. 3.2 × 2.4 μm; and the opercula granular. The sexine is punctitegillate, ca. 1.1 μm thick; the nexine as thick as the sexine except thickened at the colpi margins; and the intine ca. 1.2 μm thick, thickened to 2.4 μm beneath the pores.

OLEACEAE*

Olive Family

The Oleaceae are a diverse family of trees and shrubs widely distributed in temperate and tropical regions with several genera of economic and horticultural value, such as *Chionanthus* (fringe tree), *Forsythia* (golden bells), *Fraxinus* (ash), *Jasminum* (jasmine), *Ligustrum* (privet), *Olea* (olive), and *Syringa* (lilac) (Fig. 90).

The leaves are opposite, simple or pinnately compound, without stipules. Their hairs often give them a grayish or silvery appearance. The flowers are bisexual or unisexual (*Fraxinus*), with perianth parts typically in fours, two stamens and carpels, and a compound pistil having a capitate or bifid stigma. The fruit is various—samara (*Fraxinus*), capsule, drupe, or berry.

Except for the Rocky Mountains and adjacent regions, one or more of the 13 indigenous species of *Fraxinus* is found throughout the United States and southern Canada (Fig. 91). Species are particularly frequent in the Midwest, eastern North Carolina, and northern Florida.

Flowering and Pollen Aerobiology

The only strictly wind-pollinated genus in North America is *Fraxinus*, although a few species are predominantly entomophilous (*F. cuspidata*, introduced

Fig. 87. Morphology of representative Myrtaceae. a: *Callistemon citrinus* (crimson bottlebrush) branch with inflorescence showing exposed anthers, × ½; b: *Eucalyptus torquata* (coral gum) leaves and flowers, × ⅗.

F. ornus). There are many native and introduced species used for timber, landscaping, and street trees. The majority shed abundant pollen in the spring (March–May). In the southcentral region many species flower as early as January or February, although *F. cuspidata* flowers from May to June and *F. berlandierana* as late as August. In the Southwest flowering occurs from January to June, with February and March the predominant months.

While primarily entomophilous, olives and privets are facultatively anemophilous and as such they may shed considerable pollen. Olives flower in the Southwest typically from April to June. Privets in the South begin flowering in March or April, more commonly May and June in the Northeast, and in the Southwest flowering is characteristically during the summer months.

Airborne pollen of these three genera is readily sampled from the ambient air in regions where plants are common (Table 9). These results closely parallel the flowering times described above. In the southwestern region, for example, oleaceous aeropollen is found from January to July, the season beginning

Fig. 86. Pollen of representative Myricaceae. a–d: *Comptonia peregrina* (sweet fern), a = LM(s) polar views showing 3 onci, thickened intine below each pore ×720, b = SEM(ad) polar view ×3000, c = SEM(ad) equatorial view ×3800, d = SEM(a) showing aspidate pore ×8200; e–f: *Myrica californica* (California bayberry), e = LM(s) ×720, f = LM(a) ×1830; g–h: *Myrica cerifera* (southern bayberry), g = LM(a) ×720, h = LM(s) showing 3 onci ×1830.

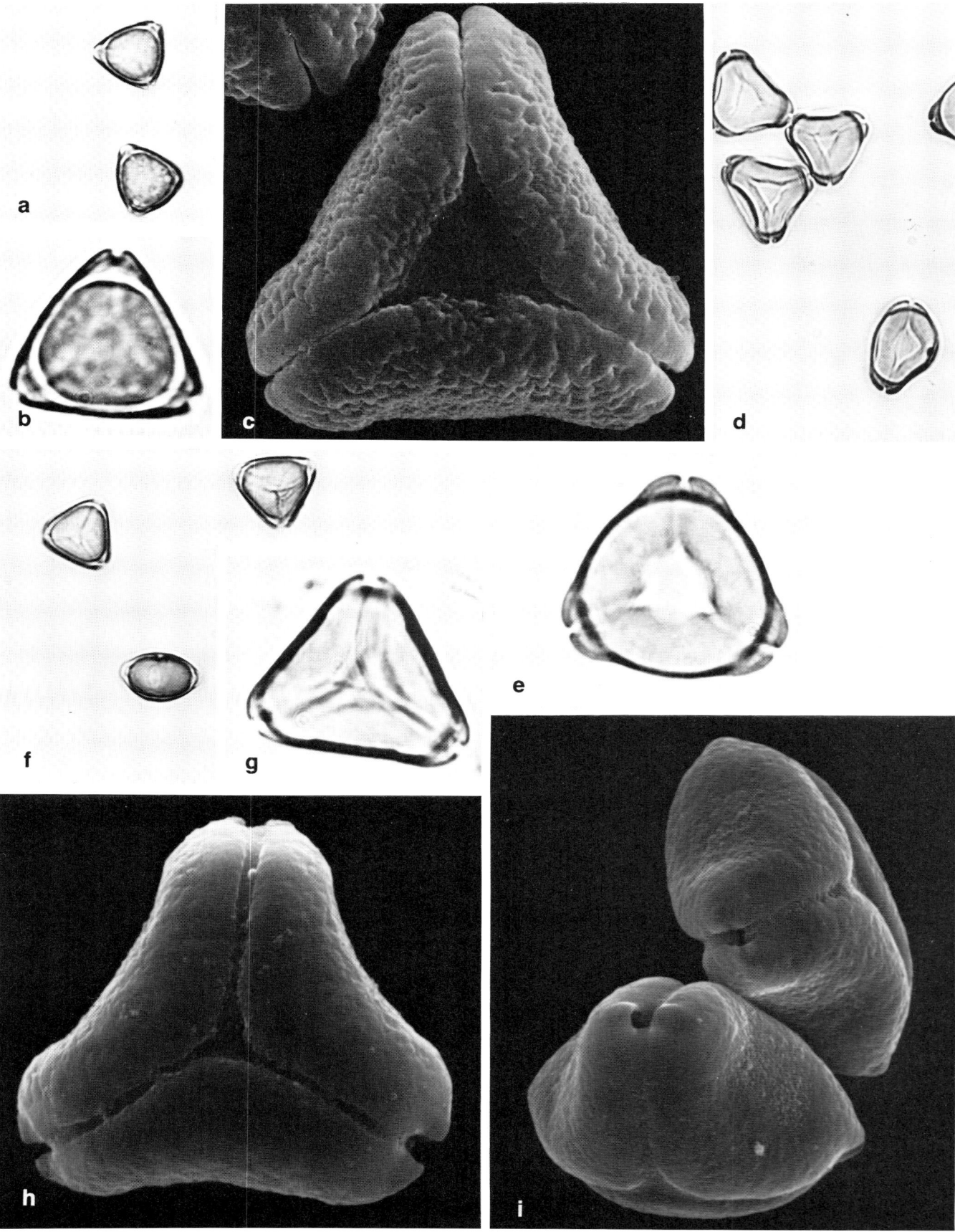

with *Fraxinus*, which flowers commonly during February and March, followed by *Olea*, from April to June, and concluding with *Ligustrum*, particularly in July. Airborne pollen of these plants has also been found widely elsewhere (Lewis & Vinay 1979).

Pollen of the other genera found in North America is only rarely encountered in aerosamples; nonetheless, the entomophilous *Forsythia*, *Jasminum*, and *Syringa* incidentally and occasionally release pollen that may be trapped by aerosamplers (Lewis & Vinay 1979).

Pollen Morphology (Figs. 92–93)

Grains are prolate or suboblate, 19–32(–38) × 13–28 μm, to spheroidal, 15–33 μm; the amb triangular, circular, or quadrangular and 3–4(–6)-colpate (*Chionanthus*, *Fraxinus*), 3–4-colporate, or 3–4-colporoidate; the colpi long and narrow, 13–26 × 0.5–2.5 μm, somewhat shorter in *Jasminum*; and the ora when definable, usually lolongate (3.6–4.8 × 0.8–2.4 μm) or occasionally lalongate (ca. 0.8 × 3.5 μm). The sexine is typically reticulate, occasionally in some areas of the grain striato-reticulate or ±tectate, the reticulations fine with small, polygonal lumina (ca. 0.5 μm) and narrow muri (<0.5 μm) as in *Chionanthus* and *Fraxinus*, or coarse with larger lumina (to 3.2 μm), polygonal to irregular, and broader muri to 1 μm as in *Ligustrum* and *Olea*, 1.0–2.4 μm thick; the nexine as thick as the sexine or often thinner; and the intine 0.5–1.2 μm thick, much thicker (to 2.5 × 13.0 μm) below the apertures.

Although similar by having principally 3-colpi and finely reticulate grains, *Chionanthus* grains are smaller (ca. 19 × 13 μm) than those of *Fraxinus* (ca. 26 × 28 to ca. 33 μm in diameter). *Ligustrum* and *Olea* grains are alike in their size (22–31 × 22–28 μm), with coarsely reticulate sculpturing.

Allergenicity

Fraxinus nigra (black ash) and *F. pennsylvanica* (green ash) were considered major offenders in spring pollinosis in Minnesota (Ellis & Rosendahl 1933), but these species were reported in Oklahoma as only moderately allergenic (Levetin & Buck 1980). In Phoenix, Arizona, however, ashes were among the chief tree offenders during February and March (Randolph & McNeil 1944; Walkington 1960). In St. Louis skin tests using *F. americana* (white ash) extracts showed

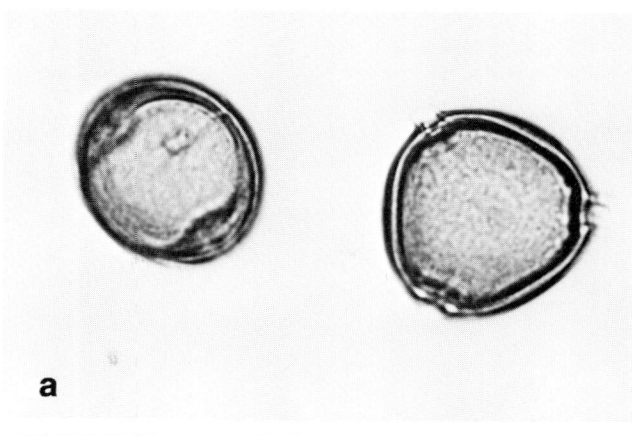
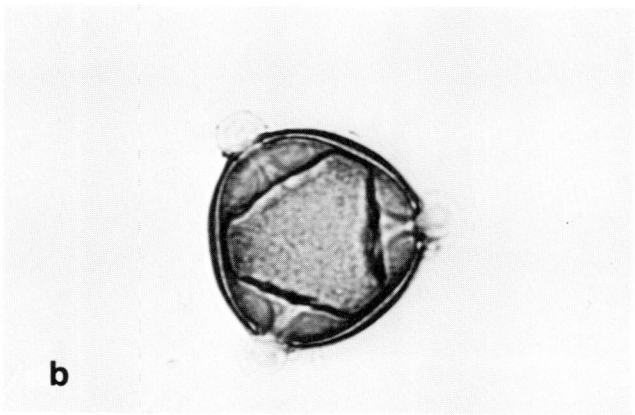

Fig. 89. Pollen of *Nyssa sylvatica* (black, or sour, gum). a = LM(s) polar and equatorial views ×720, b = LM(s) optical polar view ×720.

more moderate reactions than for any other tree extracts sampled and there were few cross-reactions between *Fraxinus* and other trees (Lewis & Imber 1975b).

In our survey of ashes involving responses from 13 allergists in seven states (Alabama, California, Connecticut, Iowa, New Jersey, Oregon, and Texas) approximately 1,190 allergic patients examined annually responded positively (2+–4+) to ash pollen extracts. The majority elicited 2+ and 3+ reactions, the overall average being moderately positive, 2.6+, suggesting that in general the ashes possess allergens that do not elicit the severest of responses.

Allergists from four states (Alabama, California, Florida, and Louisiana) reported a total of 1,489 patients annually suffering from pollinosis who responded strongly positive (average 3.9+) to skin test extracts of *Ligustrum* pollen (Lewis & Vinay 1979). Cross-reactions with a second offender, *Olea* pollen, were common. In response to olive extracts, 2,015 patients annually suffering from pollinosis in California reacted strongly positive (average 3.9+) on skin testing (Lewis & Vinay 1979). The latter data are corroborated by 395 additional patients with inhalant

Fig. 88. Pollen of representative Myrtaceae. a–b: *Callistemon citrinus*. a = LM(s) ×720, b = LM(s) ×1830; c: *Eucalyptus camaldulensis* (Murray red gum), SEM(a) polar view ×2100; d–e: *Eucalyptus rudis* (desert gum), d = LM(a) ×720, e = LM(a) ×1830; f–g: *Melaleuca quinquenervia* (cajeput), f = LM(a) ×720, g = LM(a) ×1830; h–i: *Melaleuca decora*, h = SEM(a) polar view ×5100, i = SEM(a) oblique-equatorial views ×4400.

Fig. 90. Morphology of representative Oleaceae. a: *Ligustrum vulgare* (common privet) branch with flowers, ×1½; b: *Olea europaea* (common olive) branch with flowers, ×¾; c: *Fraxinus pennsylvanica* (green ash) leafless branch showing staminate flowers, ×¾; d: *Fraxinus americana* (white ash) branch with fruit (samara), ×⅓; e: *Forsythia* x *intermedia* (forsythia) branch with flowers, ×1.

TREES AND SHRUBS

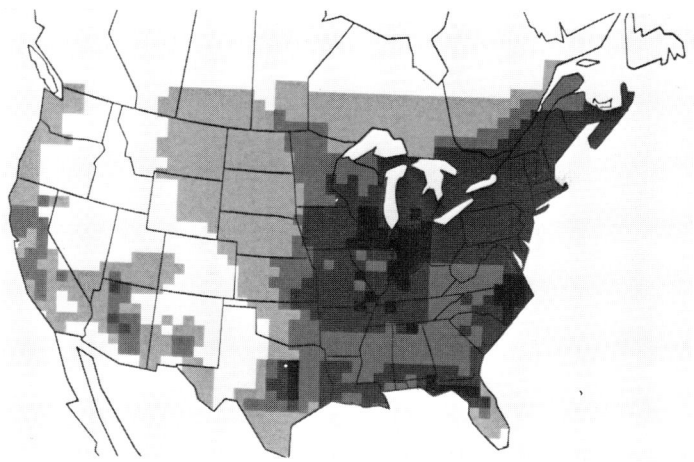

Fig. 91. Generalized composite distribution of 13 indigenous species of *Fraxinus* (☐ 1 sp., ▨ 2 spp., ▨ 3 spp., ■ 4 spp.).

allergy from different areas of California who also responded strongly positive (average 3.9+) to skin tests from olive pollen. Similarly, olive was considered the most allergenic tree in Palm Springs, California (Unger 1977). Response elsewhere to these allergens is equally severe. In Argentina, for example, pollen of *L. lucidum* was considered one of the most important causes of hay fever (Giscafre & Ragonese 1946). Allergenicity to olive pollen has also been reported from Arizona (Phillips 1932), Spain (Jiminez-Diaz 1932), Portugal (Silva 1960), and Israel, where it was second only to grass pollen in importance as a cause of hay fever (Kessler 1958).

In Israel Gutmann (1950) noted that even though not typically wind-pollinating, olive pollen is light. Bouyancy is an important consideration among insect-pollinated plants, for if their usually sticky pollen is light, opportunities to remain airborne some distance from the plant become more frequent. Pollen of *Olea europaea* weighs only 4.9 ± 0.2 (s.e.) ng, compared to 9.2 ± 0.3 ng for wind-pollinated *Fraxinus americana* pollen and 21.7 ng for the strictly entomophilous pollen of *Syringa vulgaris* (common lilac). Its unusual lightness, therefore, is one reason why olive pollen is so successful in remaining aloft and thus exposing, sensitizing, and eliciting severe inhalant allergic reactions among atopic individuals.

The other genera in North America are allergenically insignificant. In Missouri *Forsythia* is reported rarely as a cause of pollinosis, and, similarly, in Connecticut and Iowa *Syringa* and in California *Jasminum* have been implicated. In these instances reactions may have been the direct consequence of sniffing pollen onto mucous membranes by sensitized individuals (Lewis & Vinay 1979). Sensitivity to *Chionanthus*, *Forestiera* (some species wind-pollinated and dioecious), and *Osmanthus* pollen, while expected occasionally, is not known.

PLATANACEAE*
Planetree Family

The planetrees, or sycamores, are widely grown as street and shade trees in the United States, the most

TABLE 9. Oleaceae (*Fraxinus*, *Ligustrum*, *Olea*) aeropollen frequency in regions of North America based on percentage of pollen captured, by month (1974–78).

REGION	MONTH									
	1	2	3	4	5	6	7	8	9	10
Fraxinus (ash)										
nNe			0.4	2.0	3.7					
sNE		0.5	0.3	2.8	0.8					
SE			0.5							
NC				5.8	2.7					
SC	2.1	7.7	1.8	2.4	1.7					
NW			0.6	7.5	4.0					
SW	0.3	5.0	4.4	1.1	0.2	1.1	0.4			
sCA	1.2	10.9	0.3							
Ligustrum (privet)										
nNe						0.1	0.2			
SC			0.2	1.4						
SW						0.9	3.3	0.7	0.3	
sCA						0.4	0.7	0.4		0.4
Olea (olive)										
SW			0.1	7.0	7.1	8.2	0.9	0.6		
sCA			0.1	4.6	3.3	2.7	0.4			

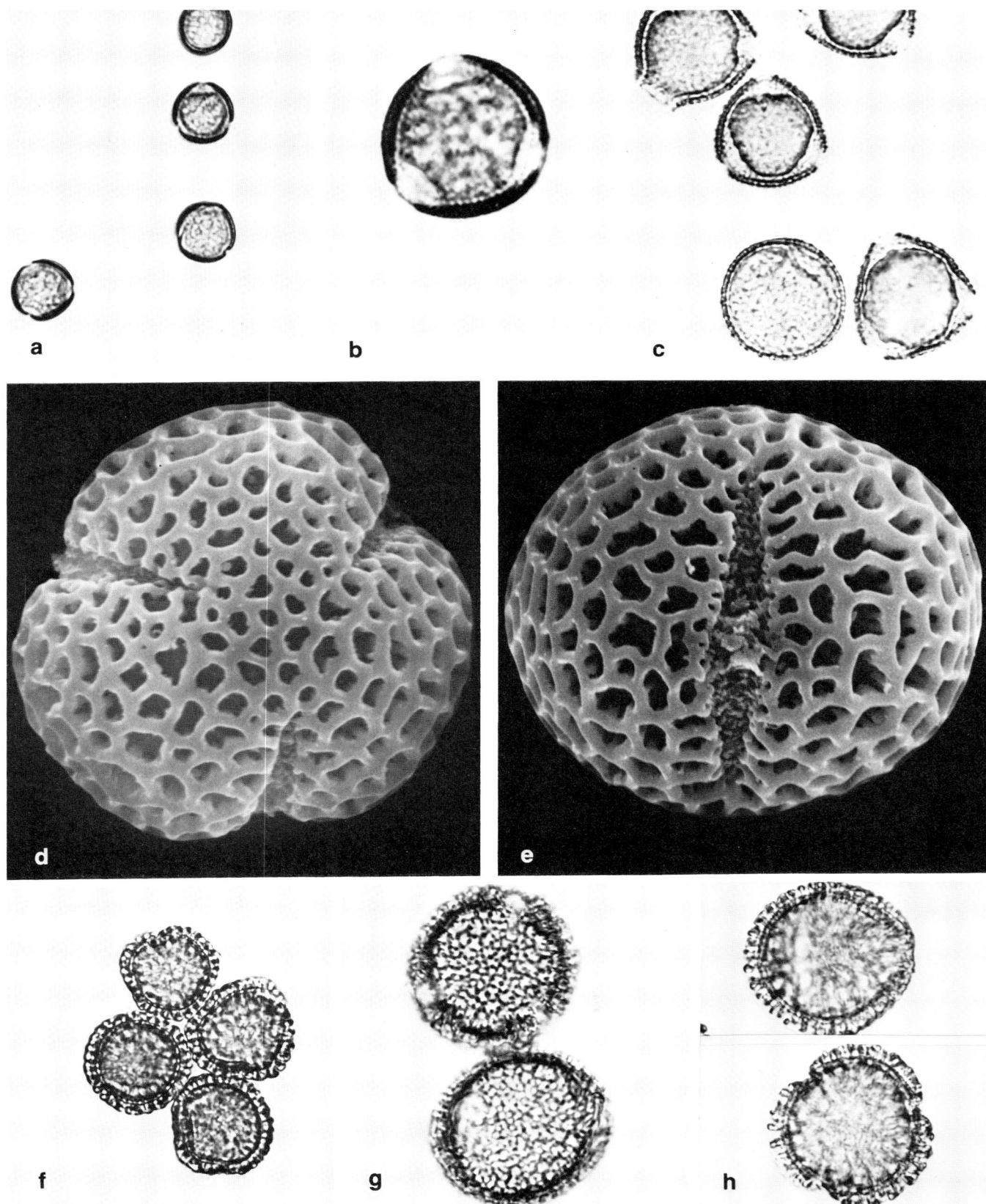

Fig. 92. Pollen of representative Oleaceae (*Chionanthus, Forsythia, Ligustrum*). a–b: *Chionanthus virginicus* (fringe tree), a = LM(s) ×720, b = LM(s) polar view ×1830; c: *Forsythia suspensa* (forsythia), LM(s) 1 equatorial and 4 polar views ×720; d–e: *Ligustrum vulgare* (common privet), d = SEM(a) polar view ×3400, e = SEM(a) equatorial view showing 1 aperture ×3400; f: *Ligustrum japonicum* (japanese privet), LM(a) ×720; g–h: *Ligustrum sinensis* (Chinese privet), LM(a) surface (g) and optical (h) views ×720.

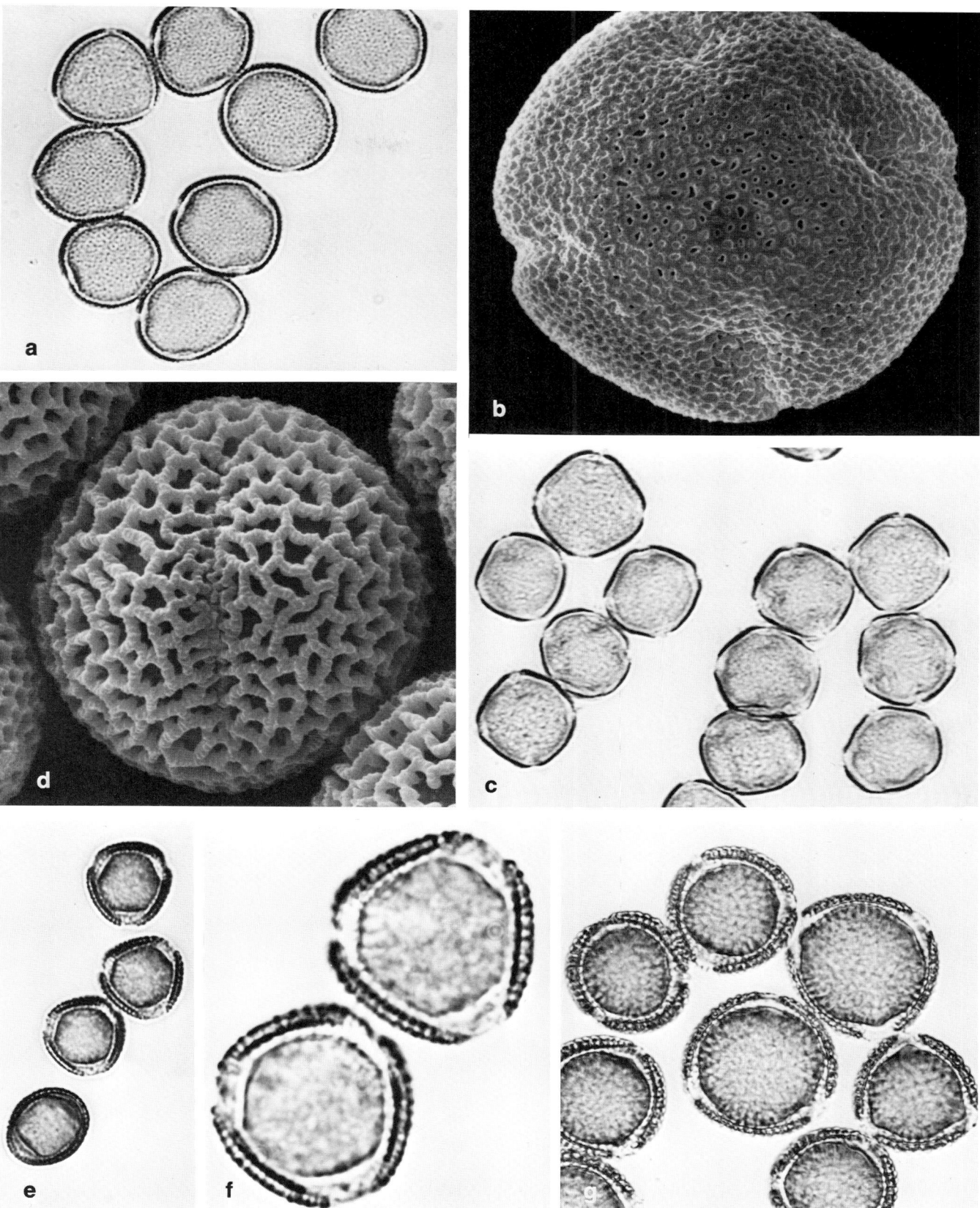

Fig. 93. Pollen of the Oleaceae (*Fraxinus, Olea, Syringa*). a: *Fraxinus nigra* (black ash), LM(s) ×720; b: *F. americana* (white ash), SEM(a) polar view showing 4 colpi ×3300; *F. pennsylvanica* (green ash), LM(s) ×720; d–f: *Olea europaea* (common olive), d = SEM(a) equatorial view showing 1 aperture ×3350, e = LM(s) with 3 polar views and 1 equatorial view ×720, f = LM(s) polar views ×1830; g: *Syringa reticulata* (Japanese tree lilac), LM(s) ×720.

common being *Platanus occidentalis* (American sycamore) (Fig. 94) and *P.* × *acerifolia* (*P. occidentalis* × *P. orientalis*), a hybrid called London planetree. They are large trees typified by grayish scaling bark, simple leaves with stellate hairs, and unisexual flowers in globose heads on long peduncles. The staminate flowers have three to eight subsessile stamens with three to eight small, hairy sepals, and the pistillate flowers usually have six to nine free, superior carpels. The fruit is a globose head of top-shaped achenes.

Flowering and Pollen Aerobiology

Flowering in the southcentral region begins in March (until April), April in the southeastern and (southern) northeastern regions, and May in the more northerly areas of the Northeast. Anthesis ends in June. In California *Platanus racemosa* flowers from February to April and *P. wrightii* of Arizona during April and May.

All species are wind-pollinated and shed large amounts of bouyant pollen that is readily caught on atmospheric slides (Wodehouse 1971).

Fig. 94. *Platanus occidentalis* (American sycamore) branches showing globose heads of staminate flowers (right) and pistillate flowers (left) in both a (×1) and b (×½).

Pollen Morphology (Fig. 95)

Grains are suboblate, 17–19 × 18–20 μm; the amb triangular and lobed and 3-colpate; the colpi moderately short and broad, ca. 11 × 5 μm; and the opercula with dense microverrucae. The sexine is finely reticulate; the lumina small (ca. 0.8 μm) and polygonal, and 0.8 μm thick; the nexine as thick as the sexine; and the intine ca. 0.8 μm thick, thickened to 1.4 μm below the colpi.

Allergenicity

Even though the tree is common throughout the St. Louis area, pollen of *Platanus occidentalis* proved only moderately allergenic based on skin tests (Lewis & Imber 1975b). This correlates with our survey of sycamore allergenicity throughout the United States, for in five states (Alabama, California, Kansas, New Jersey, Oregon), based on reports from 13 allergists, a total of 514 atopic patients averaged only 2.0+ in skin tests, a generally moderate response of sensitized individuals to pollen extracts, particularly where sycamores were common and popular. These trends are consistent with other reports of allergenicity, e.g., moderate in Oklahoma (Levetin & Buck 1980) and of secondary importance in the Southwest (Sellers 1935).

RHAMNACEAE
Buckthorn Family

A large family of principally insect-pollinated trees and shrubs, the buckthorns nevertheless may shed pollen locally and be a contributing factor in endemic pollinosis. *Ceanothus* (Fig. 96) is a potentially limited offender; its species, which are particularly common in the West, flower in the late spring and early summer but begin earlier (March) in the southcentral region.

Pollen Morphology (Fig. 97)

Grains are prolate to subspheroidal, 17–22 × 20–23 μm; the amb triangular and the sides convex and 3-colporate; the colpi long and narrow, 14–22 × 1.6 μm, occasionally syncolpate at one pole; the margins uneven; the ora lolongate to lalongate with tapering ends when transversely elongate; and the opercula granular. The sexine is rugulate, scabrate, 0.8 μm thick; and the nexine as thick as the sexine, slightly thickened adjacent the apertures.

ROSACEAE
Rose Family

The rose family is large and predominantly north temperate in distribution and is highly valued for its fruit crops and many popular horticultural ornamen-

TREES AND SHRUBS

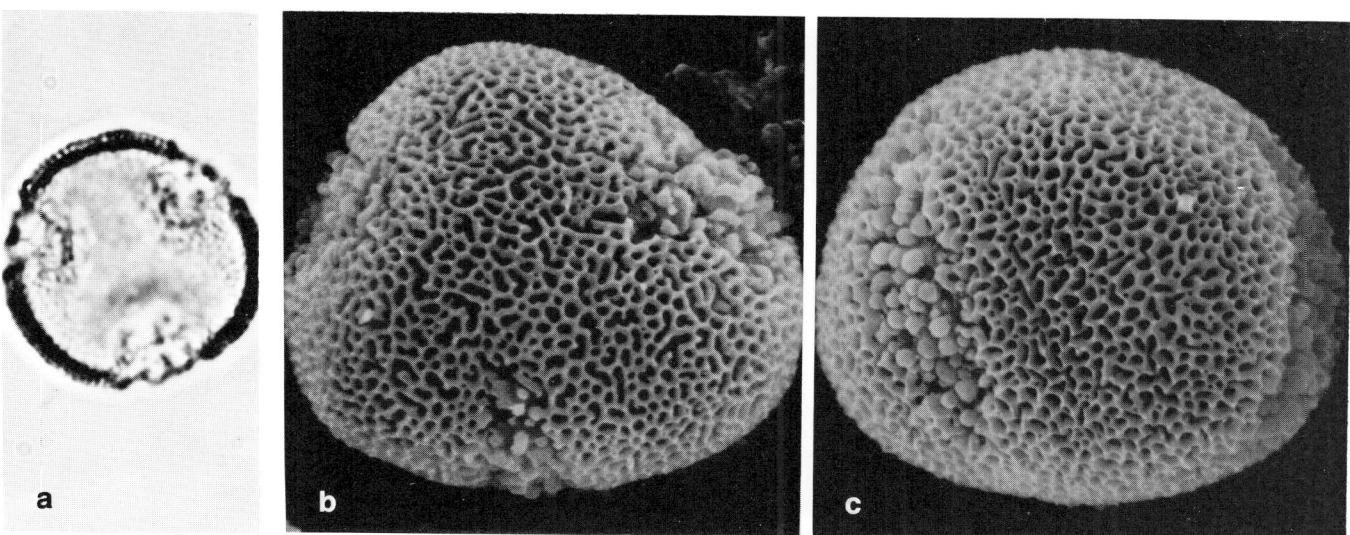

Fig. 95. Pollen of *Platanus occidentalis* (American sycamore). a = LM(a) ×1830, b = SEM(a) polar view ×4700, c = SEM(a) equatorial view showing two colpi.

tals (Fig. 98). Most plants are insect-pollinated, although abundant pollen is produced and many are facultatively or incidentally wind-pollinated, such as *Cercocarpus*. A few, including the uncommon *Acaena* and *Sanguisorba*, are primarily wind-pollinated.

Allergenic groups may reflect the subfamilial classification that includes:

1. subfamily Spiraeoideae—*Holodiscus*, *Spiraea* (bridal wreath);
2. subfamily Rosoideae—*Acaena*, *Cercocarpus* (mountain mahogany), *Cowania* (cliffrose), *Sanguisorba* (burnet), *Rosa* (rose);

Fig. 96. *Ceanothus velutinus* (tobacco brush) branch with inflorescence, ×2½.

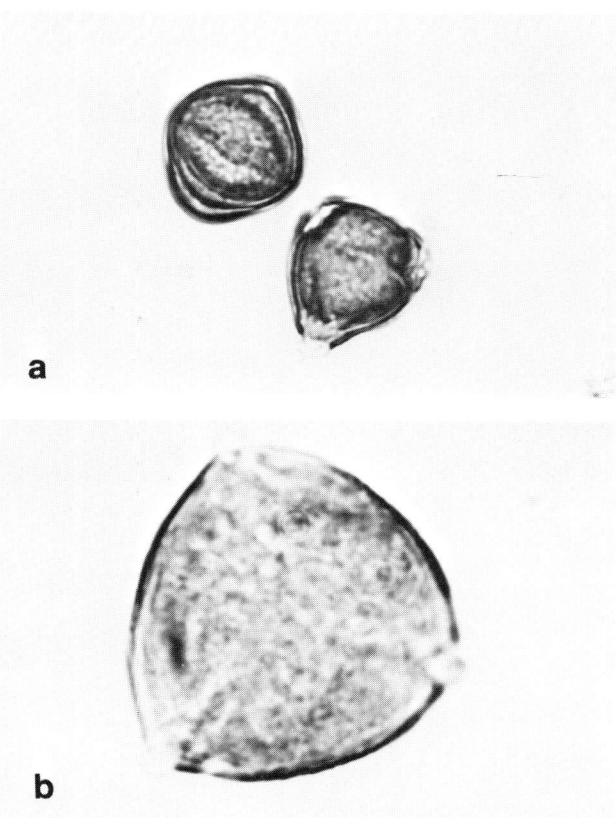

Fig. 97. Pollen of *Ceanothus*. a: *C. velutinus*, LM(s) polar and equatorial views ×720; b: *C. sanguineus* (wild lilac), LM(s) polar view ×1830.

Fig. 98. Morphology of representative Rosaceae. a: *Crataegus crus-galli* (cockspur thorn) branch with flowers, ×1; b: *Rosa eglanteria* (sweetbriar) branch with flower showing numerous stamens, ×1; c: *Prunus caroliniana* (Carolina laurelcherry) branch with inflorescence, ×¾.

3. subfamily Prunoideae—*Prunus* (apricot, cherry, plum);
4. subfamily Maloideae—*Amelanchier* (serviceberry), *Crataegus* (hawthorn, thorn), *Malus* (apple), *Pyrus* (pear), *Sorbus* (mountain ash).

The last subfamily is particularly isolated in the family.

Flowering and Pollen Aerobiology

Members of the rose family flower characteristically in the spring and early summer. *Holodiscus*, *Spiraea*, and *Rosa* often continue flowering during the summer, but generally the peak of anthesis ends by June.

Although pollen of insect-pollinated genera has been collected airborne, such as *Prunus* in Colorado, Pennsylvania, Australia, and Europe, *Rosa* in Massachusetts and *Spiraea* in Britain (Lewis & Vinay 1979), *Filipendula* and *Potentilla* in Britain (Tinsley & Smith 1974), and undetermined rosaceous pollen in California, Missouri, British Columbia, Australia, Britain, and France (Lewis & Vinay 1979), only occasionally does pollen become airborne.

Pollen Morphology (Figs. 99–100)

Grains are prolate to subspheroidal or oblate-spheroidal; the amb triangular; the sides either lobed or convex, and 3-colpate (*Cercocarpus*, *Cowania*), 3-colporate, or 3-colporoidate; the colpi long and narrow, 13–34 × 0.5–2.5 μm, with margins often conspicuously thickened; the ora, when present, lalongate to lolongate, to 5 × 7 μm; and the opercula granular. The sexine is tectate and although the striate condition is widespread in the family, sculpturing also includes reticulate, striato-reticulate, and striato-rugulate, 0.7–1.6 μm thick; the nexine as thick as the sexine or slightly thinner, but often thickened adjacent the apertures; and the intine generally thin (<1 μm), slightly thicker below the apertures.

Grains of *Holodiscus* and *Spiraea* are small with their polar axis 15–18 μm, whereas those of *Rosa* are much larger (up to 46 μm).

Allergenicity

Pollen extracts of *Crataegus*, *Rosa*, *Spiraea*, and *Sorbus* elicited some significantly positive skin test reactions among atopic individuals (Lewis & Vinay 1979). These were limited, however, because by and large exposure to the pollen was restricted to the immediate vicinity of the orchards or street and garden plantings. Of the few reported cases of pollinosis, only *Spiraea* pollen elicited strongly positive skin test reactions (average 3.5+), suggesting that not only is incidence low but allergenic activity is also moderate or low for pollen of most genera.

RUTACEAE

Citrus Family

The Rutaceae are a large, predominantly tropical family valued for their citrus fruit and oils for perfumery and medicine. The flowers are insect-pollinated, the pollen only incidentally becoming airborne.

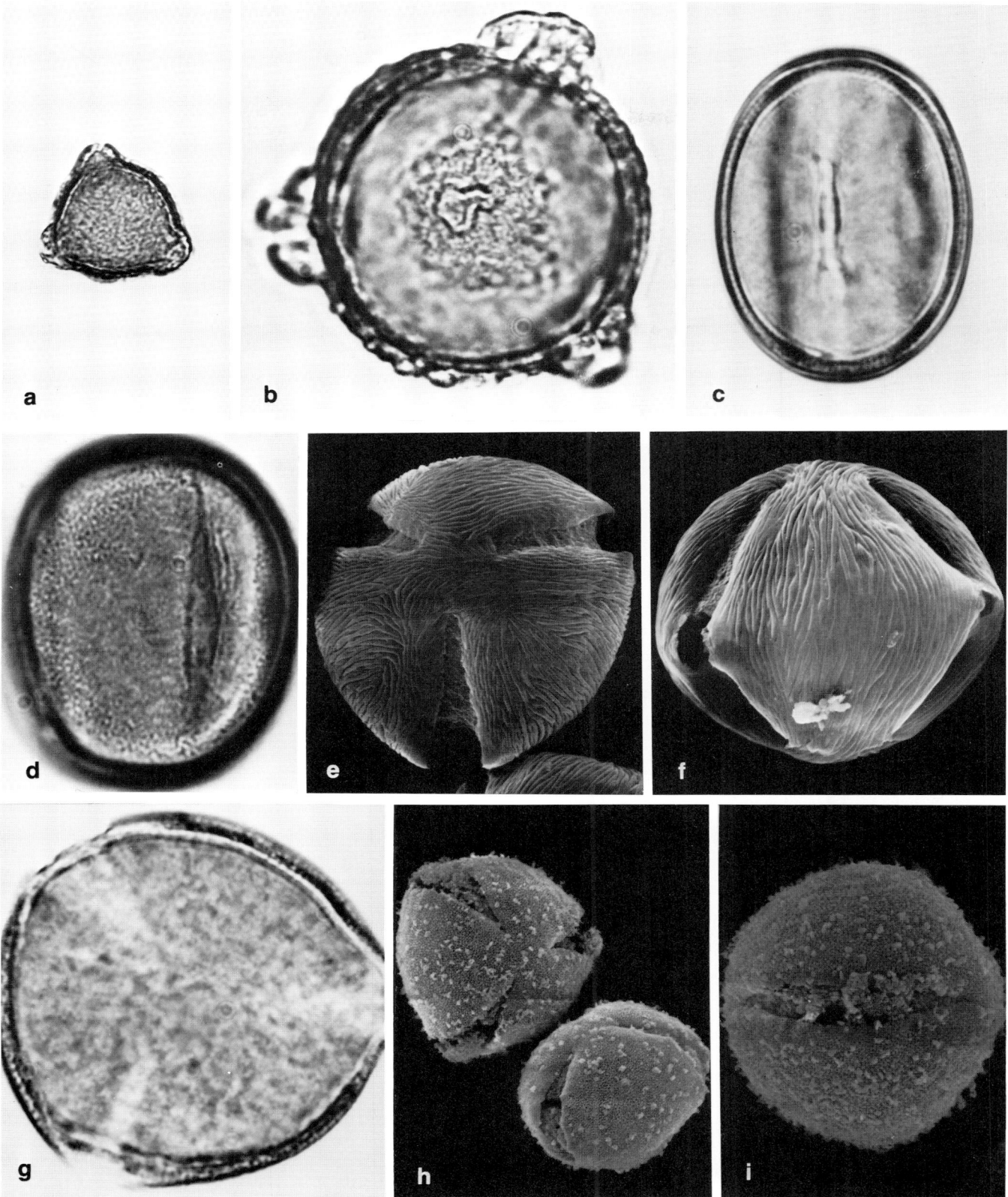

Fig. 99. Pollen of representative Rosaceae (*Acaena-Cowania*). a–b: *Acaena californica*, a = LM(s) ×720, b = LM(a) polar view ×1830; c: *Amelanchier canadensis* (Canada serviceberry), LM(s) equatorial view ×1830; d: *Cercocarpus montanus*, LM(s) equatorial view ×1830; e–g: *Crataegus viridis* (green hawthorn), e = SEM(a) polar view ×2100, f = SEM(a) equatorial view showing 2 apertures ×2000, g = LM(s) polar view ×1830; h–i: *Cowania mexicana* (cliffrose), h = SEM(a) oblique views ×1600, i = SEM(a) equatorial view ×2400.

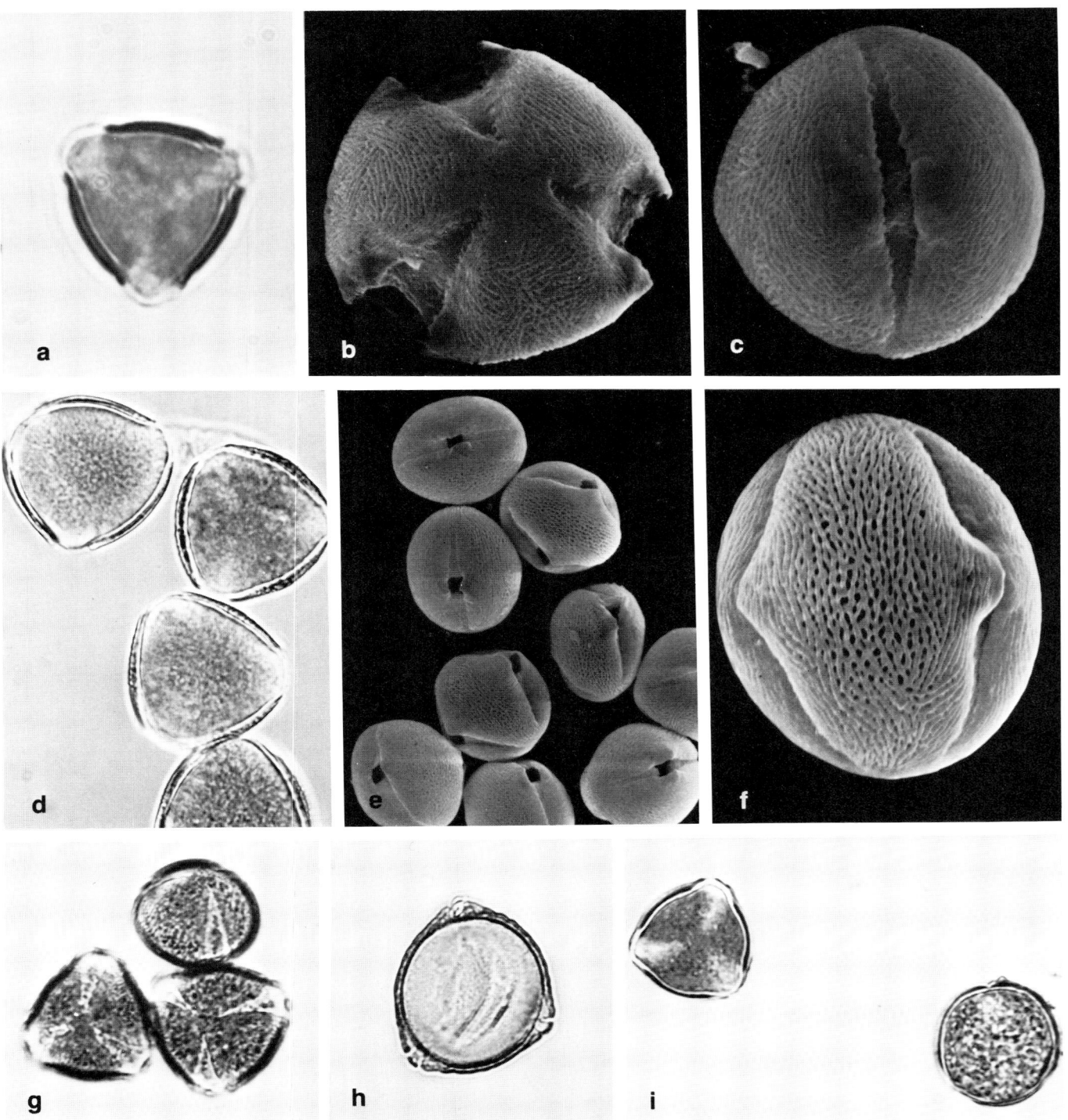

Fig. 100. Pollen of representative Rosaceae (*Holodiscus-Sorbus*). a–c: *Holodiscus discolor* (creambush), a = LM(s) polar view ×1830, b = SEM(a) polar view ×4000, c = SEM(a) equatorial view showing 1 aperture; d: *Prunus × cistena* (purple-leaved sandcherry, cultivated hybrid), LM(s) ×720; e–f: *Prunus serotina* (black cherry), e = SEM(a) ×1100, f = SEM(a) equatorial view showing 2 apertures ×3200; g: *Rosa* hybrid tea, LM(s) ×720; h: *Sanguisorba minor* (burnet), LM(a) polar view ×720; i: *Sorbus decora* (showy mountain ash), LM(s) ×720.

Citrus (Fig. 101) flowers in Florida from January to April, and in California from March to May; *Phellodendron* (cork tree) mostly in June; *Ptelea* (hop tree) from April to June; and *Zanthoxylum* (prickly ash) very early in southern latitudes (December in Texas, February in California, March in the Southeast) and April or May in northeastern and northcentral regions. Limited airborne pollen has been reported in the vicinity of citrus groves in California, Florida, and Israel (Lewis & Vinay 1979).

Fig. 101. X *Citrofortunella mitis* (*Citrus* × *Fortunella*) (calamondin) branch with flowers and immature fruit, × ⅗.

Pollen Morphology (Fig. 102)

Grains are prolate to prolate-spheroidal, 23–37 × 17–32 μm; the amb triangular, lobed or with convex sides and 3–4(–5)-colporate; the colpi long and narrow; and the ora generally lalongate. The sexine is often reticulate or faintly striato-reticulate; the lumina various (0.5–4.0 μm in diameter in *Phellodendron*), 0.8–1.6 μm thick; the nexine as thick as or thinner than the sexine; and the intine ca. 0.8 μm thick, about twice as thick below the apertures.

Citrus pollen is 4(–5)-colporate, having a broad lalongate os and reticulate sculpturing, and pollen of *Phellodendron* has a thick, coarsely reticulate sexine.

Allergenicity

As expected, there are only limited cases of pollinosis. Several allergists reported positive skin test reactions to citrus pollen extracts among atopic patients in the vicinity of citrus groves (Lewis & Vinay 1979), and in Israel hay fever caused by citrus pollen is rare and occurs only among people working in the citrus groves or in their vicinity (Gutmann 1950). Among native rutaceous plants *Zanthoxylum* should be considered a potential localized offender where it occurs in abundance.

SALICACEAE*

Willow Family

The Salicaceae are a family of two major genera of mostly northern temperate trees and shrubs, the aspens, cottonwoods, and poplars (*Populus*), and the willows (*Salix*). The plants are dioecious and flowers are in catkins (Fig. 103). *Populus* is wind-pollinated, whereas *Salix* is principally insect- and only facultatively wind-pollinated. Both genera are common in North America, with many native and several introduced species from Eurasia grown as ornamentals.

Leaves are deciduous, simple, and alternate, with small or sometimes foliaceous and persistent stipules. Flowers are unisexual, the staminate flowers with one to two nectariferous glands and two to thirty stamens. The pistillate flowers also have glands and consist of a compound pistil of two carpels, numerous ovules, superior ovary, and two to four styles. Fruit is a capsule with numerous hairy seeds; the hair is not allergenic (Phillips 1932).

Both *Populus* and *Salix* are widely distributed in North America. The 9 important native species of *Populus*, however, are most frequent around the Great Lakes, with a concentration in the northcentral Midwest (Fig. 104). The 25 important species of *Salix* sampled also show a significant center of distribution in the Great Lakes region, particularly the southern part that extends to Maine and New Brunswick in the East and to Saskatchewan and Alberta in the West (Fig. 105). Few species of *Salix* occur in southern latitudes.

Flowering and Pollen Aerobiology

Flowering is predominantly during the spring, from March to May, with that of *Populus* beginning somewhat earlier than *Salix*. In the northwestern region, in particular, flowering may extend to June and July (Table 10).

The wind-pollinated flowers of poplars are unscented and do not produce nectar. They generally have more stamens than do flowers of willows and produce much more pollen, which is not sticky and is known to be carried great distances by the wind (Wodehouse 1971). Even so, the primarily vector-transported willow pollen is frequently caught in abundance on atmospheric slides.

Pollen Morphology (Figs. 106–7)

Grains are prolate to subprolate (16–27 × 13–17 μm) or spheroidal (25–40 μm in diameter); the amb triangular, 3(–4)-lobed, and 3(–4)-colpate or 3-colporoidate (*Salix*); the colpi 12–19 × 2–4 μm; the ora, when present, lolongate and faint, 3–8 × 2.0–3.2 μm; and the opercula granular or inaperturate (*Populus*). The sexine is thin in *Populus* (<1 μm) and granulate or somewhat thicker in *Salix* (to 1.2 μm) and reticulate with lumina 0.5–1.6 μm in diameter having free piloid elements, smaller toward the colpi; the nexine thinner than the sexine; and the intine 1.0–1.2 μm thick in *Populus* and ca. 0.8 μm in *Salix*, slightly thickened (to 1.2 μm) at the apertures.

Allergenicity

Pollen extracts of two species of willow, together with box-elder and hickory, elicited the greatest allergic response of trees in skin tests administered to adults (Lewis & Imber 1975b). In contrast, extracts of *Populus*

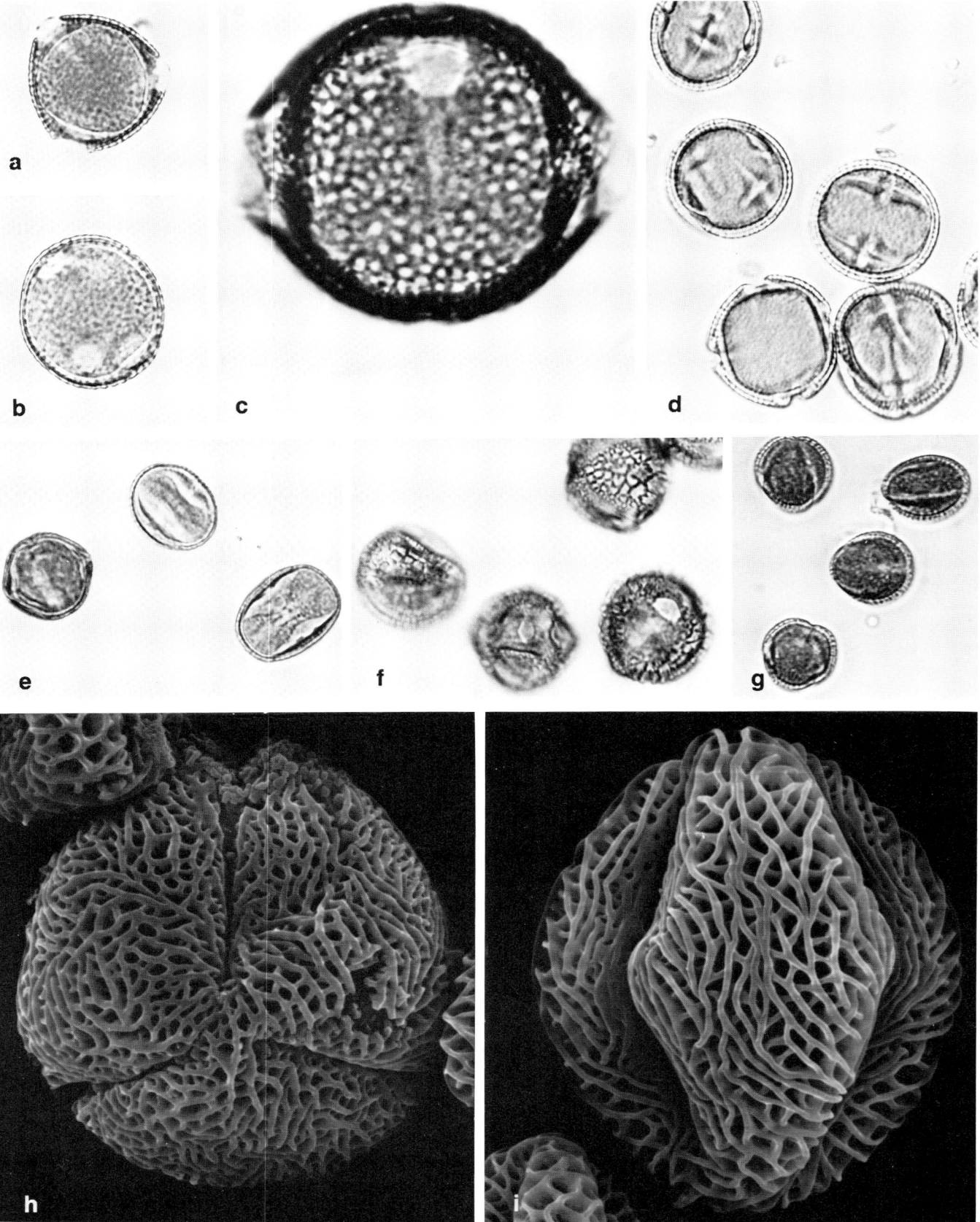

Fig. 102. Pollen of representative Rutaceae. a–d: Citrus *aurantiifolia* (lime), a = LM(s) polar view ×720, b = LM(s) oblique-equatorial view ×720, c = LM(s) oblique-equatorial view showing part of colporate aperture ×1830, d = LM(a) ×720; e: *Fortunella marginata* (oval kumquat), LM(s) ×720; f: *Phellodendron amurense* (cork tree), LM(s) ×720; g–i: *Zanthoxylum pterota*, g = LM(s) ×720, h = SEM(a) polar view ×5200, i = SEM(a) equatorial view showing 2 apertures ×4600.

deltoides elicited only moderate responses and those of *P. nigra* even lower. Nationwide, however, no distinction was observed between average levels of response, for in our survey, which included results from seven states (Alabama, California, Iowa, Kansas, Missouri, New Jersey, and Texas), positive skin tests averaged a moderate 2.4+ for poplar extracts and 2.3+ for willow extracts. What did vary markedly was the number of atopic patients (1,355) responding positively to these extracts—67.5% to poplar extracts, and 32.5% to willow extracts. This greater incidence of sensitivity to poplar pollen by a frequency two to one undoubtedly reflects more patient exposure to the abundant wind-dispersed poplar pollen compared to much less exposure to the insect-pollinated willow pollen, even if allergens of willow tended occasionally to be more allergenic, as found in St. Louis.

In Minnesota important offenders are *Populus balsamifera, P. deltoides,* and *P. tremuloides,* whereas all species of *Salix* are considered of secondary importance (Ellis & Rosendahl 1933). In the Southwest several species of *Populus* are also considered important offenders and those of *Salix* secondary (Sellers 1935). *P. alba* is a significant offender in Yugoslavia (Pujevic 1959).

Although poplar species are considerably more important as incitants of inhalant allergy than are willow species, the pollen of both is allergenic. Less pollen may be shed by willows, however, even when they are as common as poplars and may consequently elicit fewer cases of pollinosis.

High correlations of cross-reactivity on skin testing were reported by Dalen & Voorhorst (1981) between *Populus* and *Salix,* suggesting important common allergens.

SAPINDACEAE
Soapberry Family

Largely a tropical family of trees, shrubs, and woody vines, the soapberries are represented in North America by only a few genera, the most important being *Koelreuteria* (golden-rain tree) (Fig. 108) and *Sapindus* (soapberry). The former is widely cultivated as a street tree. Although insect-pollinated, anthers are exposed in the flower and pollen may incidentally become windborne. Flowering is primarily in June and July, but *Sapindus* also flowers in the spring in warmer regions.

Pollen Morphology (Fig. 109)

Grains are subprolate to oblate-spheroidal; the amb triangular with ±straight or convex sides and 3-colporate; the colpi 12–19 × 1.5–2.5 μm; the ora lalongate, 1.5–2.5 × 4.8–5.6 μm, with uneven and sometimes faint margins; and the opercula granular. The sexine is 0.8–1.0 μm thick, finely striate (*Koelreuteria*) or tegillate (*Sapindus*); the nexine about as thick as the sexine, thickened at the apertures; and the intine 0.8–1.0 μm thick, somewhat thicker below the apertures.

Sapindus pollen is smaller (ca. 15–18 μm) than that of *Koelreuteria* (22–26 × 24–28 μm).

SAXIFRAGACEAE
Saxifrage Family

A family closely related to the Rosaceae, the Saxifragaceae are important for their fruits (currants, gooseberries) and cultivated plants (Fig. 110). Pollination is largely by vectors, and only incidentally does the pollen become airborne. Flowering occurs primarily from late spring to midsummer.

Only one case of moderately positive (2+) skin test response to *Philadelphus* (mock-orange) pollen associated with pollinosis was reported by Lewis &Vinay (1979). An earlier report associated *Philadelphus* and *Hydrangea* pollen with hay fever (Scheppegrell 1922), but allergenic potential of their pollen and that of *Ribes* is unknown.

Pollen Morphology (Fig. 111)

Grains are prolate or subprolate (15–17 × 10–14 μm) to spheroidal (*Ribes*) (20–28 μm); the amb rounded-triangular, sometimes with convex sides, and 3-colporate (*Hydrangea, Philadelphus*) or 8–9-pantoporate (*Ribes*); the colpi ca. 14.0 × 0.1–2.0 μm, syncolpate in *Hydrangea*; the ora faint and lalongate (ca. 1.5 × 1.8 μm in *Hydrangea,* ca. 2.4 × 8.0 μm in *Philadephus*); the many pores of *Ribes* circular, ca. 1.6 × 3.0 μm, and densely granulate around them. The sexine is reticulate; the lumina polygonal, ca. <0.5–1.0 μm, smaller toward the apertures in *Hydrangea* and *Philadelphus,* or tectate and granulate in *Ribes,* 0.6–0.7 μm thick; the nexine often thicker or slightly thinner than the sexine; and the intine ca. 0.8 μm thick and to 1.6 μm thick beneath the aperture.

SIMAROUBACEAE
Quassia Family

Of the members of the Simaroubaceae, primarily a tropical family of trees and shrubs, the genus *Ailanthus* introduced from China in the eighteenth century is undoubtedly the best-known representative in North America. *A. altissima* (tree-of-heaven) (Fig. 112) is widely used as a yard and street tree from midlatitudes south, and it has also become an aggressive and common weed of waste places.

Pollination is primarily entomophilous but also facultatively anemophilous. Flowering and windborne

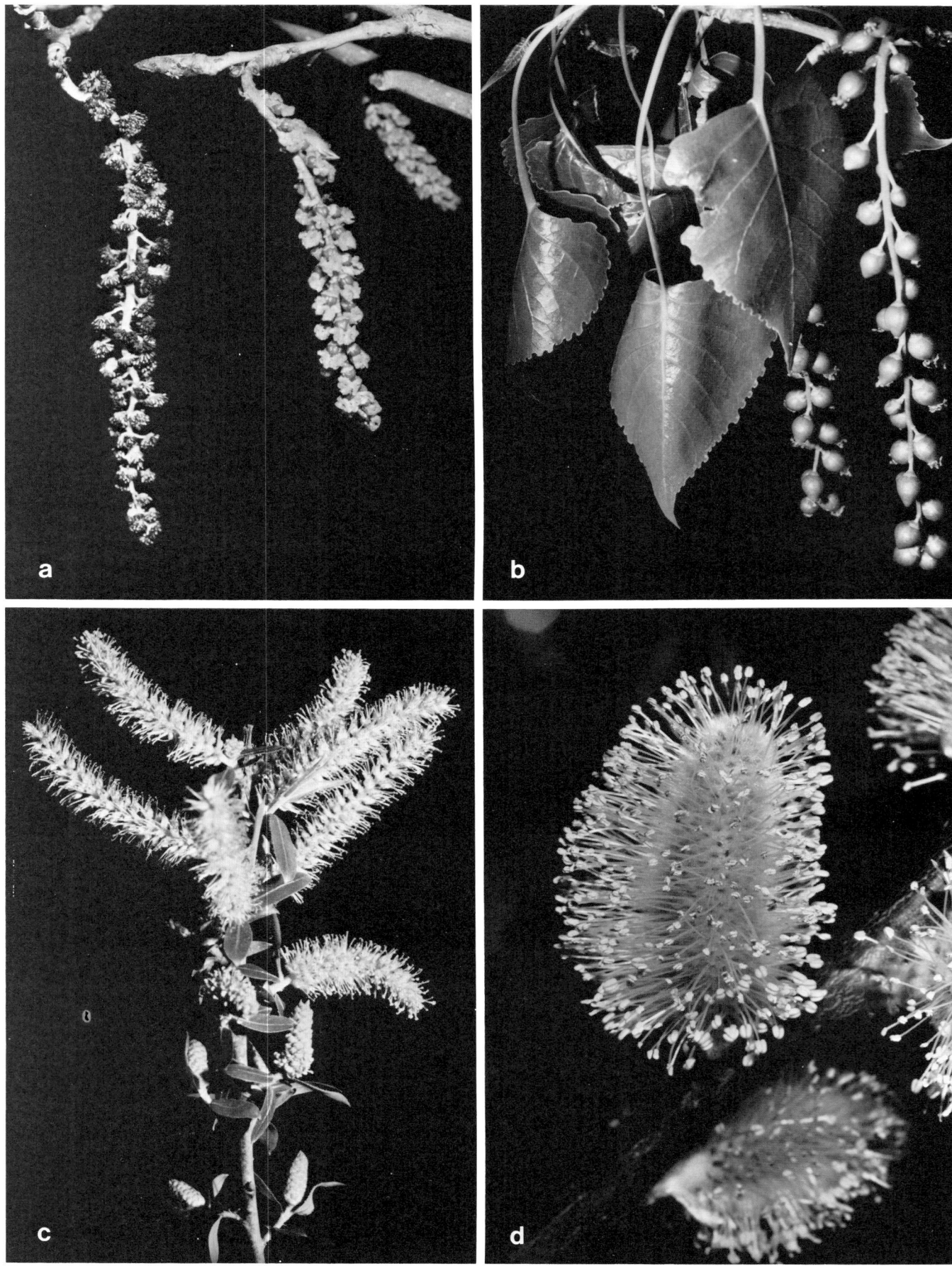

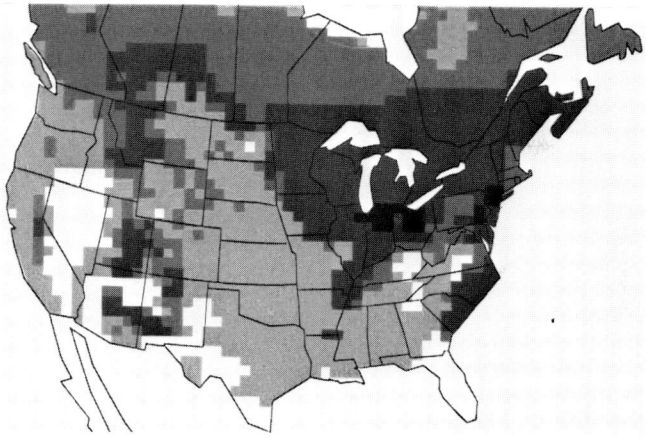

Fig. 104. Generalized composite distribution of nine indigenous species of *Populus* (☐ 1 sp., ▨ 2 spp., ▩ 3–4 spp., ■ 5–6 spp.).

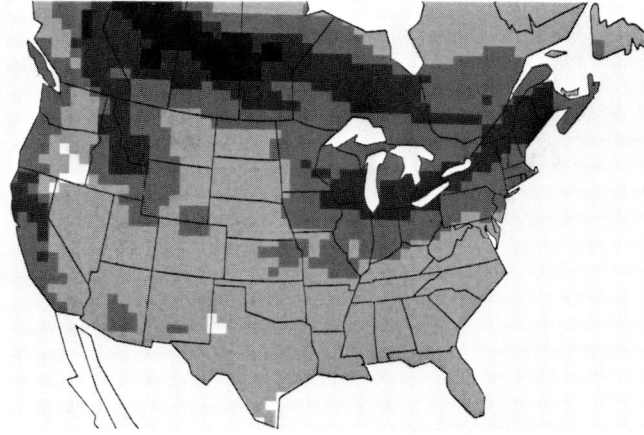

Fig. 105. Generalized composite distribution of 25 indigenous species of *Salix* (☐ 1–3 spp. ▨ 4–6 spp., ▩ 7–8 spp., ■ 9–11 spp.).

pollen can be expected mostly in May and June in the East, somewhat earlier in the southern extremes, and in June in the Southwest. However, the subtropical *Alvoradoa* of southern Florida is probably wind-pollinated (Tomlinson 1980).

Pollen Morphology (Fig. 113)

Grains are prolate to subprolate, 28–34 × 20–21 μm; the amb subtriangular and 3-colpate; the colpi long and narrow with tapering ends, ca. 27 × 1.5–2.0 μm; the ora lalongate, 3.5–4.0 × 5.6 μm; and the opercula granular. The sexine is reticulate; the lumina polygonal to elongate, <0.5–1.2 μm, smaller toward the apertures; the muri striate to striato-reticulate towards the apertures, ca. 1 μm thick; the nexine as thick as the sexine, slightly thickened adjacent to the apertures; and the intine ca. 1.2 μm thick, thickened beneath the apertures.

Allergenicity

Even though pollen is readily caught on atmospheric slides (Wodehouse 1971), there have been few reports of allergenicity in North America. However, we report 181 atopic patients from California who responded an average of 1.8+ to skin test pollen extracts. This modest level of positive reactions and moderate incidence of sensitivity is also apparent in the eastern United States, where only local pollinosis has been reported (Blumstein 1943). In Israel *Ailanthus* pollen was unquestionably responsible for four cases of inhalant allergy (Tas 1956).

TAMARICACEAE

Tamarisk Family

This small family of old-world, heathlike shrubs and trees is represented almost exclusively in North America by *Tamarix* (tamarisk, salt cedar). Several species are grown as ornamentals and for windbreaks and sand-binding; additionally, they have become widely naturalized along streams, salt flats, coastal areas, and waste places generally.

Tamarisks often flower from spring to autumn, and in the Southwest *Tamarix ramosissima* (Fig. 114) may

TABLE 10. Flowering by month of *Populus* and *Salix*, where native.

GENUS	REGION						
	nNE	sNE	SE	NC	SC	NW	SW
Populus	4–5	3–5	3–4(–5)	4–5	2–4(–7)	4–5	2–4(–6)
Salix	4–6(–7)	(2–)3–6	3–4	4–5	3–5	(3–)5–7	(2–)3–6(–6)

Fig. 103. Morphology of representative Salicaceae. a–b: *Populus deltoides* (cottonwood) leafless branch showing staminate catkins (a, ×1⅔) and a branch with leaves showing pistillate catkins developing fruit (b, ×¾); c: *Salix caroliniana* (coastal plain willow) branch with staminate catkins, ×¾; d: *Salix bebbiana* (long-beaked willow) branch with staminate catkins, ×4.

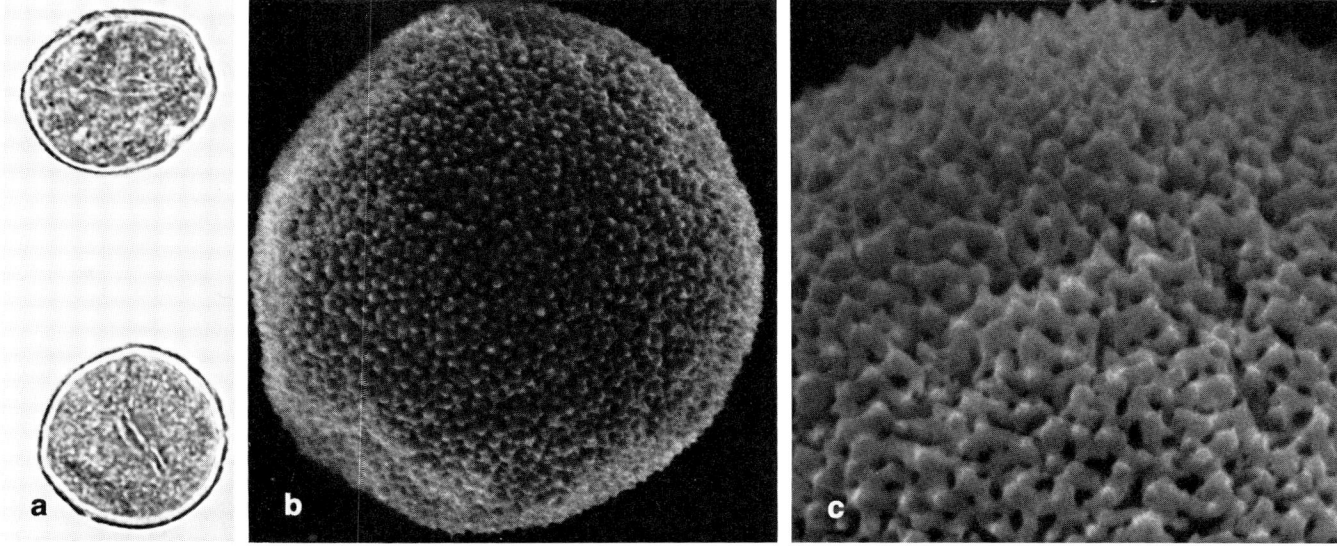

Fig. 106. Pollen of *Populus deltoides* (cottonwood). a = LM(s) ×720, b = SEM(a) ×3800, c = SEM(a) sexine surface ×11000.

flower much of the year. Pollen is disperse by vectors, facultatively by wind.

Pollen Morphology (Fig. 115)

Grains are subprolate-spheroidal, 17–18 × 14–16 µm; the amb triangular, lobate, and 3(–4)-colpate; the colpi 16 × 2.0–2.5 µm; the margins slightly uneven; and the opercula granular. The sexine is ca. 0.9 µm thick, reticulate; the lumina small (<0.5–0.7 µm) and even smaller toward the colpi and polygonal; the nexine slightly thinner than the sexine; and the intine only 0.6–0.8 µm thick.

Allergenicity

Pollen of *Tamarix* may become atmospheric. Allergenicity has occasionally been reported from Arizona (Wodehouse 1971) and California (Small & Small 1946). We report only four patients with positive skin tests (2+ to 3+) from California who suffered pollinosis on exposure to tamarisk allergens.

TILIACEAE

Linden Family

Primarily a tropical woody family, the lindens, or basswoods, are represented in eastern North America by native and introduced species of *Tilia* (Fig. 116). Flowering occurs for about five weeks, principally during June and July, but species in the southcentral region may flower from March to August. Flowers often secrete a sweet scent and nectar; entomophily is typical by bees and flies diurnally and by moths nocturnally, with anemophily secondary (Anderson 1976). *Tilia* pollen is produced in large quantities, however (Hyde and Williams 1945), and may be found in the pollen rain (Hyde 1950), often at considerable distances from the flowering trees (Wodehouse 1971).

Pollen Morphology (Fig. 117)

Grains are slightly paraisopolar, peroblate, 20–28 × 35–47 µm; the amb oblate or circular and 3-colporate; the brevicolpi 10–16 × 2–3 µm, the margins irregular, surrounded by a thickened nexinous rim ca. 3 µm wide; and the ora circular to lolongate, ca. 7 × 4 µm. The sexine is ca. 1.2 µm thick, finely reticulate to pitted; the nexine as thick as or slightly thinner than the sexine, and much thickened adjacent the apertures (to 3 µm); and the intine thickened beneath the apertures to form an oncus.

Allergenicity

There is no question of the allergenic potential of *Tilia* pollen. As reported in Virginia it can be a major inhalant offender where common and where more or less close proximity to the flowering trees is possible (Derbes 1941). *Tilia* pollinosis should receive greater consideration than it does east of the Rocky Mountains during June and July.

ULMACEAE*

Elm Family

A family of tropical and temperate trees and shrubs, the elms have leaves that are alternate, simple, and usually deciduous, with blades often oblique at their base and stipules that are paired and shed as the leaves unfold. Flowers are bisexual or unisexual, typically green, much reduced, and in clusters. There

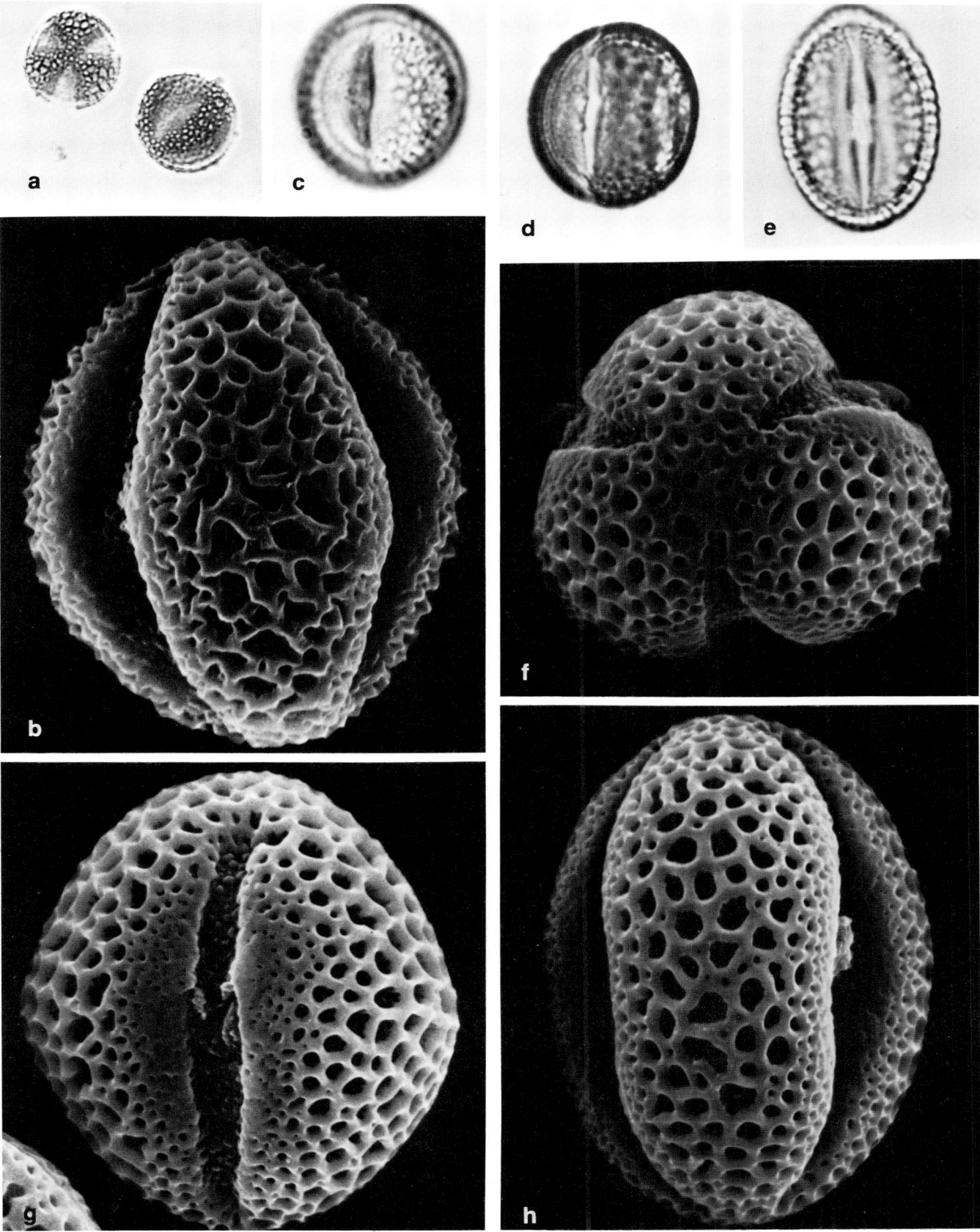

Fig. 107. Pollen of *Salix*. a: *S. sitchensis* (Sitka willow), LM(s) polar (upper left) and equatorial views ×720; *S. bebbiana* (long-beaked willow), SEM(a) equatorial view showing 2 apertures ×5200; c–h: *S. caroliniana* (coastal plain willow), c–e = LM(s) equatorial views ×1830, f = SEM(a) polar view ×5400, g–h = SEM(a) equatorial views showing 1 (g) and 2 (h) apertures ×5800.

Fig. 108. *Koelreuteria paniculata* (golden-rain tree) branch with inflorescence, ×⅖.

Fig. 110. Morphology of representative Saxifragaceae. a: *Philadelphus pubescens* (mock-orange) branch with flowers, ×⅘; b: *Ribes viscosissimum* branch with flowers, ×1.

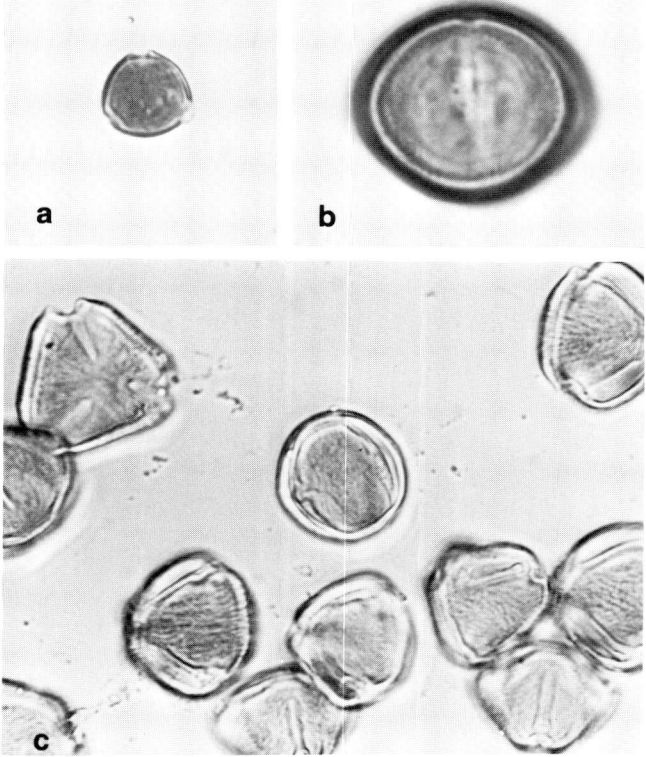

Fig. 109. Pollen of *Sapindus* and *Koelreuteria*. a–b: *S. drummondii* (western soapberry), a = LM(s) ×720, b = LM(s) equatorial view ×1830; c: *K. paniculata* (golden-rain tree), LM(a) ×720.

are five to eight united sepals, no petals, four to eight stamens, and a pistil of two carpels having a superior ovary and two styles. Fruit is a samara or drupe.

There are two major genera in North America, *Celtis* (hackberry, syn. *Momisia*) and *Ulmus* (elm) (Fig. 118), with two minor representatives in the Southeast, *Planera* (planer tree or water elm) and *Trema* (Florida trema). Many species are native, but some are also introduced from Europe and Asia. Species of *Zelkova* are being planted in the east as a replacement for elms. The family is wind-pollinated.

Indigenous species of *Celtis* and *Ulmus* occur primarily in the eastern half of the continent, although the range of *Celtis* extends in a continuous southern distribution to Arizona. *Celtis* is most common in a narrow midlatitude belt from Tennessee to Oklahoma (Fig. 119). *Ulmus* is frequent from Ontario to Georgia and west to southern Minnesota and central Texas, with a concentration of species from Tennessee to Oklahoma and central Texas (Fig. 120). Unfortunately, *U. americana*, one of the six native species, is being destroyed over vast regions of its range by the Dutch elm disease, and as a consequence frequency of elm pollinosis has been markedly reduced.

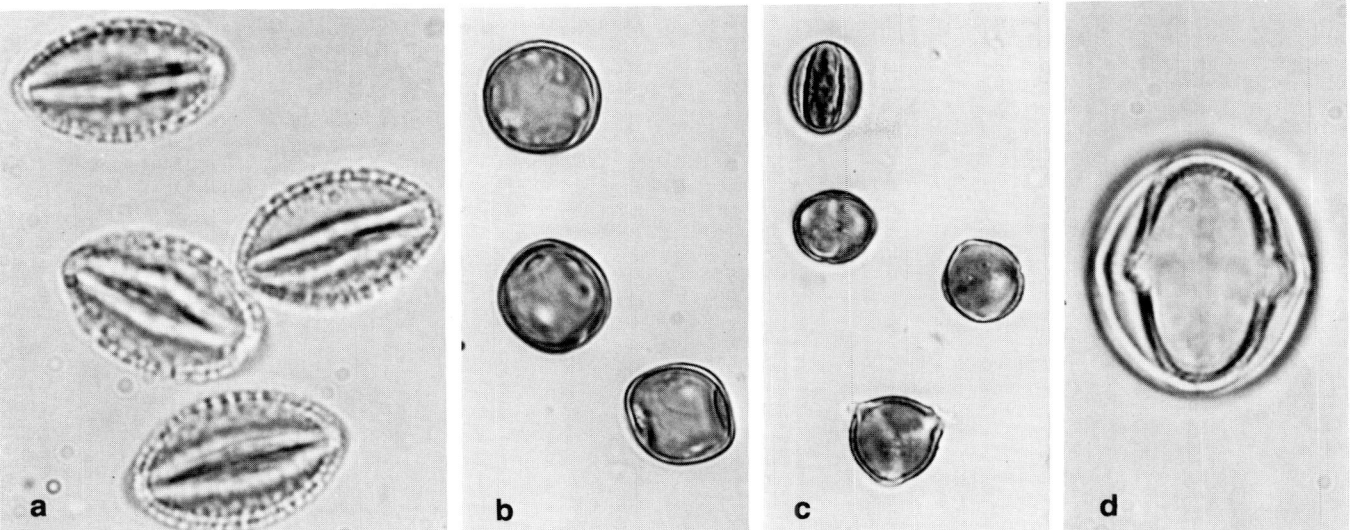

Fig. 111. Pollen of representative Saxifragaceae. a: *Hydrangea quercifolia* (oak-leaved hydrangea), LM(a) ×1830; b: *Ribes sativum* (common currant), LM(s) ×720; c–d: *Philadelphus × cymosus* (mock-orange cultivated hybrid), c = LM(s) ×720, d = LM(a) equatorial view ×1830.

Flowering and Pollen Aerobiology

Elms flower at two distinct times. The majority are among the early elements of tree pollinosis, often beginning to flower in January or February in southern latitudes and as late as April in northern regions. Peak months in the East are February and March. Two native species (*Ulmus crassifolia, U. serotina*) and the Chinese *U. parvifolia*, which is widely planted in the South and in California, flower during the late summer to October and occasionally to November. *U. crassifolia* in its native range has a long flowering period, from July to October.

The hackberries flower strictly in the spring, usually during April and May, and about one month earlier in Florida. Planer trees flower from March to April and *Trema* often flower year-round in southern Florida.

Atmospheric frequencies of *Celtis* and *Ulmus* pollen captured on slides are summarized in Table 11. In the southeastern, northeastern, and southcentral regions anthesis occurs principally from January to March, although extending beyond this time in the North, and in the northcentral and northwestern regions pollen is released primarily from February to April. Pollen shed is abundant during these periods. In the Southwest the peak spring shed is in March. Dispersal of late-flowering pollen is most significant in the southcentral and southwestern regions during August and September with reduced frequencies in July and October.

Fig. 112. *Ailanthus altissima* (tree-of-heaven) branch showing compound leaves and inflorescence (a, ×⅙), flowers (b, ×1), and fruit (c, ×⅙).

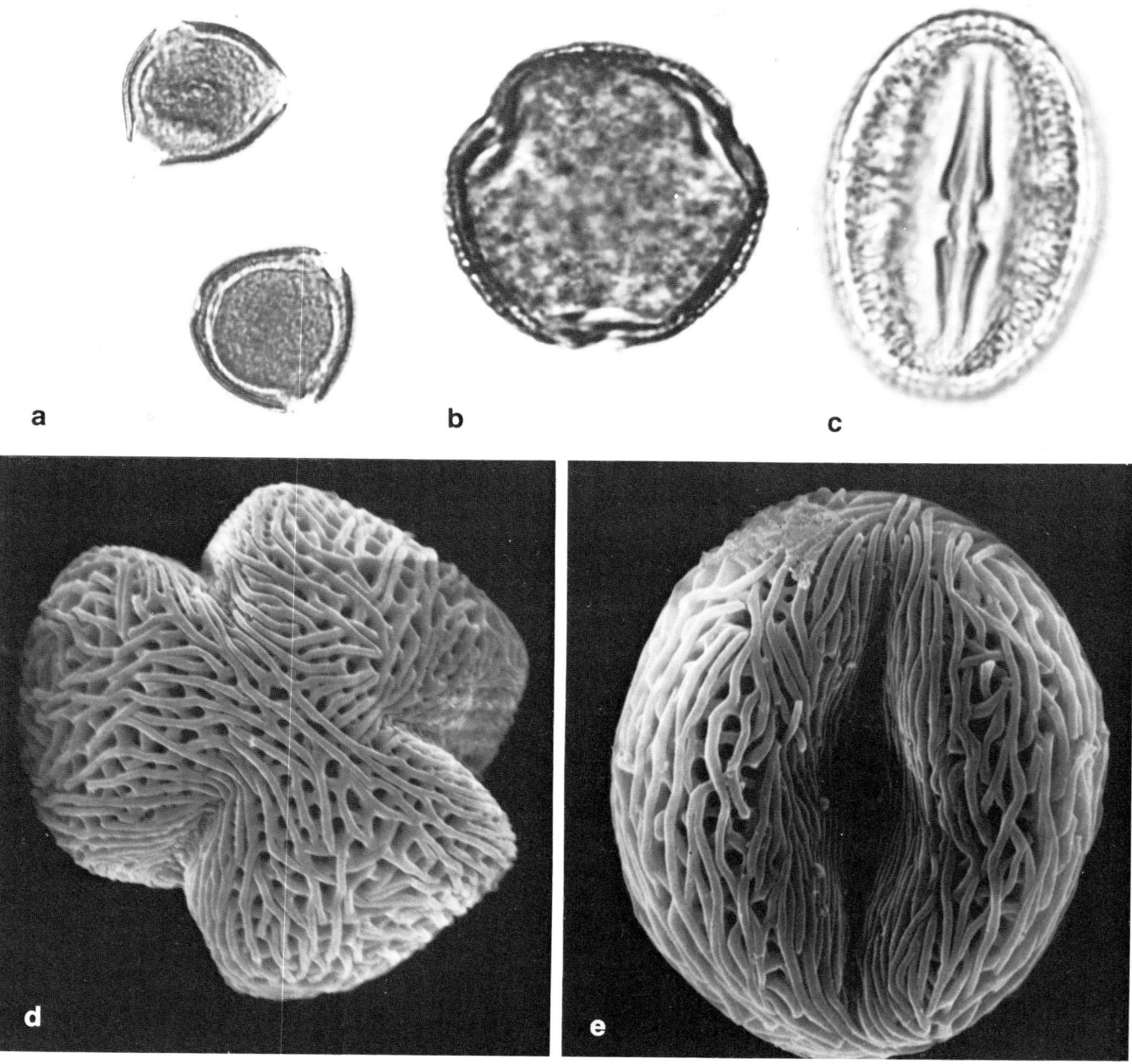

Fig. 113. Pollen of *Ailanthus altissima*. a = LM(s) ×720, b = LM(s) polar view ×1830, c = LM(a) equatorial view ×1830, d = SEM(a) polar view ×4200, e = SEM(a) equatorial view showing 1 aperture ×3800.

Pollen Morphology (Figs. 121–22)

Grains are suboblate to spheroidal, sometimes paraisopolar, 16–50 × 18–50 μm, those of *Trema* being the smallest (16–18 × 18–19 μm); the amb from oblate-spheroidal and spheroidal to angular and square, or polygonal; 2–20-porate and stephanoporate (in *Planera* and *Ulmus*), with the pores sometimes unequally spaced equatorially, lalongate to lolongate, typically 2–3 μm in diameter, but smaller (1.6 × 1.2 μm) in *Trema* and annulate; the annulus to 2.5 μm thick; and the opercula granular or flecked. The sexine is 1.2–1.5 μm (*Celtis, Planera*) or 0.4–0.7 μm (*Trema, Ulmus*) thick, often thickened toward the pores, tegillate, undulating and granulate (*Celtis*), rugulate and undulating (*Planera, Ulmus*), or microverrucate (*Trema*); the nexine thinner (ca. 0.6 μm) or thicker (0.6 μm in *Trema*) than the sexine; and the intine 1.0–1.6 μm thick, considerably thickened, forming an oncus be-

TREES AND SHRUBS

Fig. 114. *Tamarix ramosissima* (tamarisk) branch with flower and mostly floral buds, ×1½.

Fig. 116. *Tilia americana* (American basswood or linden) branches showing leaves and flowers subtended by a ligulate bract, ×1.

neath the pores, especially in *Celtis* (to 4.8 × 12 μm) and *Trema* (to 3.2 × 11 μm).

Grains of *Ulmus* are (4–)5–6(–7)-stephanoporate; those of *Celtis* are 3–10-porate.

Allergenicity

In our survey nationwide, allergists reported 1,453 atopic patients who on the average responded mod-

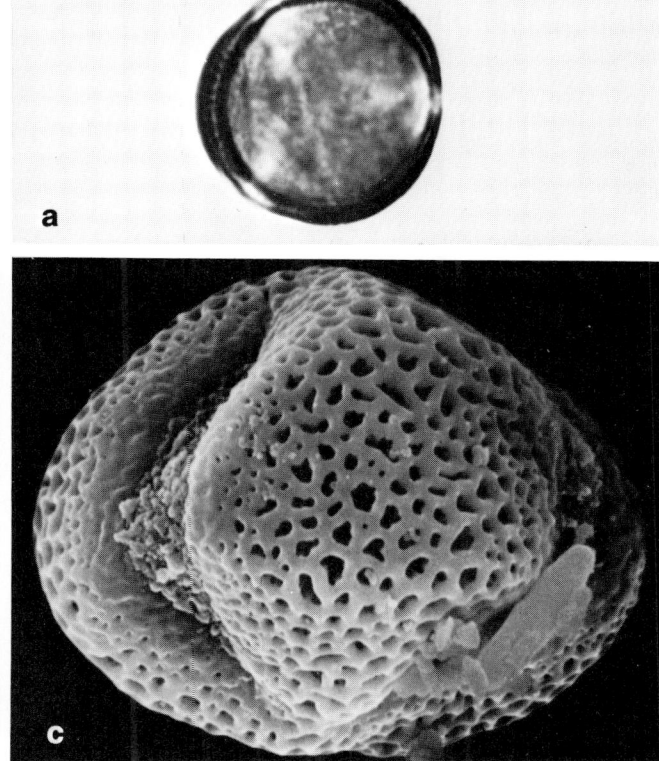

Fig. 115. Pollen of *Tamarix ramosissima*. a = LM(s) ×1830, b = SEM(a) oblique-polar view ×6000, c = SEM(a) equatorial view showing 2 apertures ×6200.

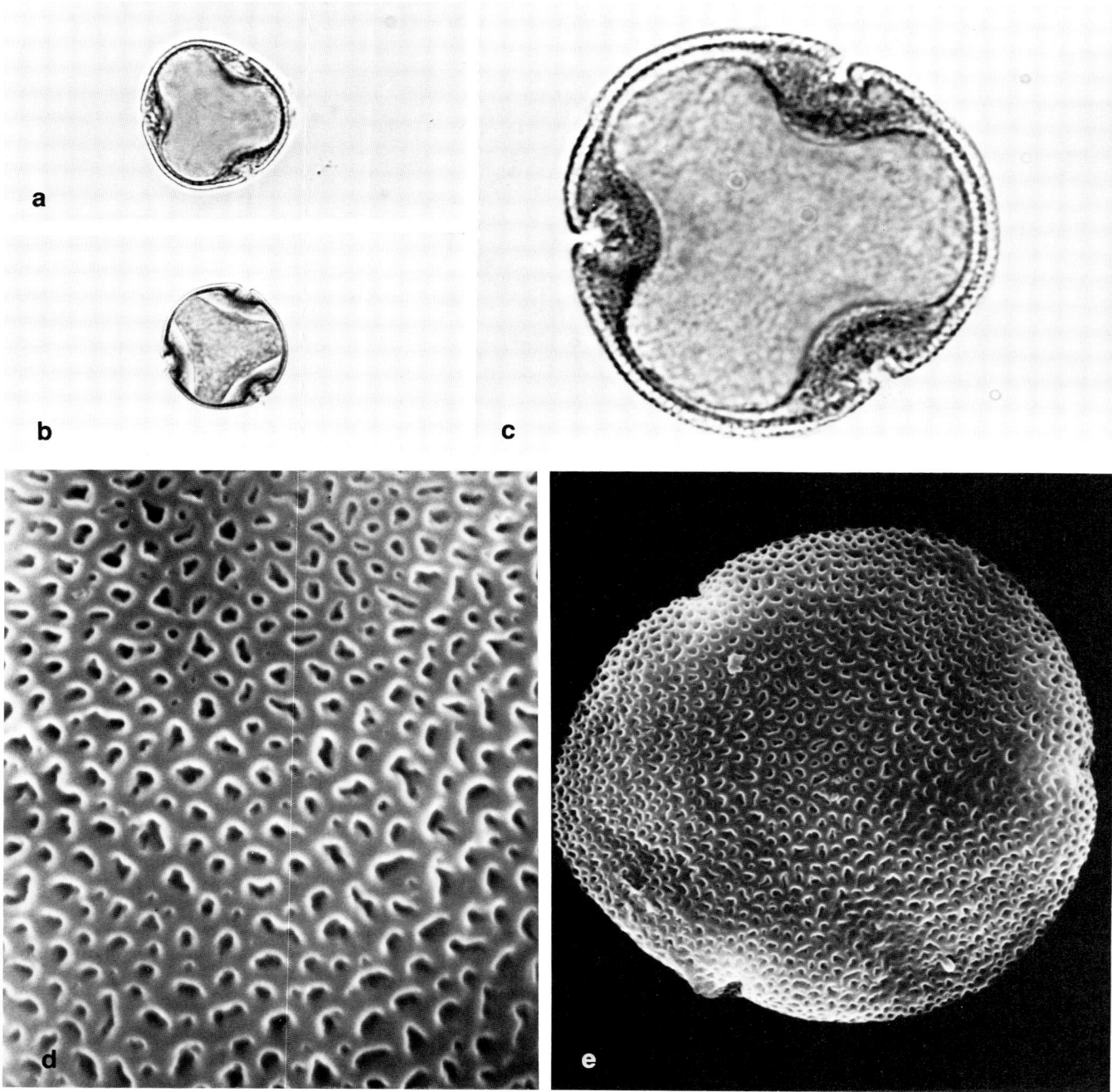

Fig. 117. Pollen of *Tilia americana*. a–c = polar views, a = LM(a) ×720, b = LM(s) showing 3 onci below the apertures ×720, c = LM(a) ×1830, d = SEM(ad) showing sexine surface ×8800, e = SEM(ad) polar view ×3600.

erately high (3.0+) to skin tests using spring elm pollen extracts; response with autumn elm pollen among 997 patients averaged somewhat more positive (3.4+). In southern California spring-flowering elms are sometimes considered unimportant and only autumn testing is undertaken. Reports of many patients severely allergic to autumn-flowering elm pollen were common from southern California.

Only 30 patients were reported as responding positively in skin tests to hackberry pollen extracts. Their response averaged a moderate 2.5+. In the Southwest, including Texas, *Celtis* has been reported of secondary importance in pollinosis (Sellers 1935). This divergence between hackberries and elms in allergenic levels and incidences, and the absence of reports of cross-sensitivies, implies some important allergens not shared by the two genera. They are classified in separate tribes of the family.

Data for *Planera* and *Zelkova* (both tribe Ulmeae) and *Trema* (tribe Celtideae) are wanting. Allergenicity

Fig. 118. Morphology of representative Ulmaceae. a: *Celtis occidentalis* (hackberry) branch with leaves and flowers, ×1½; b: *Ulmus parvifolia* (Chinese elm) branch with flowers, ×⅗; c–d: *Ulmus americana* (American elm) branch with leaves and fruit (c, ×½) and flowers showing anthers (d, ×1½).

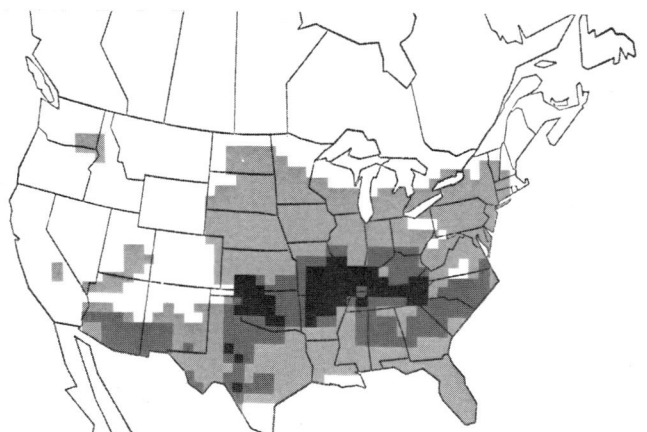

Fig. 119. Generalized composite distribution of 5 indigenous species of *Celtis* (☐ 1 sp., ▦ 2 spp., ▪ 3 spp.).

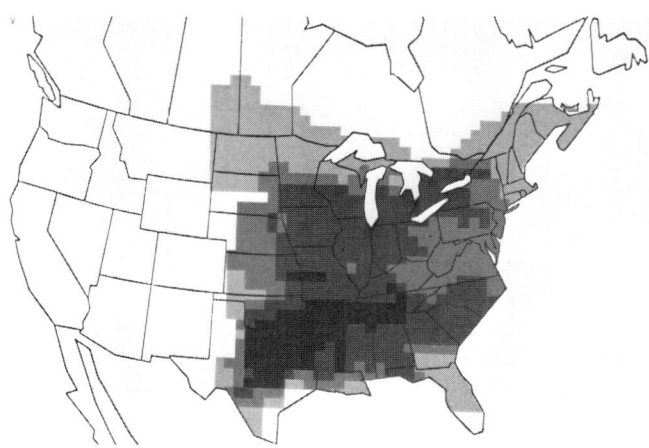

Fig. 120. Generalized composite distribution of 6 indigenous species of *Ulmus* (☐ 1 sp., ▦ 2 spp., ▪ 3 spp., ■ 4–5 spp.).

TABLE 11. *Ulmus* and secondarily *Celtis* aeropollen frequency in regions of North America based on percentage of pollen captured, by month (1974–78).

REGION	MONTH										
	1	2	3	4	5	6	7	8	9	10	11
nNE	9.6	61.2	54.8	23.2	2.6	1.6					
sNE	12.2	43.8	29.3	7.4	6.2	4.3					
SE	7.0	2.6	7.3	0.2							
NC		45.1	61.1	16.1	4.5						
SC	24.6	61.3	24.0	7.7	0.7			3.2	7.7	1.0	
NW		28.1	7.1	12.3	2.1	0.5					
SW	1.2	0.9	5.6	2.1	0.4			9.6	7.9	2.1	
sCA	0.3	0.2	0.2				2.2	10.2	9.7	2.8	0.6

can be expected, perhaps following the tribal classification, whereby pollen of the first two along with *Ulmus* will be of more significance than that of *Trema* and *Celtis*. This will depend on frequency of exposure and, compared to other plants in the family, elm pollen exposure in North America will continue to be paramount. In Argentina, however, where *Celtis* is common, *C. tala* is the chief offender in 44% of patients with hay fever (Carron & Malvarez 1941).

VISCACEAE

Mistletoe Family

The mistletoes are green parasites of shrubs and trees anchored to the host by suckers (modified adventitious roots). They are mainly tropical but extend into temperate areas, being particularly frequent in southern and western regions.

Dwarf mistletoes (*Arceuthobium*) consistently liberate pollen that is dispersed by wind from well-exposed anthers (Player 1979). Dispersal distance is not known, but as the plants are often found high in trees, pollen conceivably could be carried some distance depending on convection currents. Flowering occurs during the spring and summer. We are not aware of reports of pollinosis due to *Arceuthobium* or *Phoradendron* (mistletoe).

Pollen Morphology (Fig. 123)

Grains are subprolate, 26–29 × 22 μm; the amb triangular with slightly concave sides and 3-colpate (or colporoidate); the colpi ca. 16 × 0.5 μm, slightly constricted equatorially; and the ora very faint, lolongate. The sexine is ca. 1.6 μm thick, with spinules ca. 1 μm high; the nexine ca. 0.8 μm thick; and the intine thinner, ca. 0.6 μm thick.

VITACEAE

Grape Family

Mostly climbing plants, the Vitaceae are best known in North America by grapes (*Vitis*) and by Boston ivy and Virginia creeper (*Parthenocissus*) (Fig. 124). The Vitaceae flower from May to July depending on latitude, but occasionally grapes flower in April. Pollination is by insects. Pollen is secondarily wind-dispersed (Knuth 1908) and atmospheric pollen is infrequently collected on exposed slides (Allessio & Rowley 1966). Allergenicity is not known.

Pollen Morphology (Fig. 125)

Grains are prolate to subprolate; the amb triangular with ±straight sides, angulaperturate, and 3-colporate; the colpi 23–38 × 0.5–2.4 μm with tapering ends and somewhat thickened margins; the ora lalongate to lolongate, 1.8–3.2 × 1.6–4.8 μm; the margins ±thickened; and the opercula granular. The sexine is 0.6–0.9 μm thick, reticulate (*Parthenocissus*) with polygonal to elongate lumina, 0.5–1.2 μm in diameter, or punctate (*Vitis*); the nexine as thick as or thinner than the sexine; and the intine ca. 0.8 μm thick, to ca. 1.7 μm thick beneath the apertures.

Grains of *Parthenocissus* are much larger (40–48 × 27–39 μm) than those of *Vitis* (26–31 × 19–22 μm).

ZYGOPHYLLACEAE

Lignum Vitae Family

A largely tropical family, the Zygophyllaceae are represented in the United States by one, potentially allergenic, species, *Larrea divaricata* (creosote bush, greasewood) (Fig. 126). The species flowers from February to August in the southcentral regions and from March to May in the Southwest. It is insect-polli-

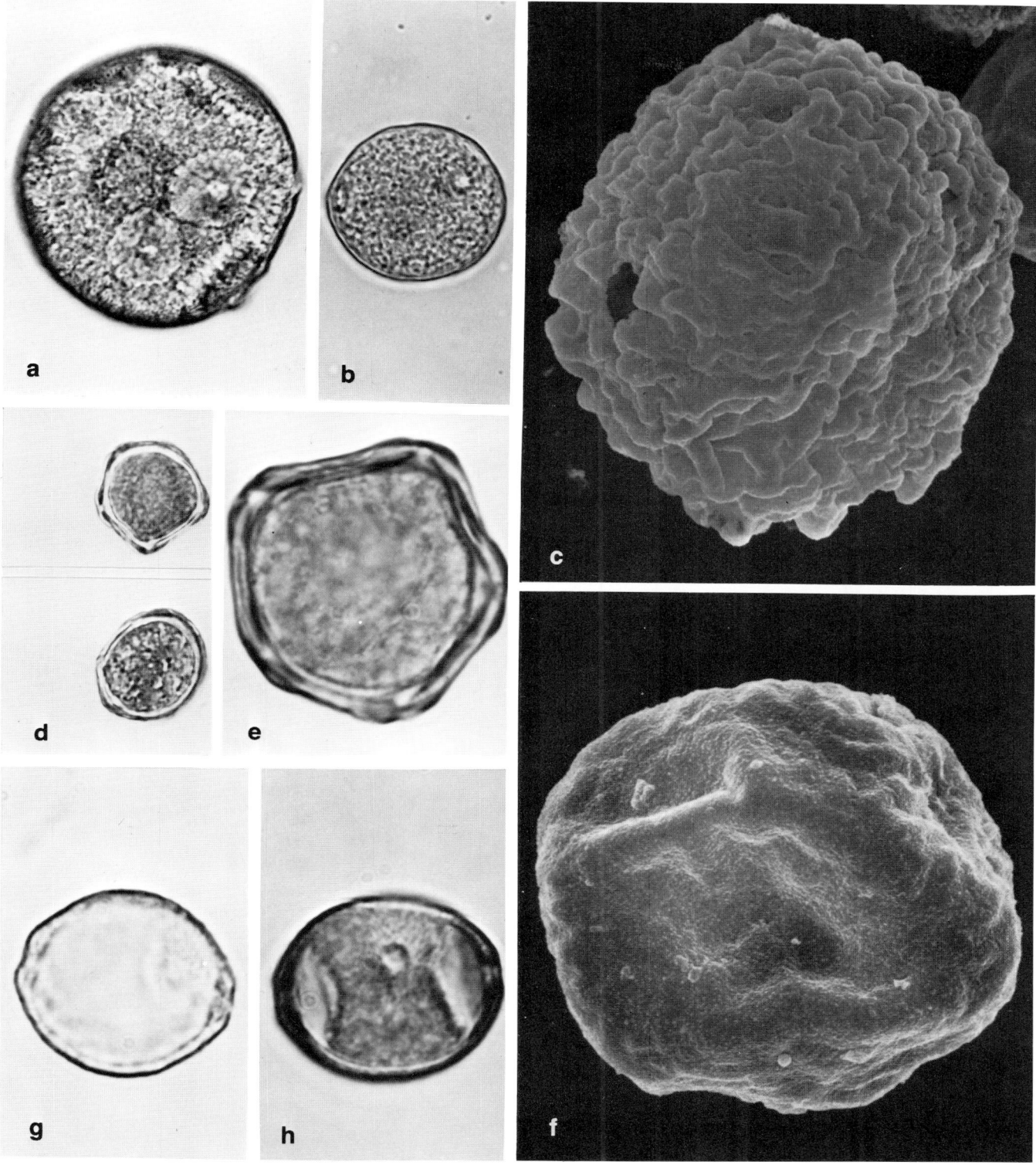

Fig. 121. Pollen of representative Ulmaceae (*Celtis-Trema*). a: *Celtis tenuifolia* (Georgia hackberry), a = LM(s) ×720; b–c: *Celtis occidentalis* (hackberry), b = LM(s) ×720, c = SEM(ad) ×2300; d–f: *Planera aquatica* (planer tree), d = LM(s) ×720, e = LM(s) polar view ×1830, f = SEM(ad) ×3400; g–h: *Trema micrantha* (Florida trema), g = LM(a) equatorial view ×1830, h = LM(s) equatorial view showing thickened intine ×1830.

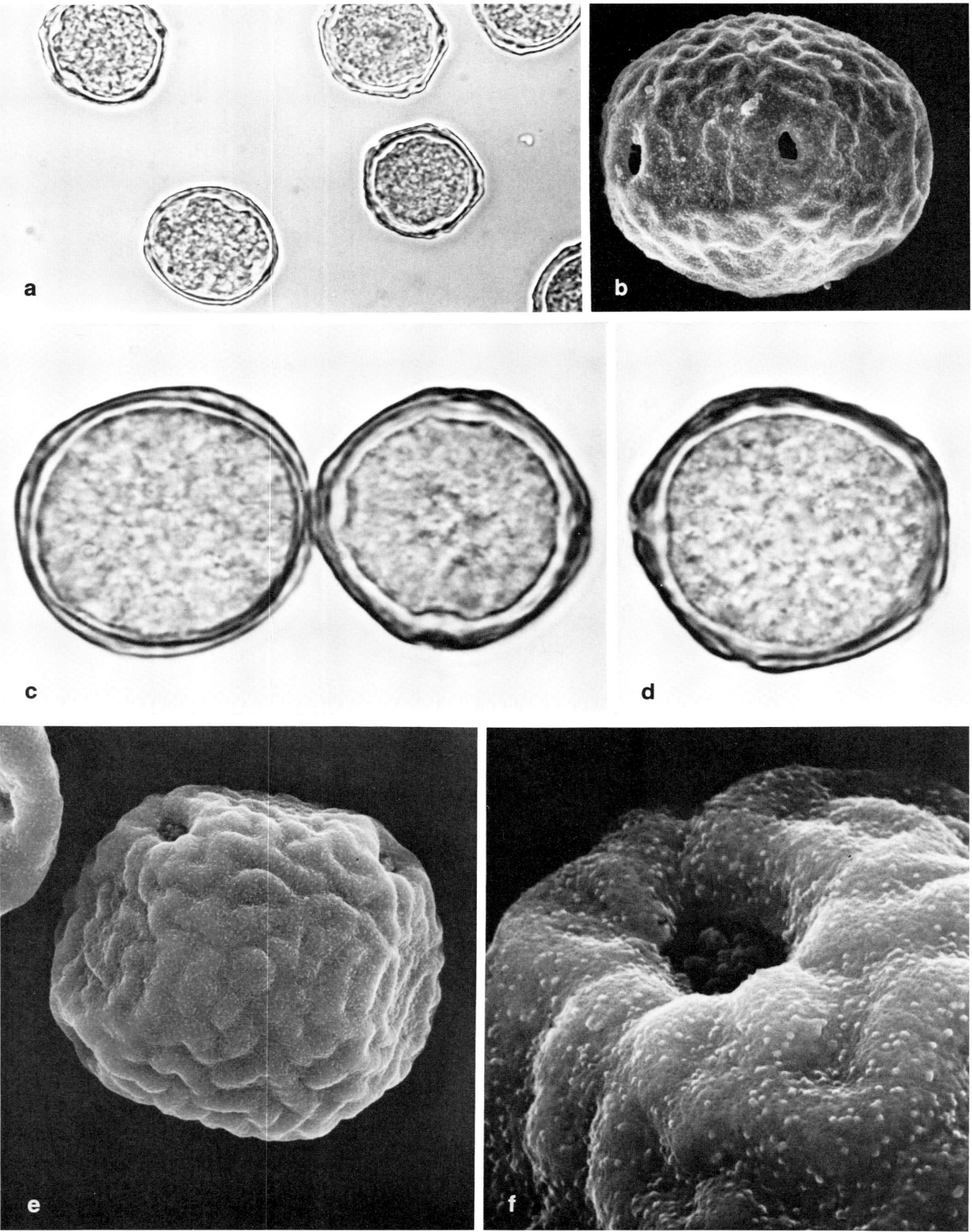

Fig. 122. Pollen of *Ulmus*. a–b = *U. americana* (American elm), a = LM(s) ×720, b = SEM(a) equatorial view ×2000; c–f: *U. parvifolia* (Chinese elm), c–d = LM(s) ±polar views ×1830, e = SEM(a) near polar view ×2800, f = SEM(a) pore morphology and sexine surface ×11000.

Fig. 123. Pollen of *Arceuthobium vaginatum* (dwarf mistletoe), LM(s) ×720.

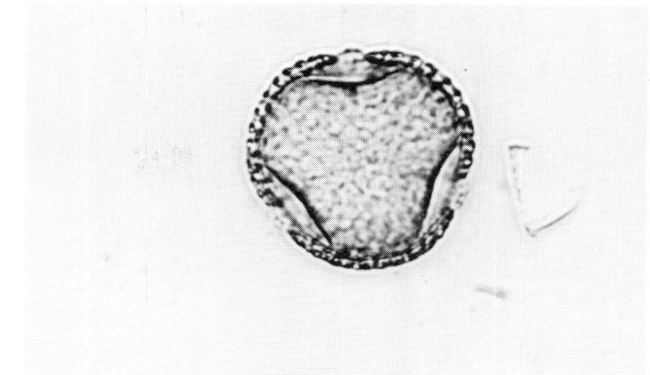

Fig. 124. Morphology of representative Vitaceae. a–b: *Vitis aestivalis* (pigeon or summer grape) branch with leaf and inflorescence (a, ×1⅓) and flowers (b, ×3⅓); c–d: *Parthenocissus tricuspidata* (Boston ivy) branches on wall with leaves and inflorescences (c, ×4) and flowers (d, ×4).

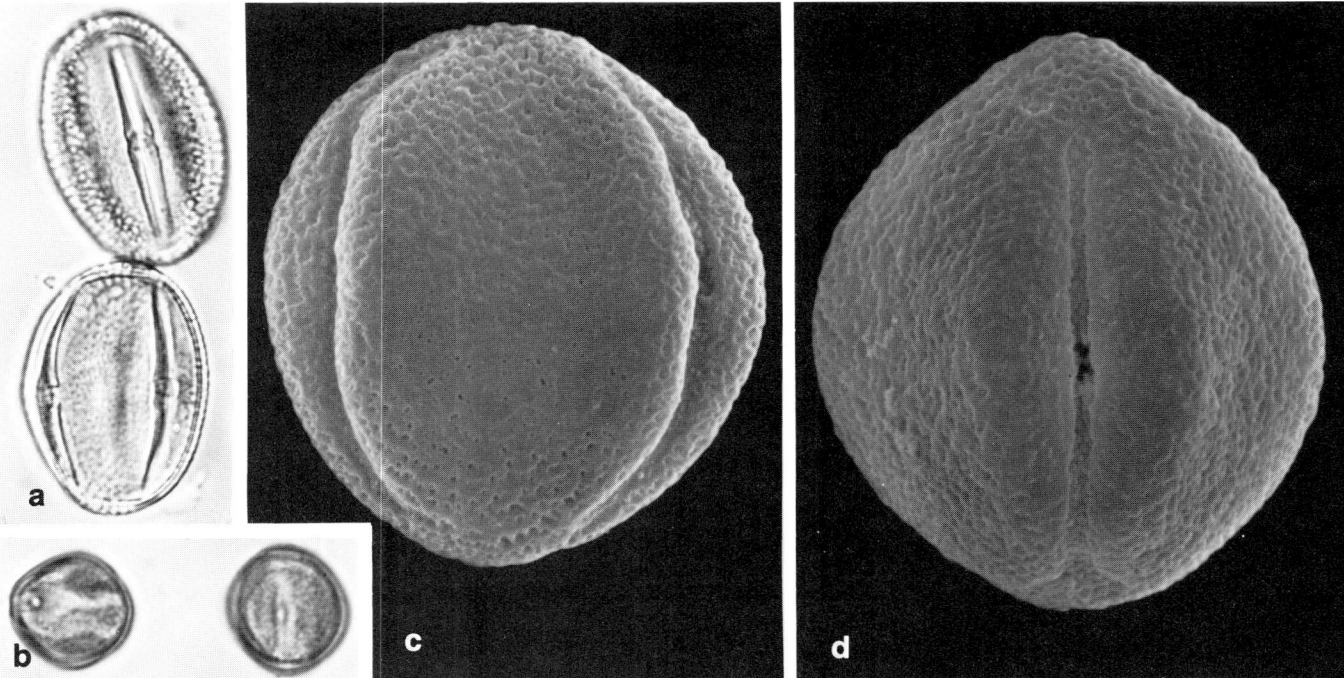

Fig. 125. Pollen of representative Vitaceae. a: *Parthenocissus tricuspidata* (Boston ivy), LM(a) equatorial views ×720; b–d: *Vitis labrusca* (fox grape), b = LM(s) ×720, c–d = SEM(a) equatorial views showing 1 (c) and 2 (d) apertures.

Fig. 126. *Larrea divaricata* (creosote bush) in a New Mexican habitat (a) and having leaves and flowers (b, ×1).

nated, but presumably the pollen may incidentally become airborne. The report of possible allergenicity is from California (Small & Small 1946).

Pollen Morphology (Fig. 127)

Grains are prolate to prolate-spheroidal, 19–20 × 15–17 μm; the amb triangular with convex sides and 3-colporate; the colpi 16 × 1.5–2.5 μm, slightly constricted equatorially; the ends pointed; the ora lalongate, 2.4 × 3.5–7.0 μm, or occasionally lolongate, faintly demarked; and the opercula granular. The sexine is ca. 0.8 μm thick, reticulate; the lumina small (0.5–0.7 μm) and polygonal; the nexine as thick as the sexine; and the intine also thin (ca. 0.8 μm).

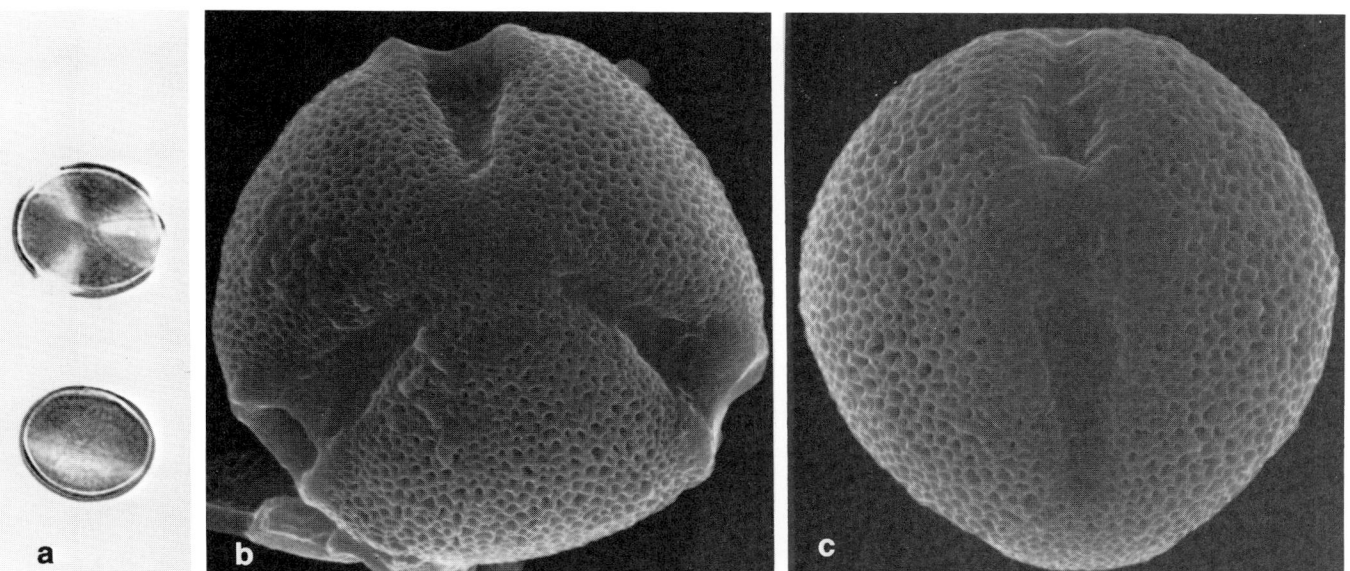

Fig. 127. Pollen of *Larrea divaricata*. a = LM(s) polar (upper) and equatorial views ×720, b = SEM(a) polar view ×4800, c = SEM(a) equatorial view showing 1 aperture ×4800.

Chapter 2

Grasses and Grasslike Plants

POACEAE (Gramineae)

Grass Family

A large family of about 10,000 species, of which 1,200 occur in North America, the grasses are ecologically the most dominant and economically by far the most important in the world. They are cosmopolitan in distribution, representing the principal component in about 20% of the earth's vegetational cover. Grasses provide all the world's cereal crops, most of the sugar, grazing for domesticated and wild animals, building materials in many parts of the world (bamboos, thatch, matting), aromatic oils for soaps and perfumes, and make a major contribution to much of the world's landscape including urban and suburban environments. Many are, however, allergenic and because of their wind-dispersed pollen, cause serious hardship in many parts of the world where they are often the primary cause of pollinosis.

Grasses are annual or perennial herbs with fibrous roots and rhizomes, often from upright stems that are cylindrical, hollow, or sometimes pithy. Leaves are in two rows on the stem consisting of a sheath and a blade, the sheath tightly enveloping the stem and the blade parallel-veined, flat, long, and narrow. At the junction of the sheath and blade is the ligule, a membranous or ciliate appendage perhaps functioning to prevent rain from entering the sheath. Inflorescences usually surmount the stem and consist of several to many spikelets (analogous to flowers), each with rows of scales alternating along the axis, the lowest scales (glumes) being empty, the others (lemmas) each forming a floret that in turn usually consists of two small scales, ± 3 stamens, and two feathery stigmas surmounting the superior ovary, each with one ovule. Fruit is 1-seeded, with copious endosperm.

Bisexual spikelets are usual, but occasionally distinct staminate and pistillate spikelets or plants are found in the family. The florets remain open for a few hours to allow exposure of the floral parts to wind-pollination, cross-pollination usually being affected by the earlier maturation of anthers than stigmas. Wind-pollination, though characteristic, may be limited by self-pollination or rarely by facultative insect-pollination (as in *Paspalum dilatatus*, Adams et al. 1981). Therefore, little pollen becomes airborne from the largely selfing, cultivated species of *Avena* (oats), *Oryza* (rice), and *Triticum* (wheat), but as *Secale* (rye) and *Zea* (corn) are cross-pollinated, pollen is widely dispersed by them.

Classification

The Poaceae (Figs. 128–29; see Pls. VII and VIII) are so large and diverse a family that an internal classification is imperative. Not only is it important to botanists, but as classification reflects relationship, it is significant also to allergists. For example, pollen of *Cynodon dactylon* (Bermuda grass) is allergenically distinct from the majority of common grasses found in North America and many patients respond favorably to hyposensitization based on this grass alone. They do not need extracts from the more distantly related grasses. However, they may be sensitive to and cross-react with those grasses closely related to Bermuda grass because of partially shared allergens. When distributions and flowering periods overlap with *Cynodon*, species of *Bouteloua* (grama), *Buchloe* (buffalo grass), *Chloris* (fingergrass), and *Spartina* (cordgrass), ought to be considered for skin testing and, if positive, for desensitization, since each, though sharing allergens with *Cynodon*, may have unique ones to which patients are allergic. Reference to the best classification available gives the allergist the opportunity to make these judgments.

Those grasses classified in the same tribe are expected to have more antigens/allergens in common than those of adjacent tribes. Similarly, those grouped in the same subfamily should possess more antigens/allergens in common than grasses classified in other subfamilies. There may be imperfections in this ideal natural classification because of incomplete research, but by and large the classification may be viewed as a reasonable reflection of antigenic properties of pollen. A summary follows of the botanical classification of grasses insofar as genera are represented in North America with known or suspected allergenic properties. The most significant in pollinosis are indicated by an asterisk.

Subfamily Bambusoideae (bambusoids): tribe Oryzeae—*Oryza* (rice), *Zizania* (wild rice).

Subfamily Arundinoideae (arundinoids): tribe Arundineae—*Arundo* (giant reed), *Phragmites* (common reed).

Subfamily Chloridoideae (chloridoids): tribe Eragrostideae—*Eragrostis* (lovegrass); tribe Chlorideae—*Bouteloua** (grama), *Buchloe* (buffalo grass), *Chloris* (fingergrass), *Cynodon** (Bermuda grass), *Spartina* (cordgrass); tribe Sporoboleae—*Muhlenbergia* (muhly), *Sporobolus* (dropseed); tribe Zoysieae—*Hilaria* (galleta), *Zoysia* (Japanese lawngrass or zoysia).

Subfamily Panicoideae (panicoids); tribe Paniceae—*Digitaria* (crabgrass), *Echinochloa* (barnyard grass), *Panicum* (panicum including switchgrass, witchgrass, millet), *Paspalum* (Dallis grass, knotgrass), *Pennisetum* (fountain grass, elephant grass), *Rhynchelytrum* (Natal grass), *Setaria* (bristlegrass), *Stenotaphrum* (St. Augustine grass); tribe Andropogoneae—*Andropogon* (beardgrass, bluestem), *Eremochloa* (centipede grass), *Saccharum* (sugarcane), *Schizachyrium* (little bluestem), *Sorghastrum* (Indian grass), *Sorghum** (Johnson grass, Sudan grass, sorghum), *Tripsacum* (gamagrass), *Zea* (corn or maize).

Subfamily Pooideae (pooids): tribe Poeae—*Briza* (quaking grass), *Chasmanthium* (syn. *Uniola* in part, sea oats), *Cynosurus* (dogtail), *Dactylis** (cocksfoot or orchard grass), *Diarrhena*, *Distichlis* (saltgrass), *Festuca** (fescue), *Lolium** (darnel or ryegrass), *Melica* (melicgrass), *Poa** (blue, June, or spear grass), *Tridens* (purpletop); tribe Aveneae—*Anthoxanthum** (vernalgrass), *Arrhenatherum* (tall or tuber oatgrass), *Avena* (oats), *Deschampsia* (hairgrass), *Holcus** (velvetgrass) *Koeleria* (crested hairgrass), *Phalaris* (canary grass), *Trisetum* (trisetum); tribe Agrostideae—*Agrostis** (bentgrass, redtop), *Alopecurus* (foxtail), *Ammophila* (beachgrass), *Calamagrostis* (reedgrass), *Cinna* (woodreed grass), *Oryzopsis* (ricegrass), *Phleum** (timothy); tribe Bromeae—*Bromus** (bromegrass, chess, rescue grass); tribe Triticeae—*Agropyron* (couchgrass, quackgrass, wheatgrass), *Elymus* (wild rye), *Hordeum* (barley), *Hystrix* (bottlebrush), *Secale* (rye), *Sitanion* (squirreltail), *Triticum* (wheat); tribe Stipeae—*Stipa* (feather or needle grass).

The largest and most important of the temperate subfamilies, the pooids, includes many allergenically significant grasses such as *Dactylis, Festuca, Lolium, Poa, Anthoxanthum, Holcus, Agrostis, Phleum,* and *Bromus*. These grasses illustrate a high degree of immunological similarity, but diffusable antigens/allergens from pollen of members of other subfamilies differ substantially from the pooids. Moreover, the arundinoids, chloridoids, and panicoids show relatively little cross-reactivity outside of their taxonomic groups and even less between any one of them and the pooids (Watson and Knox 1976). These data parallel those of Lewis & Imber (1975a) who found patient response to skin test pollen extracts significantly correlated among the pooids *Anthoxanthum, Dactylis, Phleum, Agrostis,* and *Poa,* but below average in cross-reactivities among these grasses and the chloridoid *Cynodon* and the panicoid *Sorghum*. Also, *Cynodon* and *Sorghum* were below average in cross-reactivity. In all instances, however, the level of cross-reactivity was insufficient to justify the clinical use of a single grass as a dependable indicator of allergenicity, as already concluded in European (Heijer & Goransson 1965) and Australian (Wright & Clifford 1964) studies, although the correlation of reaction between *Dactylis, Phleum,* and *Poa* was sufficiently great to suggest that any one may be a useful gauge of common allergens among them (Lewis & Imber 1975a).

Flowering

In the temperate zone grasses flower predominantly during the summer months, although a number begin flowering in the spring and some extend well into autumn. Some summer grasses are early flowering; others do not begin flowering until late in the summer. However, in the southern part of the temperate zone and in the subtropical zone flowering for some grasses may be very prolonged so that anthesis is perennial rather than seasonal.

A series of histograms (see Figs. 130–37) illustrates by regions the annual flowering periods of the most relevant grass genera compiled over many years. During any single year the length of flowering may be reduced somewhat from that shown because of varying environmental conditions annually or locally, but the times given do illustrate the potential for annual grass flowering. Flowering peaks are not indicated, but usually flowering is greatest during the first or second month.

Because of their association with human environments, lawngrasses are an important consideration, yet they are difficult to categorize by flowering time. Watering and mowing techniques vary greatly throughout the continent and so do preferred lawngrasses. If water is provided throughout the year in the warmer regions, grasses may flower continuously, and if the popular Bermuda grass or *Poa annua* (annual bluegrass) are used, their short inflorescences will escape the mower's blade and will release pollen as long as reproductive potential exists (Bookman 1978). Combined with the unkept lawns in the neighborhood and the lawngrasses that have escaped and become established in nearby waste areas, these and some other grasses may flower in a well-cut lawn, so it is difficult to prevent grass pollen from being part of the ambient air when the growing season is extended by watering.

In the northern part of the northeastern region 41 genera are significant sources of pollen that may cause pollinosis (Fig. 130). Of these, 11 genera are consid-

Fig. 130. Histogram of monthly flowering of important grasses found in the northern part of the northeastern region (see map, Fig. 3).

the others continue to do so depending on local conditions.

In the southern part of the northeastern region, flowering of the grasses is more prolonged than in the northern part, and there are more potential incitants. Anthesis of the early-flowering species begins in April, and the later-flowering species extends to November, with *Poa* spanning the entire period (Fig. 131). Prior to June, 47.9% of the grass flora may be flowering, though perhaps only half in large numbers, but of the 12 genera significant in grass pollinosis, 58.3% have already initiated flowering prior to summer. Nevertheless, during June 79.1% of the

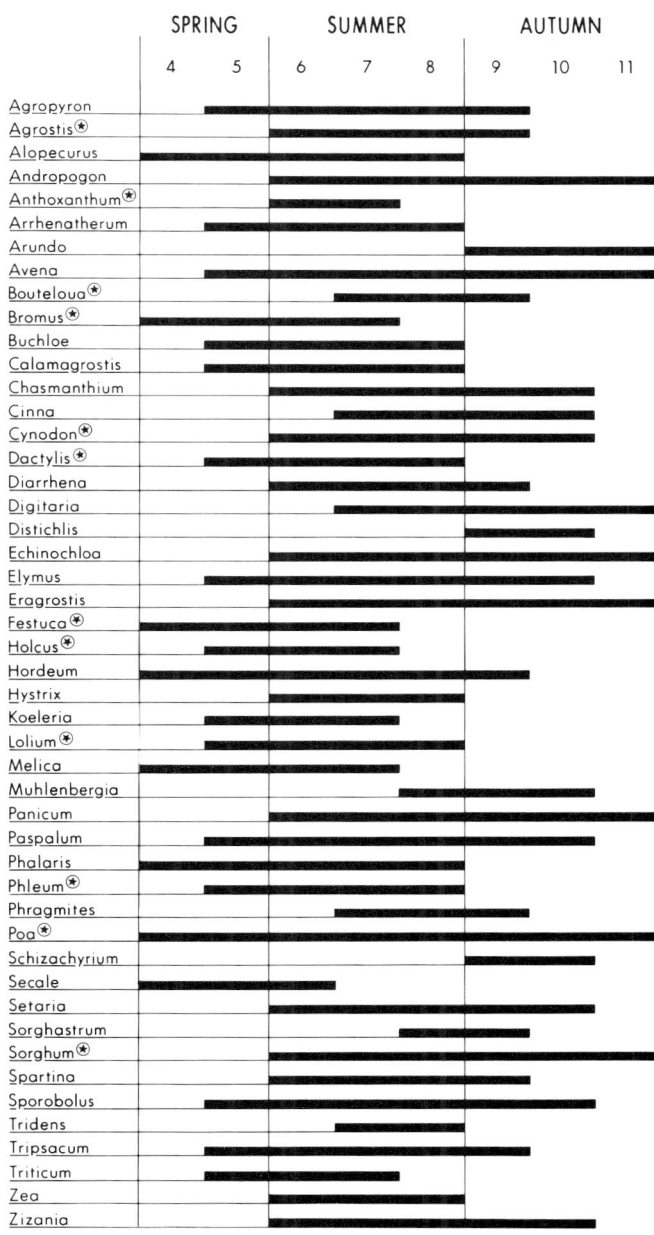

Fig. 131. Histogram of monthly flowering of important grasses found in the southern part of the northeastern region (see map, Fig. 3).

ered most important allergenically. In May, very few grasses flower, by June 43.9%, and July 75.6%; but by August and September, 80.5% and 61.0%, respectively, are still in flower. Of the most significant genera, much the same pattern exists: only 9.1% in May, 81.8% in June, but 45.5% by September. These data illustrate far longer flowering periods than anticipated, and they also show, depending on the environment any single year, that pollen can become atmospheric all summer and for particular genera even into the early fall. Important June-flowering genera are *Agrostis, Anthoxanthum, Bromus, Dactylis, Festuca, Holcus, Lolium,* and *Poa,* whereas *Cynodon* and *Sorghum* flower later. By August, *Anthoxanthum, Dactylis, Festuca,* and *Holcus* are no longer flowering, but

grasses flower, July 87.5%, August 77.1%, September 60.4%, October 41.7%, and November 18.8%. The important early-flowering grasses beginning in April-May are *Bromus, Dactylis, Festuca, Holcus, Lolium, Phleum,* and *Poa,* the somewhat later-flowering ones in June include *Agrostis, Anthoxanthum, Cynodon,* and *Sorghum,* and still later in July, *Bouteloua.* The later-flowering grasses ought to be considered seriously as causes of a secondary flare of grass pollinosis, for when initiating flowering they can shed abundant pollen for short periods of time. This group includes *Arundo, Bouteloua, Cinna, Digitaria, Distichlis, Muhlenbergia, Phragmites, Schizachyrium, Sorghastrum,* and *Tridens.*

Flowering of grasses in the southeastern region (excluding the subtropical zone) parallels closely that of the southern part of the Northeast: 49 genera, some with protracted flowering times from April to November, typify the region. Flowering by April is initiated for 18.8% of the grasses, including the important *Agrostis, Anthoxanthum, Bromus, Festuca,* and *Poa,* and by May 40.1% of the genera are flowering, including *Cynodon, Dactylis, Holcus,* and *Sorghum,* depending on latitude. By June 63.3% of grasses flower, July and August 71.4%, September 61.1%, October 53.1%, and November 8.2% (Fig. 132).

As for the Northeast, the northcentral region of the continent has a more limited grass-flowering period (May-October) and any single grass is often restricted in flowering time. The majority flower during the summer—52.4% in June, 83.3% in July, and 54.8% in August. A grass flora of 42 genera is important (Fig. 133), principally, the early-flowering (May) *Agrostis, Bromus, Dactylis, Festuca,* and *Poa* and the later-flowering *Bouteloua* (July), *Phleum* (June), and *Sorghum* (July).

In the southcentral region, grasses flower typically from spring (March) through autumn (November), but in lower latitudes of Texas, winter flowering extends this anthesis when rains occur during winter months. Thus, pollen shed may be year-round and pollinosis perennial. Even without winter rains many grasses have prolonged flowering as, for example, *Buchloe, Chloris, Cynodon, Paspalum,* and *Rhynchelytrum,* spanning a three-season period; and a span of two seasons is not unusual. As shown in Fig. 134, 42.6% of the genera may have flowering representatives in March and April, 55.3% in May, 74.5% in June, 70.2% in July, 74.5% in August, 60% in September, 55.3% in October, and as many as 46.8% are still flowering in November. The most important genera have typical flowering periods as follows: *Agrostis* March-October, *Bouteloua* May-November, *Bromus* March-September, *Cynodon* March-November, *Dactylis* March-August, *Festuca* March-August, *Lolium* March-June, *Phleum* June-August, *Poa* March-June, and *Sorghum* May-November.

In the northwestern region (Fig. 135), grass an-

Fig. 132. Histogram of monthly flowering of important grasses found in the southeastern region (see map, Fig. 3).

thesis is concentrated in the summer months: 84.4% of the potentially allergenic grasses flower in June, 88.9% in July, and 66.7% in August. Of the 12 most important genera, *Festuca* and *Poa* flower March-August; *Anthoxanthum* April-July; *Lolium* May-July; *Bromus, Dactylis,* and *Phleum* June-August; *Agrostis, Cynodon,* and *Holcus* June-September; *Bouteloua* July-September; and *Sorghum* July-October.

Based largely on flowering reported in *A California Flora and Supplement* (Munz 1959), which includes few cultivated species, Fig. 136 is probably a conservative estimate of grass anthesis in the southwestern region. Although important, this overview nevertheless fails to consider irrigated crops and lawns where year-

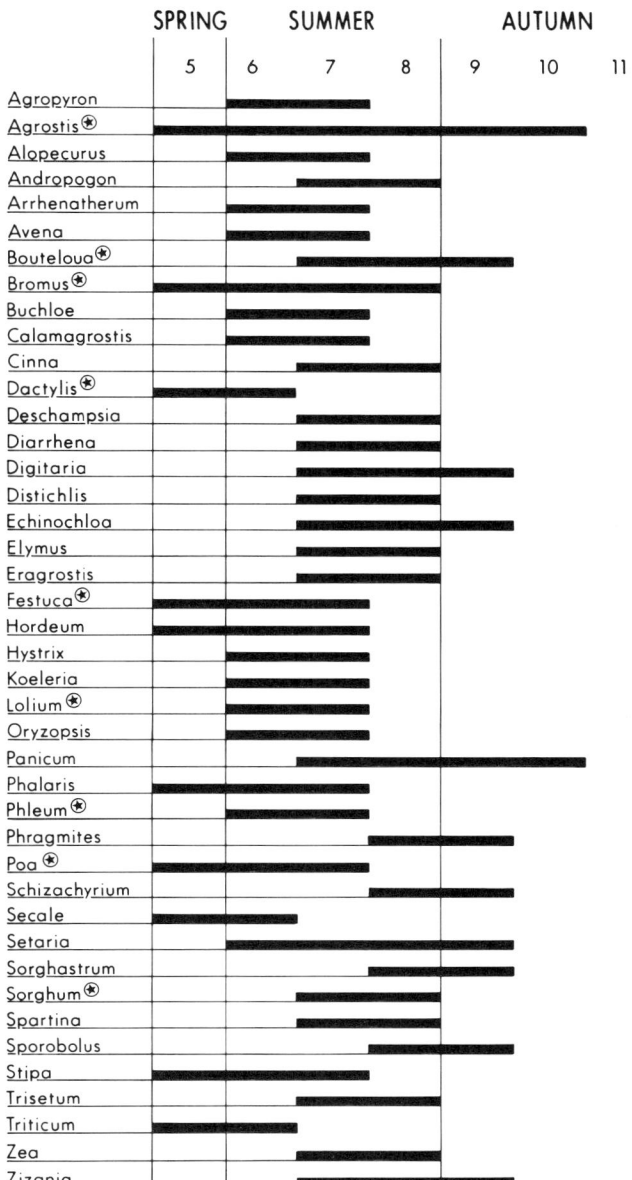

Fig. 133. Histogram of monthly flowering of important grasses found in the northcentral region (see map, Fig. 3).

Fig. 134. Histogram of monthly flowering of important grasses found in the southcentral region (see map, Fig. 3).

round flowering of some genera can be expected in certain urban, suburban, and agricultural areas. With this qualification, the allergenically troublesome *Anthoxanthum, Bouteloua, Dactylis, Phleum,* and *Sorghum* begin flowering in May, whereas *Bromus, Festuca, Lolium,* and *Poa* begin a month earlier and *Agrostis, Cynodon,* and *Holcus* a month later. The majority of these genera have long, overlapping flowering periods well into the summer and autumn, but another group, which includes *Andropogon, Chloris, Deschampsia, Echinochloa, Oryza, Panicum, Pennisetum, Phragmites,* and *Trisetum,* begins flowering during this time and compounds the problem of autumnal exposure for those allergic to grasses in the southwest.

Of the three subtropical areas recognized in the continental United States, that of subtropical Florida, from approximately Tampa east and south throughout the southern peninsula and Keys, is the least similar in grass flowering to its associated temperate region in the Southeast. Strikingly different (Fig. 137) are those grasses that shed pollen the whole year or nearly so, such as *Cynodon, Rhynchelytrum, Setaria, Sporobolus,* and *Zea.* Many introduced tropical grasses are common only in southern Florida and some are of unknown allergenicity, but all are potentially significant. Of special consideration are the alien lawn-grasses, including St. Augustine grass (*Stenotaphrum*), which is now escaping and may be collected in moist areas from South Carolina to Texas, and centipede grass (*Eremochloa*), which, like Bermuda grass, has an inflorescence short enough to miss the mower's blade and will shed pollen even after cutting.

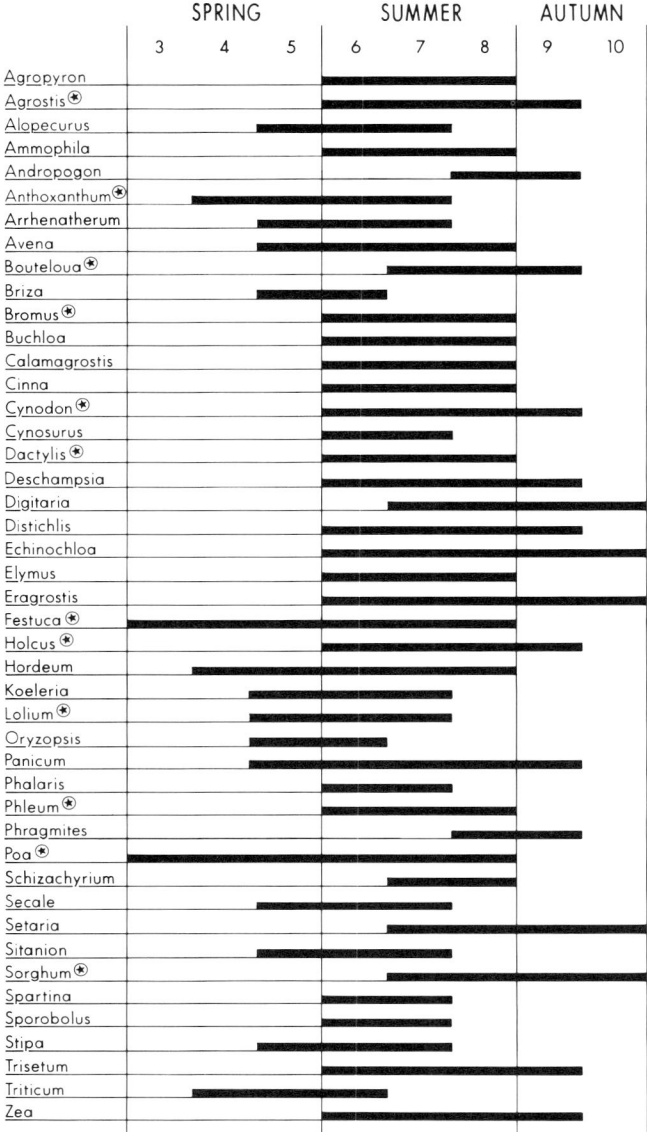

Fig. 135. Histogram of monthly flowering of important grasses found in the northwestern region (see map, Fig. 3).

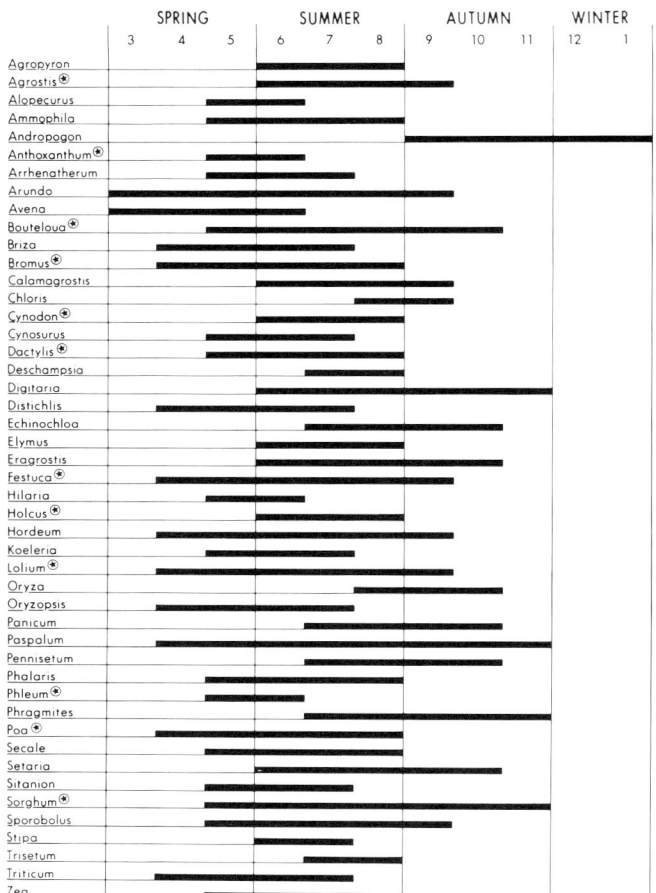

Fig. 136. Histogram of monthly flowering of important grasses found in the southwestern region (see map, Fig. 3).

Pollen Aerobiology

Pollen captured by gravitational techniques is presented as a frequency of all pollen found by month in each region of the continent (Table 12). As shown by flowering data, grass pollination was significant from May through October in the northern Northeast. The principal months of trapped pollen were June and July, but September and October were important enough to contribute to a secondary episode of pollinosis. In the southern Northeast a similar trend is apparent, although grass pollen was more abundant earlier (in April and especially May), less frequent in July, but more common in August and September.

Ambient pollen of the southeastern region is not well documented, but apparently it was frequent in May and particularly in June and July, very low in September (perhaps because of the proportionately higher frequency of *Ambrosia*) but up to 8.9% of all pollen in October.

The majority of grass pollen was collected from the air in May, June, and July in the northcentral region in correlation with flowering times. Some pollen was recovered in August and September, but comparatively little compared with other pollen (ragweed) even though many species of grass flower then.

May through August were the primary months for capturing grass pollen in the southcentral region, with frequencies in June and July being particularly great.

High grass pollen frequencies were apparent in the northwestern region from May to July, with June being the primary month. Trapped pollen extended from April to September.

In the southwestern region, grass pollen was recovered throughout the year, though particularly from April to December. Fewer aerosamples were recovered from January to March.

Many factors affect the shed and dispersal of grass pollen. Incidence is correlated with meteorological

GRASSES AND GRASSLIKE PLANTS

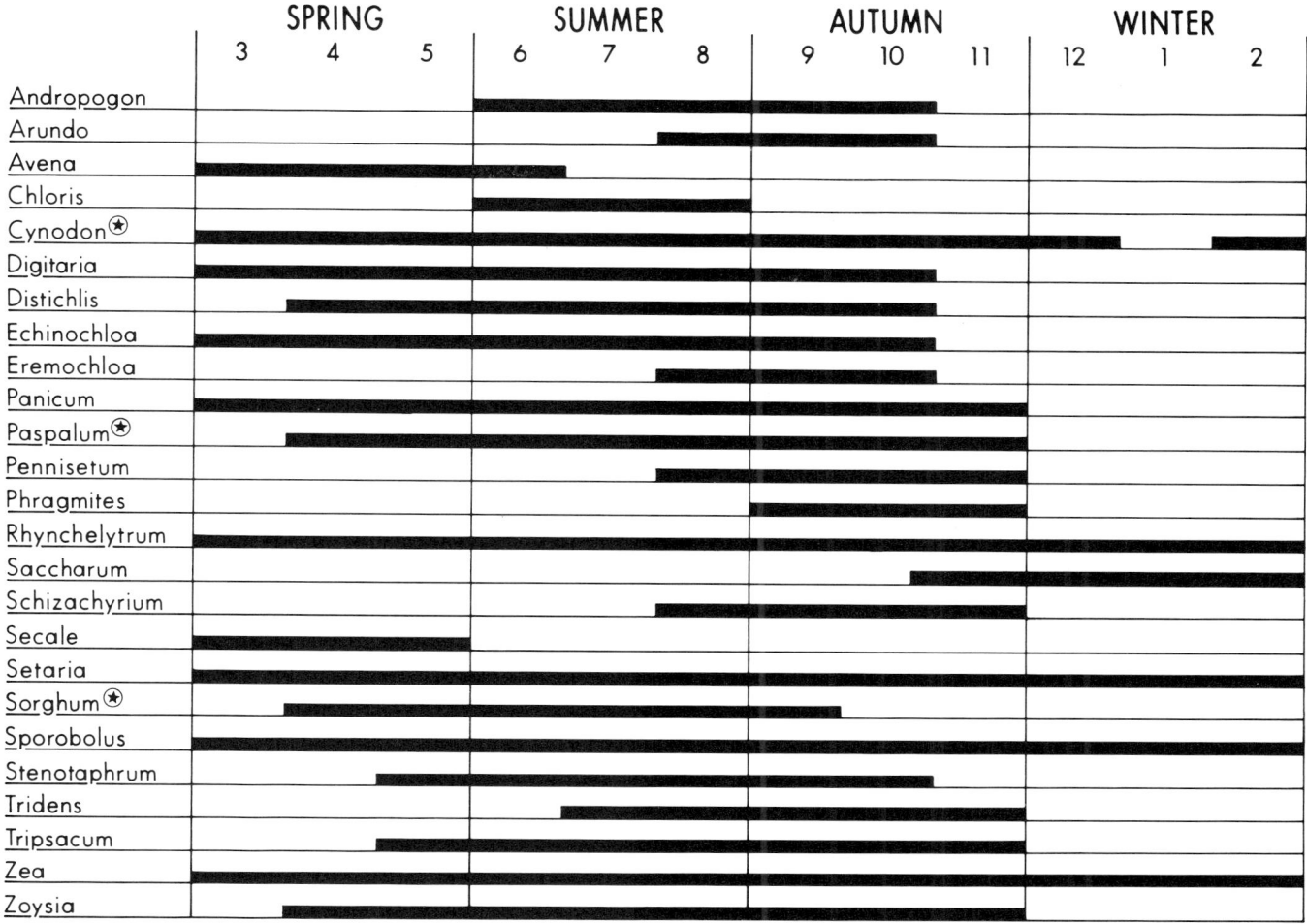

Fig. 137. Histogram of monthly flowering of important grasses found in subtropical Florida (see map, Fig. 3).

factors: greatest frequency with high temperatures, lowest frequency on days of high humidity or rainfall, and high frequency during times of increased wind. Diurnal patterns vary according to species, environmental conditions, and locality. Some *Poa* species shed pollen between 3 A.M. and 8 A.M., some *Festuca* species between 3 P.M. and 8 P.M., whereas *Phleum pratense* (timothy), *Dactylis glomerata* (orchard grass), and *Phragmites communis* (common reed) release pollen during the morning and *Agropyron repens* (quackgrass) during the afternoon (Ogden et al. 1974). A recent report from Australia (Smart & Knox 1979) indicated that grass pollen concentrated over Melbourne at night during the height of the pollination season, a sustained incidence due to environmental factors that may prove generally important. Distance

TABLE 12. Grass aeropollen frequency in regions of North America based on percentage of pollen captured, by month (1974–78).

REGION	MONTH											
	1	2	3	4	5	6	7	8	9	10	11	12
nNE				0.1	4.7	40.8	63.3	2.4	5.2	5.6		
sNE				3.4	21.9	40.7	33.5	14.2	6.1	1.6		
SE				1.6	22.1	80.9	93.8	?	0.8	8.9		
NC				3.3	10.7	13.1	10.4	2.2	0.8	3.9		
SC			1.2	1.8	18.3	42.4	45.7	17.4	4.8	1.9	1.0	
NW				1.2	20.3	83.5	46.4	5.6	3.0			
SW	1.5	0.7	1.9	4.2	14.8	28.8	19.2	18.4	36.6	25.0	37.3	6.7
sCA	0.5	1.5	2.8	10.9	21.9	42.0	59.5	51.7	38.9	29.6	11.6	?

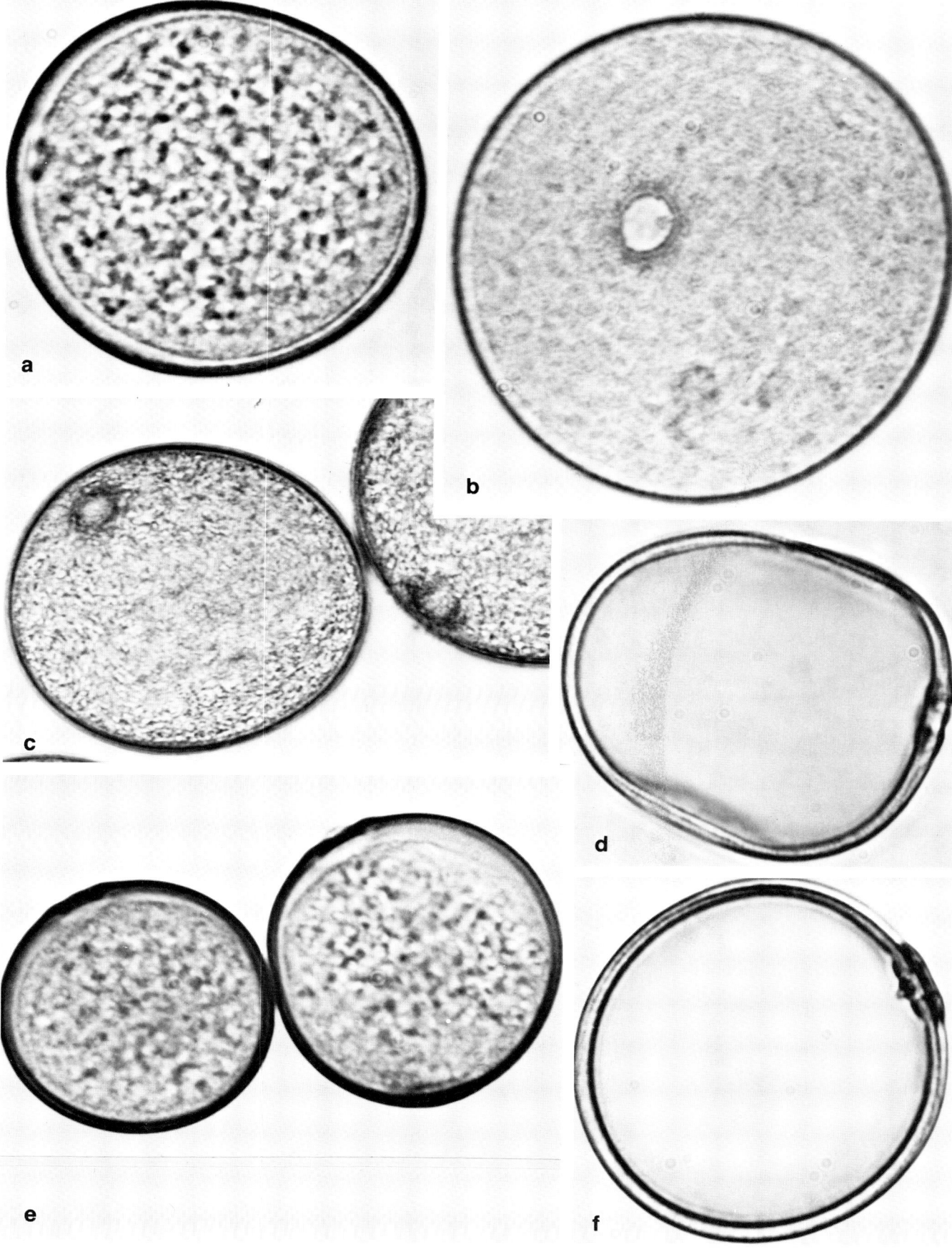

of dispersal involving both environmental and pollen characteristics is also a consideration. Long-distance distribution may occur, but more usually pollen is dispersed nearby. For example, Raynor et al. (1972) showed that more than 99% of the pollen released by a stand of *Phleum pratense* was deposited within 1 km of the source.

The larger pollen of *Zea mays* (corn), which is reputed to be of little significance in pollinosis because of its size, consequent lack of bouyancy, and poor dispersal, may be found some distance from the source. A review of corn pollination illustrates the significance of diurnal and seasonal periodicity, number of pollen shed with time, and dispersal. Anthers of corn usually begin shedding pollen at sunrise, a process that is completed in a few hours. Pollen remains viable for no more than 24 hours, the corn tassel continuing to shed pollen typically for five to eight days, with maximum shedding during the third day. Pollen is produced in large amounts: a single anther sheds about 2,500 grains, a spikelet 15,000 grains, and a tassel 2 to 5 million grains. Distribution is by wind and gravity, and with a high wind even the large grains of corn can be carried distances of up to 0.5 km from the plants (Purseglove 1972).

Pollination occurs after the filaments of the stamens elongate and the anthers exert and hang outside the spikelet and dehisce, i.e., they split and release the pollen. Grasses are essentially anemophilous, for their pollen is mainly transferred from floret to floret and from plant to plant by wind; but insects, mainly pollen-collecting bees, also incidentally transfer pollen during the process of collecting it for food.

Pollen Morphology (Figs. 138–39)

Grains are spheroidal to ovoidal, sometimes ellipsoidal to subellipsoidal, and 22–122 μm in diameter; the exine notably thin, 1.0–1.5 μm; the sexine as thick as or slightly thicker than the nexine; and the intine usually much thicker, 1.5–5.0 μm, particularly below the apertures. The surface is granular to finely reticuloid, forming a continuum between these extremes and of limited taxonomic significance (Page 1978). The apertures are 1-porate; the pore is 2–8 μm in diameter, usually circular to ovoidal, occasionally transversely ellipsoidal surrounded by a thickened protruding ring (annulus), and covered by an operculum that is either smooth or granular; the pore margin smooth or wavy, with the largest grains provided with the largest pores. Ovoidal grains usually have the pore at the broader end.

Small grains (22–38 μm) with ±granular surfaces typify *Agrostis, Alopecurus, Cynodon, Holcus,* and *Koeleria,* and those with reticulatelike surfaces include *Briza, Chloris, Dactylis, Muhlenbergia, Poa,* and *Tridens.* Somewhat larger grains (32–45 μm) provided with ±granular surfaces are characteristic of *Andropogon, Anthoxanthum, Arundo, Chasmanthium, Elymus, Festuca, Lolium,* and *Zizania,* as well as those having reticulatelike surfaces, such as *Ammophila, Cynosurus, Digitaria, Echinochloa, Panicum, Paspalum, Pennisetum, Phalaris, Phleum, Saccharum, Setaria,* and *Sorghum.* The largest grains (40–122 μm), generally with the largest and most extended pores, include *Agropyron, Avena, Bromus, Hordeum, Secale, Sorghastrum, Triticum,* and *Zea,* which have ±granular surfaces, and *Hystrix* and *Tripsacum,* which have reticulatelike surfaces.

The most distinct grass pollen is the large (86–122 μm) spheroidal to subspheroidal grains of *Zea mays,* provided with a large pore (6–8 μm) surrounded by a thickened annulus that is covered by a granular operculum. Distinctive also is the moderately large (52–65 μm) subspheroidal to ovoidal or ellipsoidal grains of *Secale cereale* (cultivated rye), with an annulate pore situated near the broader end. Similar pollen is found for *Avena,* whose grains are 48–66 μm in diameter.

For additional morphological studies of grass pollen, see Andersen & Bertelsen (1972), Kohler & Lange (1979), and Rowley (1960, 1964).

Allergenicity (by genus)

Grass pollinosis is important in North America; it is usually second only in significance to the late summer effects of ragweed allergenicity. In western Oregon and Washington, and undoubtedly elsewhere in the West, grass pollen is of primary importance, greatly distressing sensitized individuals (Perlman 1955). The principal cause of inhalant pollen allergies elsewhere and particularly in tropical regions is often grasses, as reported, for example, in Brazil, Colombia, India, New Zealand, Puerto Rico, and South Africa.

Genera are alphabetically arranged, each provided with a brief general statement followed by a reference to allergenicity. Several of the more significant grasses are illustrated (see Figs. 128–29), and maps of distribution appear in Appendix 1. A list of common names with their equivalent generic names precedes this discussion.

Bahia grass	*Paspalum*
barley	*Hordeum*
barnyard grass	*Echinochloa*
beachgrass	*Ammophila*
beardgrass	*Andropogon*
bentgrass	*Agrostis*

Fig. 138. Pollen of representative Poaceae (light photomicrographs). a: *Anthoxanthum odoratum* (sweet vernalgrass), LM(s) ×1830; b: *Cynosurus cristatus* (crested dogtail), LM(s) ×1830; c: *Hordeum jubatum* (foxtail barley), LM(s) ×720; d: *Lolium perenne* (common ryegrass), LM(a) ×1830; e: *Poa pratensis* (Kentucky bluegrass), LM(s) ×1830, f: *Sorghum halepense* (Johnson grass), LM(a) ×1830.

Bermuda grass	*Cynodon*	reed, giant	*Arundo*
bluegrass	*Poa*	rescuegrass	*Bromus*
bluestem	*Andropogon*	rice	*Oryza*
bluestem, little	*Schizachyrium*	rice, wild	*Zizania*
bottlebrush	*Hystrix*	ricegrass	*Oryzopsis*
bristlegrass	*Setaria*	rye	*Secale*
bromegrass	*Bromus*	rye, wild	*Elymus*
buffalo grass	*Buchloe*	ryegrass	*Lolium*
canary grass	*Phalaris*	St. Augustine grass	*Stenotaphrum*
centipede grass	*Eremochloa*	saltgrass	*Distichlis*
chess	*Bromus*	sorghum, cultivated	*Sorghum*
cocksfoot grass	*Dactylis*	speargrass	*Poa*
cordgrass	*Spartina*	squirreltail	*Sitanion*
corn	*Zea*	Sudan grass	*Sorghum*
couchgrass	*Agropyron*	sugarcane	*Saccharum*
crabgrass	*Digitaria*	switchgrass	*Panicum*
Dallis grass	*Paspalum*	timothy	*Phleum*
darnel	*Lolium*	tridens	*Tridens*
dogtail	*Cynosurus*	trisetum	*Trisetum*
dropseed	*Sporobolus*	velvetgrass	*Holcus*
elephant grass	*Pennisetum*	vernalgrass, sweet	*Anthoxanthum*
feathergrass	*Stipa*	wheat	*Triticum*
fescue	*Festuca*	wheatgrass	*Agropyron*
fingergrass	*Chloris*	witchgrass	*Panicum*
fountaingrass	*Pennisetum*	woodreed grass	*Cinna*
foxtail	*Alopecurus*	zoysia	*Zoysia*
galleta	*Hilaria*		
gamagrass	*Tripsacum*		
grama	*Bouteloua*		
hairgrass	*Deschampsia*		
hairgrass, crested	*Koeleria*		
Indian grass	*Sorghastrum*		
Japanese lawngrass	*Zoysia*		
Johnson grass	*Sorghum*		
June grass	*Poa*		
June grass, western	*Koeleria*		
knotgrass	*Paspalum*		
lovegrass	*Eragrostis*		
maize	*Zea*		
melicgrass	*Melica*		
millet, broomcorn or common	*Panicum*		
muhly	*Muhlenbergia*		
Napier grass	*Pennisetum*		
Natal grass	*Rhynchelytrum*		
needlegrass	*Stipa*		
oatgrass, tall	*Arrhenatherum*		
oats	*Avena*		
oats, sea	*Chasmanthium, Uniola*		
orchard grass	*Dactylis*		
panicum	*Panicum*		
purpletop	*Tridens*		
quackgrass	*Agropyron*		
quaking grass	*Briza*		
redtop	*Agrostis*		
reedgrass	*Calamagrostis*		
reed, common	*Phragmites*		

AGROPYRON   *wheatgrass*

Most species furnish forage, several of which are among the most valuable range grasses in western regions of the continent, and in valleys they may grow in sufficient abundance to produce hay. *Agropyron repens* (quackgrass, couchgrass), introduced from Eurasia, is a troublesome weed in cultivated ground, particularly in eastern North America.

The common weedy *Agropyron repens* sheds little pollen and may as a consequence be of little relevance. *A. smithii* (western wheatgrass) and other species, however, have been considered significant contributors to allergenicity (Wodehouse 1935, 1971), particularly in Oklahoma (Blue 1955; Levetin & Buck 1980) and Wyoming (Scheppegrell 1917), yet Barrett (1934) reported very few patients sensitive to *Agropyron* pollen even though the genus is common in Utah. Gutmann (1950) found *A. junceum* pollen important in Israel. Apparently some, but not all, species of wheatgrass contribute to pollinosis.

AGROSTIS   *bentgrass*

Most species are forage grasses, either under cultivation or in the mountain meadows of the West. The best-known cultivated species is *Agrostis stolonifera* (syn. *A. alba* of authors, *A. gigantea*, *A. palustris*), called redtop or creeping bentgrass, which is used for meadows, pastures, lawns, and turf and is also

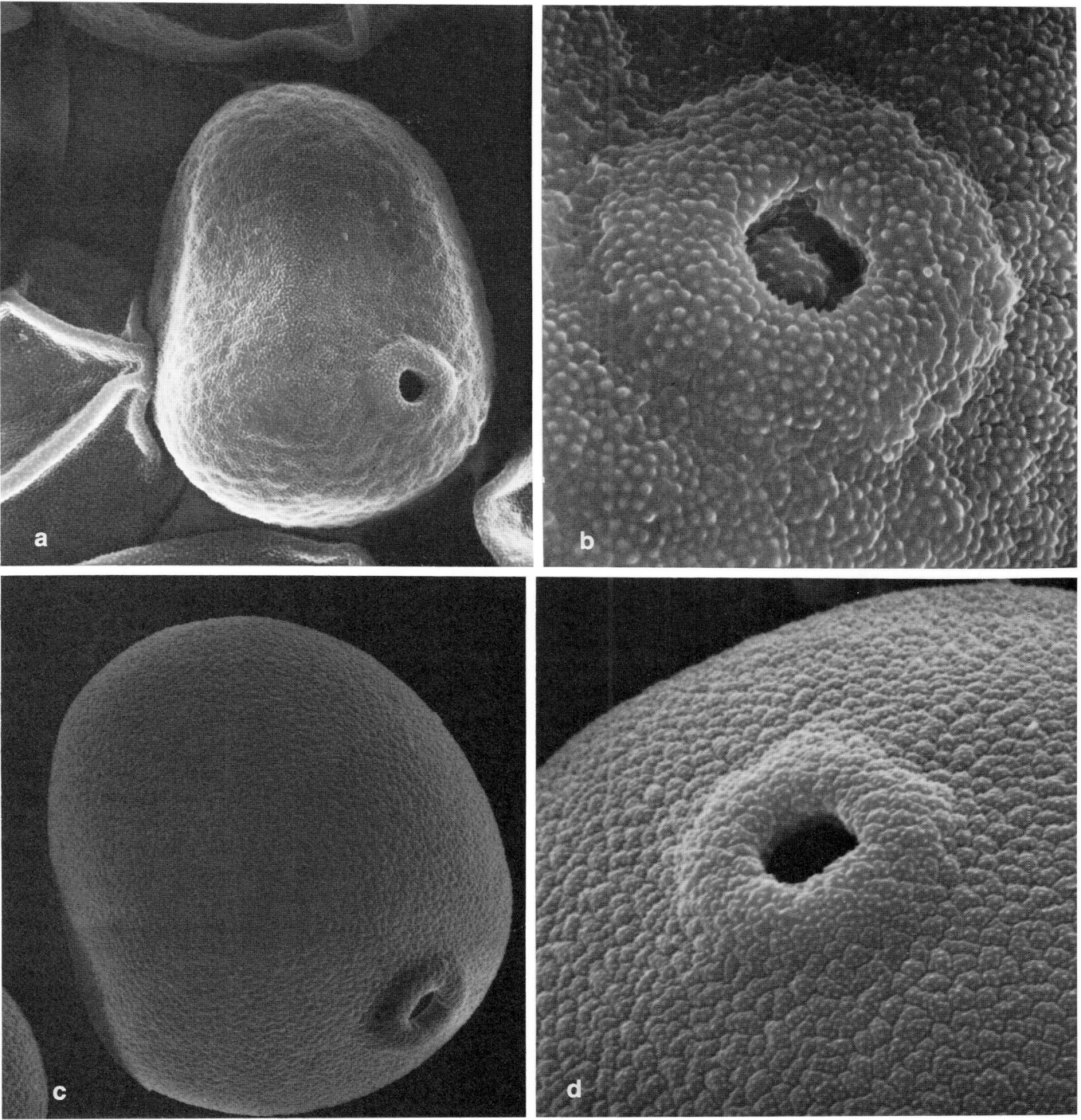

Fig. 139. Pollen of representative Poaceae (scanning electron micrographs), a–b: *Secale cereale* (rye), a = SEM(a) ×1600, b = SEM(a) showing pore morphology and sexine surface ×5800; c–d: *Tripsacum dactyloides* (eastern gamagrass), c = SEM(a) ×2000, d = SEM(a) showing pore morphology and sexine surface ×4800.

widely escaped in temperate North America. Several native species are also common.

Agrostis stolonifera sheds quantities of pollen. The species is considered, together with *Phleum pratense* (timothy), the cause of most grass pollinosis during the second wave in June and July in temperate regions (Wodehouse 1935, 1971). Indeed, all species of *Agrostis* can be implicated.

ALOPECURUS *foxtail*

The species of *Alopecurus* are all forage grasses; *A. pratensis* in particular is commonly cultivated and escaped to meadows and pastures. The species may cause pollinosis where abundant, as reported in Massachusetts (Wodehouse 1971).

AMMOPHILA *beachgrass*

Two species, the West Coast *Ammophila aremaria*, introduced from Europe, and the East Coast native *A. breviligulata*, are important sand-binding grasses. Both shed significant amounts of pollen from June to September, which probably contributes locally to summer and late season pollinosis (Wodehouse 1971).

ANDROPOGON *beardgrass*

More than 35 species are found in North America, the most important being *Andropogon gerardii* (big bluestem), a native constituent of wild hay in the prairies. It and *A. scoparium* have been implicated as offenders in Missouri (Steyermark 1963), *A. furcatus* in Minnesota (Ellis and Rosendahl 1933), *A. bicornis* in Cuba (Quintero 1955), and *A. hirtus* in Israel (Gutmann 1950).

ANTHOXANTHUM *vernalgrass*

The most important species, *Anthoxanthum odoratum* (sweet vernalgrass), introduced from Eurasia, is common in waste fields, meadows, and pastures of eastern North America and the West Coast. This species, often flowering from April, sheds enormous quantities of extremely allergenic pollen and is one of the most important causes of grass pollinosis (Wodehouse 1935). Where common, the species generally starts the grass pollinosis season (Wodehouse 1971).

ARRHENATHERUM *tall oatgrass*

Introduced from Europe as a meadow grass and now frequently escaped from cultivation in the northeastern region, *Arrhenatherum elatius* sheds relatively little pollen and is only a minor factor in pollinosis (Wodehouse 1935).

ARUNDO *giant reed*

Arundo donax is widely cultivated in the southern regions as an ornamental grass to prevent wind erosion. It should be considered a cause of local outbreaks of late-season pollinosis.

AVENA *oats*

The familiar *Avena sativa* (cultivated oats) is self-pollinating; by shedding little pollen the species is of limited importance in pollinosis. Other species, however, such as the weedy *A. barbata* and *A. fatua* of western regions, are considered causes of pollinosis in southern California (Piness 1955) and elsewhere (Wodehouse 1935, 1971). *A. barbata* and *A. sterilis* are offenders in Israel (Gutmann 1950).

BOUTELOUA *grama*

The many species of *Bouteloua* are among the most valuable of forage grasses and form an important part of grazing in the West. For example, *B. gracilis* (blue grama) is prominent in the "short grass" region of the Great Plains, and *B. curtipendula* (side-oats grama) is more widely distributed and is much used for grazing and hay.

All species of this late-flowering genus are important causes of pollinosis where abundant (Wodehouse 1935). *Bouteloua curtipendula*, for example, has large anthers with apparently abundant pollen.

BRIZA *quaking grass*

One species, *Briza minor*, may form an appreciable part of the spring forage in California and is becoming more common in the Pacific Coast region. Allergenicity is suspected, but documentation is wanting.

BROMUS *bromegrass*

The native bromegrasses, such as *Bromus carinatus*, form an important forage in the western regions but the most important agronomic species is the Eurasian introduction *B. inermis* (smooth brome) (see Fig. 128a, b). It is cultivated as a hay and pasture grass, particularly in northern areas, and for reseeding western mountain ranges.

In California bromegrass pollen is a major cause of pollinosis (Rowe 1955), particularly *Bromus carinatus*, *B. mollis*, and *B. rigidus*, which flower in the summer throughout the Pacific region (Rowe 1929; Wodehouse 1971). In Minnesota pollen of *B. inermis* is important (Ellis & Rosendahl 1933). *B. secalinus*, however, sheds little pollen and probably is of limited significance allergenically (Wodehouse 1971).

BUCHLOE *buffalo grass*

Buchloe dactyloides is dominant over large areas of the Great Plains uplands and is one of the most important grazing grasses in the region. Wodehouse (1971), however, suggested that it is unimportant in pollinosis.

CALAMAGROSTIS *reedgrass*

Several species are important native forage grasses: *Calamagrostis canadensis* (bluejoint) is a source of much of the wild hay in the northeastern region, and *C. rubescens* (pinegrass) is a leading range grass in the Pacific Northwest. Reedgrasses have been implicated in pollen allergenicity in Minnesota (Ellis & Rosendahl 1933).

CHASMANTHIUM (from *Uniola*) *sea oats*

Of value as sea-binders and ornamentals, sea oats are probably nowhere common enough to shed abundant pollen. In the southeastern and southcentral regions, however, *Chasmanthium latifolium* might be responsible for limited summer pollinosis.

CHLORIS *fingergrass*

Chloris gayana (Rhodes grass) is cultivated in the warmer regions of the continent, particularly in irrigated areas of the Southwest. Allergenicity is reported occasionally (Cuba) (Quintero 1955), but local occurrences can be expected as this promising meadow grass is more widely grown. It is a cross-pollinating plant.

CINNA *woodreed*

Although not generally abundant, the tall native woodreeds may shed considerable pollen from each plant. Allergenicity is unknown.

CYNODON *Bermuda grass*

The introduced *Cynodon dactylon* is common in the southern half of the continent, where it grows in open ground, grasslands, fields, and waste places. It is used as a lawngrass that at maturity is short enough to miss the mower's blade and may continuously shed buoyant pollen from summer through autumn.

The species is allergenically distinct from most grasses (Wodehouse 1971) as would be anticipated by its isolated botanical position. Cross-reactions with other grasses are limited but can be expected in the tribe Chlorideae, among others. Bermuda grass sensitivity is widely acknowledged as the most serious cause of pollinosis wherever common, as in Arizona (Phillips 1955), California (Piness 1955; Rowe 1955), New Mexico (Kirchner 1955), Oklahoma (Levetin & Buck 1980), Texas (French 1930; Fein & Kamin 1962), as well as Cuba (Quintero 1955), Mexico City (Salazar Mallen 1955), and Israel (Gutmann 1950). Patients respond well to hyposensitization in large part because a species-specific pollen extract can be administered.

CYNOSURUS *dogtail*

Found in fields and waste places, *Cynosurus cristatus* is a summer-flowering introduction of concern in inhalant allergy wherever common.

DACTYLIS *orchard grass*

A common introduction from Eurasia, *Dactylis glomerata* (see Figs. 128c, d) sheds enormous quantities of pollen and accounts for more positive skin test reactions in the eastern United States than any other early-summer-flowering grass (Wodehouse 1935, 1971). It has been reported of major allergenic significance in California (Hall 1922), Minnesota (Ellis & Rosendahl 1933), and Oklahoma (Levetin & Buck 1980). The species is often associated with *Poa pratensis* (June grass).

DESCHAMPSIA *hairgrass*

Often the dominant grass in mountain meadows, *Deschampsia caespitosa* is not reported as allergenic, but wherever common this summer-flowering species might be an offender.

DIARRHENA

Found in rich or moist woods, the single species *Diarrhena americana*, where occurring in abundance, could be a source of pollen that might be implicated in pollinosis.

DIGITARIA (syn. *Syntherisma*) *crabgrass*

Digitaria ischaemum and *D. sanguinalis* are common weeds of fields, roadsides, and waste places and are often troublesome in lawns, particularly in the eastern and southern United States. These late-flowering species occasionally release enough pollen to cause a late flare-up of grass pollinosis as reported in New York City (September) (Wodehouse 1935) and Minnesota (July-August) (Ellis and Rosendahl 1933), but apparently they do not produce enough pollen to be of general importance in pollinosis in spite of the abundance of plants (Wodehouse 1971).

DISTICHLIS *saltgrass*

Often forming dense mats along seashores and salt marshes, *Distichlis spicata* is considered an important offender in California (Hall 1922) and of moderate significance in Oklahoma (Levetin & Buck 1980).

ECHINOCLOA *barnyard grass*

Where common in moist open places, ditches, cultivated fields, etc., the introduced *Echinocloa crusgalli* could be responsible for local outbreaks of pollinosis as reported in Minnesota (Ellis & Rosendahl 1933). *E. colonum* (jungle-rice), also introduced to the United States, is considered an offender in Israel (Gutmann 1950).

ELYMUS *wild rye*

A large genus of good forage grasses widely distributed in North America, *Elymus canadensis*, for example, is found along river banks, in moist open or partly shaded ground, and also in dry, sandy, or gravelly soil. This species has been implicated as a minor offender in Missouri (Steyermark 1963) and others have been suspected elsewhere, particularly in California (Hall 1922, Piness 1955).

ERAGROSTIS *lovegrass*

Although *Eragrostis* is a common genus continentally, it may be of little significance allergenically because of small anthers and presumably low pollen shed. *E. cynosuroides*, however, is considered a major offender in Israel (Gutmann 1950).

FESTUCA *fescue*

Many species of fescue are important forage grasses in the grazing regions of the West. The most important cultivated species is the European *Festuca elatior* (meadow fescue) (see Fig. 128e, f) used for hay and pasture in the Midwest.

Festuca elatior, F. pratensis, F. ovina, and other species shed abundant pollen and where common are serious causes of pollinosis (Wodehouse 1935, 1971; Piness 1955; Levetin & Buck 1980). Nevertheless, representatives of the genus are often neglected in skin testing.

HILARIA *galleta*

Several species of *Hilaria* are important range grasses in Texas and the southwestern region, but their pollen has not yet been reported as allergenic.

HOLCUS *velvetgrass*

Two introduced European species, *Holcus lanatus* and *H. mollis,* are found in open ground, meadows, grassy banks, and moist places and are particularly common along the Pacific Coast. *H. lanatus* is occasionally cultivated as a meadow grass on light or sandy soil.

Holcus lanatus sheds large amounts of bouyant, allergenic pollen, which is an important cause of grass pollinosis in California (Hall 1922; Row 1928), Oregon (Wodehouse 1971), and Washington (Scheppegrell 1917), as well as England (Chamberlain 1927).

HORDEUM *barley*

Cultivated barley (*Hordeum vulgare*) is largely self-pollinating and as such disperses little pollen. However, *H. jubatum* (foxtail barley) (see Fig. 128g, h) and *H. murinum* (mouse barley) are widespread, troublesome weeds, particularly in the West. Although neither sheds much pollen (Wodehouse 1971), it is apparently enough for the first species to be considered a major offender in Minnesota (Ellis & Rosendahl 1933) and Oklahoma (Levetin & Buck 1980), and the second species in Israel (Gutmann 1950).

HYSTRIX *bottlebrush*

The allergenic significance of *Hystrix patula,* which is cultivated as an ornamental, is not known.

KOELERIA *crested hairgrass, western June grass*

Koeleria cristata is an important forage grass and constitutes much of the native pasture throughout western regions. The species sheds much pollen and is considered important in pollinosis in Oregon (Chamberlain 1927), southern California (Piness 1955), and Wyoming (Scheppegrell 1917). *K. phleoides* pollen is of allergenic importance in Israel (Gutmann 1950).

LOLIUM *darnel, ryegrass*

A widely used meadow grass in Europe, *Lolium perenne* (see Fig. 128i, j) is also cultivated in eastern and southern North America in parks, roadsides, and public grounds, where vigorous early growth is required, and for winter forage. It sheds abundant pollen and is a major contributor in pollinosis wherever common (Hall 1922; Blue 1955; Piness 1955; Rowe 1955; Wodehouse 1971; Levetin & Buck 1980). Pollen of *L. tenulentum* and other species are also offenders (Wodehouse 1971).

Lolium perenne pollen has been show to have four allergenic components, termed Groups I to IV. Group I seems to be the major allergen since 95% of grass-sensitive patients react to it by skin test and leukocyte histamine release. Group II is reactive in 60% of grass-sensitive patients, and many of them are highly sensitive to Group III. About 20% of patients allergic to Group I also react to Group IV (Lichtenstein et al. 1969; Marsh et al. 1970). All groups are immunologically distinct from one another and from *Phleum pratense* (timothy) allergen B, and have molecular weights from 11,000 daltons (Group II, isoallergenic variant B) to 50,000 daltons (Group IV).

MELICA *melicgrass*

A common genus in North America, melicgrass is not gregarious nor used much as forage. Allergenicity is not reported but should be considered a moderate factor in early-season grass pollinosis (Hall 1922).

MUHLENBERGIA *muhly*

A large genus consisting of important range grasses in western regions, *Muhlenbergia* has small anthers, however, which probably produce little pollen. Nevertheless, species of the genus ought to be considered offenders in pollinosis as were two species in Minnesota (Ellis & Rosendahl 1933).

ORYZA *rice*

Rice is cultivated in Arkansas, California, Louisiana, and Texas and is also adventive near the coast from Virginia to Texas. Self-pollination is usual, but small and varying amounts of cross-pollination by wind do occur. Thus, particularly where rice is commercially grown, instances of late-autumn pollinosis should be considered, as in California (Hall 1922) and also in Japan, where late-flowering rice coincided with severe attacks of bronchial asthma (Matsumura et al. 1969).

ORYZOPSIS *ricegrass*

Species of *Oryzopsis* are usually not found in abundance, except *O. hymenoides,* which is common in the semiarid areas of the western regions where it furnishes much feed for stock. No report of allergenicity, however, is associated with these grasses.

PANICUM *panicum, common millet*

A large genus in North America, sometimes cultivated, but more often a weed in waste places, *Panicum* is occasionally associated with pollinosis, as in Cuba (Quintero 1955) and Israel (Gutmann 1950), but it undoubtedly contributes more to summer and perhaps early-autumn symptoms than is recognized in North America.

PASPALUM *Dallis grass, Vasey grass*

Particularly frequent in the southeastern region where it flowers from May to October, *Paspalum* should be considered a significant factor in hay fever (second-most important cause in Cuba, Quintero 1955), but its role is not fully understood (Wodehouse 1971). Some species are apomictic, the common form of *P. dilatatum* is also partially insect-pollinated, and they may shed only limited pollen.

PHALARIS *canary grass*

An important constituent of lowland hay in the prairies, *Phalaris arundinacea* is suspected of contributing to pollinosis in Minnesota (Ellis & Rosendahl 1933), as is the weedy introduced *P. minor* where abundant in California, Louisiana, and Texas (Wodehouse 1971). In fact, *Phalaris* species are considered among the most important grass offenders in southern California.

PHLEUM *timothy*

Widespread in moist habitats throughout the continent and commonly cultivated for fodder, timothy (*Phleum pratense*) (see Fig. 129a, b) sheds large amounts of pollen. Apparently it is second only to *Dactylis* (orchard grass) in allergenicity (Wodehouse 1935) and is consequently one of the worst causes of early-summer pollinosis (Wodehouse 1971). Fortunately, timothy is not too successful an invader of urban areas and fewer cases are attributable to timothy than otherwise would be expected. Both *P. arenarium* and *P. subulatum* are of allergenic concern in Israel (Gutmann 1950).

Timothy pollen has been extensively studied for allergenic components (Malley et al. 1975). The major allergen is A (molecular weight 13,000 daltons); another is allergen B (MW 10,500 daltons), to which about 40% of patients who are sensitive to ryegrass (*Lolium perenne*) Group I are also sensitive.

PHRAGMITES *common reed*

Frequent in moist areas, except the southeastern, and although not yet implicated in North America as an offender, *Phragmites* should be considered in local late-summer flare-ups. *P. communis* pollen is known to be allergenic in Israel (Gutmann 1950).

POA *bluegrass, June grass, or spear grass*

The bluegrasses are of great importance because of their forage value, some being cultivated for pasture and others forming large areas of natural mountainous pasture in the West, and because of their wide use as lawngrasses.

Pollen of any member of this large genus may elicit moderate to major allergic reactions. Even the small *Poa annua*, which produces limited pollen, is nevertheless important (Gutmann 1950), but other species like *P. compressa*, *P. pratensis* (see Fig. 129c, d), and *P. trivialis* disperse abundant pollen and unquestionably are important factors in early summer pollinosis (Hall 1922; Ellis & Rosendahl 1933; Piness 1955; Rowe 1955; Levetin & Buck 1980). In fact, *P. pratensis* (Kentucky bluegrass or June grass), together with *Dactylis* (orchard grass), may account for more grass pollinosis in eastern North America during the first half of the season than all other grasses combined (Wodehouse 1971).

RHYNCHELYTRUM (syn. *Tricholaena* in part) *Natal grass*

Cultivated as a meadow grass in the southeastern region and as indoor dried decorations and found as an escape in open woods and fields, *Rhynchelytrum repens* is at least a moderate offender in pollinosis where common (Quintero 1955).

SACCHARUM *sugarcane*

Cultivated particularly in southern Louisiana and Florida, sugarcane is pollinated by wind and may be responsible for limited pollinosis among those working in and living around the sugarcane fields during anthesis in the winter months. Quintero (1955), however, found no clinical sensitivity to its pollen in Cuba, perhaps in part because sugarcane was cut before it had flowered.

SCHIZACHYRIUM *little bluestem*

Native to the prairies, open woods, and dry hills and fields, *Schizachyrium scoparium* may be responsible for limited pollinosis during late summer.

SECALE *rye*

Found throughout North America as an escape and widely cultivated in the Canadian prairies, Midwest, and southern Ontario, rye sheds prodigious quantities of pollen. Although grains are large with a short dispersal range (Heslop-Harrison 1979), they are nevertheless responsible for some pollinosis (Ellis & Rosendahl 1933; Wodehouse 1935, 1971).

SETARIA *bristlegrass*

Most species of *Setaria* are introduced from the Old World, with a few native to the southwestern region. *Setaria* pollen has been implicated in pollinosis in

Cuba (Quintero 1955) and *S. glauca* (pigeon grass) in Minnesota (Ellis & Rosendahl 1933), but its significance is undoubtedly secondary.

SITANION *squirreltail*

Found in the western regions, pollen of *Sitanion* is not known to be allergenic, but it may be a factor in late-spring and early-summer pollinosis.

SORGHASTRUM *Indian grass*

Sorghastrum nutans is the most important of these native late-flowering grasses that are constituents of wild or prairie hay. The species has been implicated in late-summer hay fever in Minnesota (Ellis & Rosendahl 1933).

SORGHUM *Johnson grass, sorghum*

Widely cultivated plants or troublesome weeds in more southern latitudes of the continent, these largely self-pollinating species produce ±large pollen in sufficient quantities to affect allergic reactions. In fact, the degree of cross-pollination is influenced by the wind and the inflorescence type, the more open heads of sorghum being more liable to have greater dispersal of their pollen.

Major allergic reactions have been reported principally to pollen of *Sorghum halepense* (Johnson grass) (see Fig. 129e, f) and secondarily to the cultivated sorghums where grown commonly in the United States (Hall 1922; French 1930; Wodehouse 1935, 1971; Howe 1955; Kirchner 1955; Phillips 1955; Piness 1955; Fein & Kamin 1962; Steyermark 1963; Levetin & Buck 1980), and in northern Mexico (Salazar Mallen 1955). Some major allergens may be shared with *Zea* (corn), *Sorghastrum* (Indian grass), and other grasses in the tribe Andropogoneae and perhaps also with those in the allied tribe Paniceae.

SPARTINA *cordgrass*

Species of *Spartina* form dense colonies in wet habitats, and although not implicated in pollinosis they may be responsible for local reactions where frequent.

SPOROBOLUS *dropseed*

Dropseeds are forage grasses that become abundant in the southwestern region, but their pollen allergenicity is little known. It may be of secondary importance where common, as in Cuba (Quintero 1955).

STENOTAPHRUM *St. Augustine grass*

Cultivated as a lawngrass in the Southeast and California, St. Augustine grass is also escaping to low waste places where pollen shed may be ample to implicate the species in local midsummer cases of pollinosis. In Texas, however, Fein & Kamin (1962) reported that it rarely causes inhalant allergy.

STIPA *needlegrass, feathergrass*

Even though *Stipa* is common in western regions, its allergenic potential is not known.

TRIDENS *tridens*

Of scattered occurrence throughout North America, pollen allergenicity of *Tridens* remains unknown.

TRIPSACUM *gamagrass*

Of limited pollen allergenicity, species of *Tripsacum* should nevertheless be considered significant where common. In Hawaii, for example, many patients suffering from grass pollinosis react strongly positive (3+–4+) on skin testing (G. Fournier-Massey, personal communication).

TRISETUM *trisetum*

Common in alpine meadows and more northern latitudes, members of the genus are not known, however, to elicit pollinosis.

TRITICUM *wheat*

Abundant pollen is shed within the floret, but wheat is largely self-pollinated, so there is only limited exposure to the pollen produced. It is considered of little importance in pollinosis (Wodehouse 1935, 1971), although some individuals are allergic to its pollen (Gutmann 1950).

ZEA *corn, maize*

One of the most important economic plants of the world, corn (*Zea mays*) is abundantly cultivated in the Midwest. Even with large pollen having a short distribution (to 0.5 km), the extremely allergenic pollen may cause pollinosis (Wodehouse 1935, 1971), particularly among individuals living in the vicinity of the crop (Hall 1922; Ellis & Rosendahl 1933; Levetin & Buck 1980).

ZIZANIA *wildrice*

Found in marshes and borders of streams and ponds, usually in shallow water, *Zizania* sheds abundant light pollen. Because of low pollen allergenicity (Wodehouse 1935) and perhaps low exposure frquency, Wodehouse (1971) considered wildrice pollen unimportant in inhalant allergy, yet Ellis & Rosendahl (1933) thought it significant in Minnesota.

ZOYSIA *zoysia*

Introduced lawngrasses for milder climates of the continent that occasionally have escaped, zoysia may be of concern as some inflorescences of these short plants are unscathed by the mower and may release pollen throughout the summer. Allergenicity is, however, unreported.

CYPERACEAE
Sedge Family

Similar to the Poaceae, the sedges and reeds differ by usually having solid, 3-angled stems, a closed leaf sheath, no ligule, and each flower subtended by a single bract (glume).

The family is large, about 4,000 species, and found throughout the world, though it is abundant in wet, marshy areas of temperate and subarctic regions. Because of this kind of habitat, generally away from population centers, exposure to the wind-dispersed sedge pollen is minimal. Reports of allergenicity are not frequent. Principal genera are *Carex*, one of the largest known genera in numbers of species, *Cyperus* (umbrella-sedge), *Fimbristylis*, and *Scirpus* (bulrush) (see Fig. 129).

Flowering and Pollen Aerobiology

Many species of *Carex* and *Scirpus* shed their pollen in the spring and early summer, but limited flowering may continue to the autumn (Table 13). Pollen of *Cyperus* and *Fimbristylis* species is shed somewhat later and may extend to the early winter in warmer regions of the continent. Often large amounts of buoyant pollen are dispersed and frequently caught on atmospheric slides (Wodehouse 1971).

Pollen Morphology (Fig. 140)

Grains are generally pyriform to less commonly ovoid, often oval to circular in polar view, 24–65 μm long and 21–40 μm across; the exine thin, ca. <1 μm; the sexine usually as thick as the nexine; the surface densely granular; the intine up to 20 μm in thickness, being particularly thick at the narrower end of the pollen and comparatively thin at the wider end below the aperture. Apertures are poorly defined, 1–4-poroid, 3–20 μm wide, somewhat sunken, elongate, elliptical or ovoidal, with a loosely granular or fragmented surface. In 4-aperturate grains one aperture is found at the wider end and three others on the sides around the grain.

Allergenicity

Considering the volume of pollen produced and the close relationship to the grass family, it is perhaps surprising that there are only a few reports of sensitivity to cyperaceous pollen. It may be simply a matter of limited exposure, for few of these plants are closely associated with population centers. This is apparently not true in Israel, however, where several sedges are important offenders (Gutmann 1950), nor in Hawaii, where about 50 atopic patients annually responded positive (3+) to skin tests using sedge pollen extracts (A.W. Neilson, Jr., personal communication).

CAREX *sedge*

Four species are said to constitute serious pollinosis in Minnesota (Ellis & Rosendahl 1933), but only two positive skin test reactions were reported from California (Hall 1922). However, as many as 60 patients annually responded 3+ to pollen extracts in Hawaii.

CYPERUS *umbrella-sedge*

Pollen of *Cyperus esculentus* (yellow nutgrass), particularly common in the eastern half of the continent, is suspected of causing pollinosis (Bassett et al. 1978). In Puerto Rico allergenicity of *C. rotundus* pollen was found to be low but needing further study (Marchand 1948). In Israel pollen of *C. antiquorum* (syn. *C. papyrus*), or papyrus, and *C. longus* was confirmed as responsible for pollinosis by direct exposure, positive skin test reactions, and successful treatment using pollen extracts (Gutmann 1950).

SCIRPUS

Three species of *Scirpus* were involved in pollinosis in Minnesota (Ellis & Rosendahl 1933), and, as described for *Cyperus*, *S. litoralis* pollen was implicated in pollinosis in Israel (Gutmann 1950).

JUNCACEAE
Rush Family

Rushes are grasslike herbs lacking spikelets but with flowers in ±clustered heads. The perianth consists of two whorls of three scales that may be green,

TABLE 13. Flowering by month of the Cyperaceae.

GENUS	REGION							
	nNe	sNE	SE	NC	SC	NW	SW	sFL
Carex	5–8	3–10	3–10	4–8	3–9	4–8	3–9	3–8
Cyperus	7–9	5–10	6–10	6–10	1–12	6–10	5–12	3–11
Fimbristylis	8–9	5–10	7–10		6–11		6–11	6–11
Scirpus	5–9	4–10	5–10	6–9	3–11	5–9	5–9	4–11

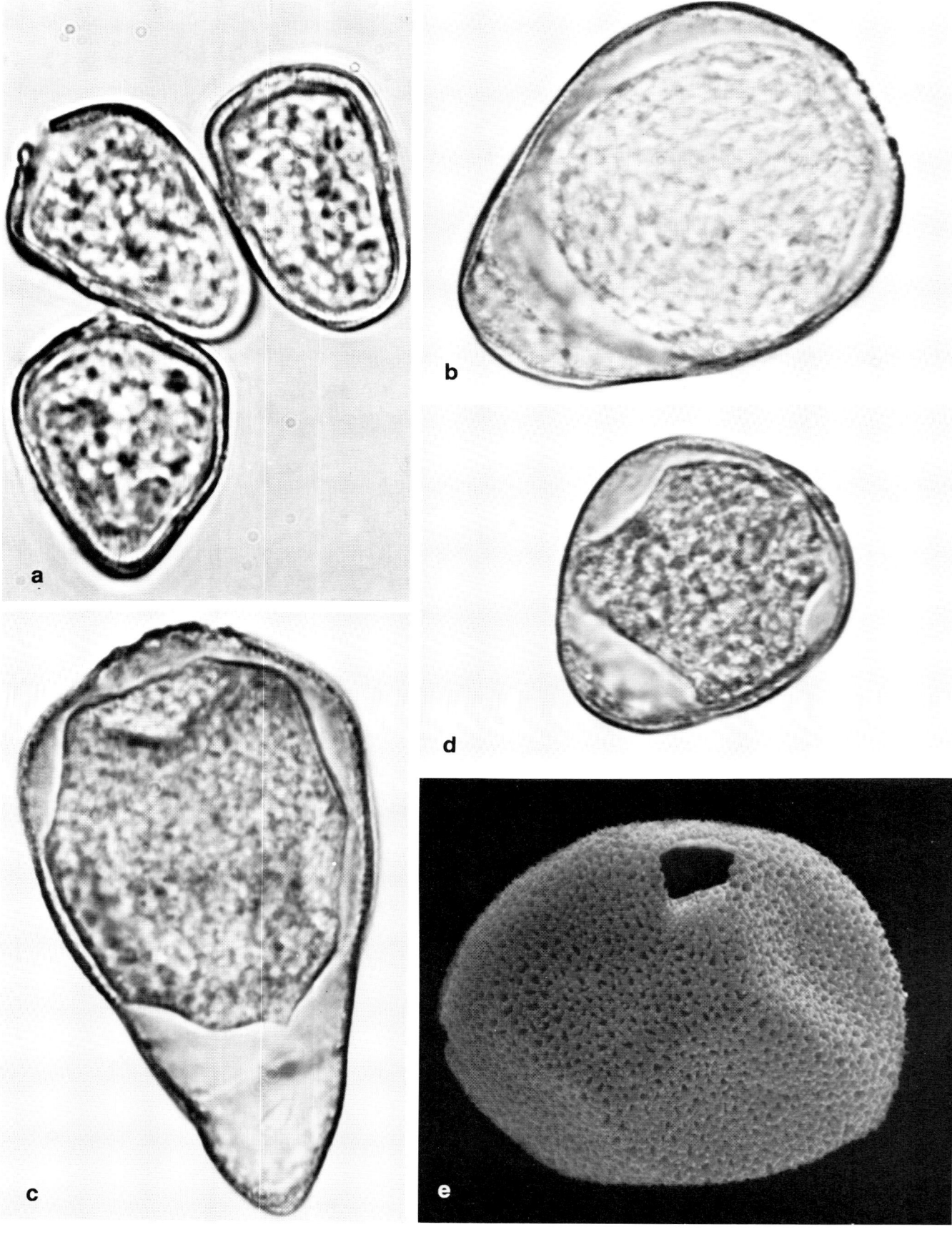

TABLE 14. Flowering by month of the Juncaceae.

GENUS	REGION							
	nNE	sNE	SE	NC	SC	NW	SW	sFL
Juncus	5–10	5–11	5–11	6–9	3–8	4–9	4–10	3–11
Luzula	5–7	4–6	3–8	6–7	3–5	4–9	3–8	

brown, or black, occasionally white or yellowish. There are three to six stamens, and the ovary is superior, 3-carpellate, and with three stigmas. The fruit is a dehiscent capsule.

This small family of nine genera is found in habitats similar to the Cyperaceae and probably few individuals are exposed to their wind-dispersed pollen. The chief genera are *Juncus* (rush or bogrush) and *Luzula* (woodrush) (see Fig. 129).

Flowering and Pollen Aerobiology

Luzula pollen is shed largely during the spring and early summer, typically somewhat longer for *Juncus* (Table 14). Large quantities are dispersed (Wodehouse 1935).

Pollen Morphology (Fig. 141)

Grains are united in tetrahedral tetrads, 20–44 μm in diameter, 1-aperturate (*Luzula multiflora*) or inaperturate (*L. pilosa*) with bandlike thickenings adjoining the four monads; the exine thin, <1 μm; the surface finely granular; the intine generally thick, up to 4 μm, but extremely thin beneath the invaginated distal region.

The tetrads of *Luzula* are almost spherical because the monads are closely appressed. The tetrads are 35–45 μm in diameter; the monads are 26–32 μm in diameter, with invaginated distal regions that are provided (when aperturate) with an ulceroid aperture having a coarsely granular sexine, which is separated by finely granulated bands.

Allergenicity

According to Wodehouse (1935), pollen is not yet implicated in pollinosis, for exposure would be limited except in low, wooded areas. However, *Juncus acutus* (great searush) is of allergenic significance in Israel based on direct exposures, positive skin test reactions, and successful treatment using pollen extracts (Gutmann 1950).

Fig. 140. Pollen of representative Cyperaceae. a: *Cyperus alternifolius* (umbrella plant), LM(s) ×1830; b: *Carex crinita* (fringed sedge), LM(s) ×1830; c–d: *Scirpus maritimus* (saltmarsh bulrush), c = LM(s) longitudinal view showing greatly thickened intine at the narrower end x1830, d = LM(s) oblique view ×1830; e: *Fimbristylis autumnalis*, SEM(a) showing a pore ×3600.

JUNCAGINACEAE
Arrowgrass Family

The arrowgrasses are sedgelike, with leaves resembling grasses. The inflorescences are few-flowered spikes (or racemes) with bisexual or unisexual flowers, and the perianth consists of two similar whorls each of three segments. There are usually six stamens; pollination is by wind. The ovary is superior, 3- to 6-carpellate, with as many often plumose stigmas. The fruit is a 1- to several-seeded follicle.

The Juncaginaceae are a small family found in marshes and bogs of the colder regions of the temperate zone, the principal genera being *Triglochin* (arrowgrass) and *Scheuchzeria*, neither of which is common in populated areas of the continent.

A few pollen grains of *Triglochin maritima* (Fig. 142) have been found on slides in Canada, but they have not been implicated in pollinosis (Bassett et al. 1978). Flowering occurs typically from May to July, or to August for *Triglochin*.

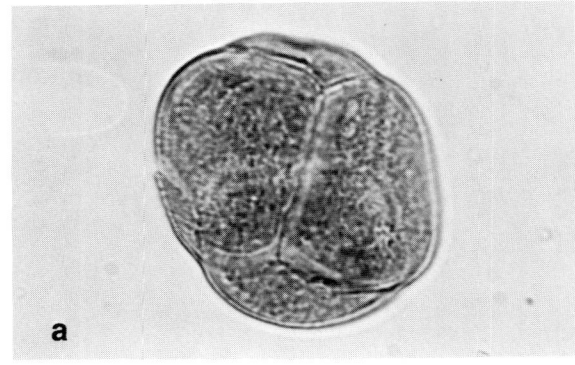

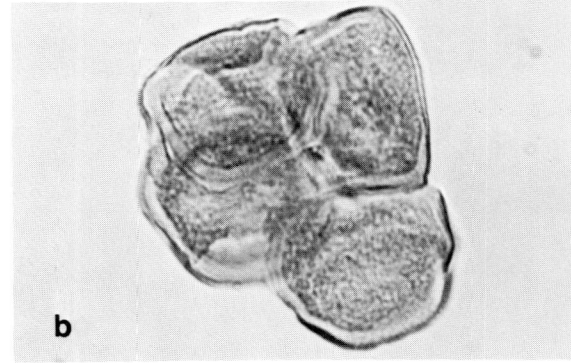

Fig. 141. Pollen tetrads of *Luzula multiflora* (many-flowered woodrush). a = LM(s) ×720, b = LM(s) showing tetrad breakup x720.

Pollen Morphology (Fig. 143)

Grains are inaperturate; spheroidal to ovoid; 22–34 μm in diameter; with a thin (ca. 1 μm) reticulated exine and an extremely thin intine (ca. <0.5 μm); and the sexine with polygonal lumina ca. 1 μm wide and muri ca. 0.5 μm thick.

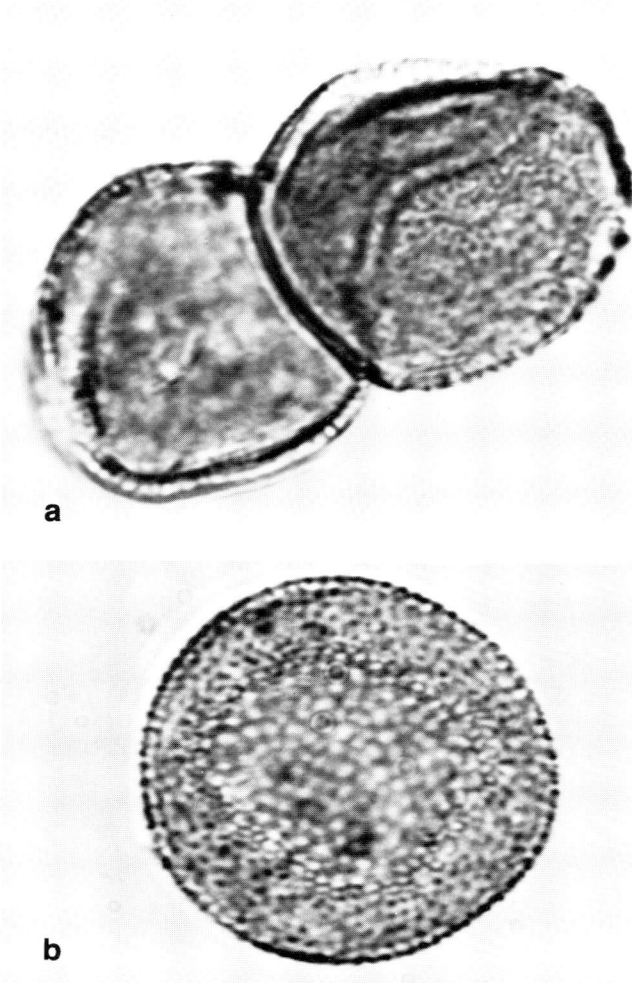

Fig. 143. Pollen of *Scheuchzeria* and *Triglochin*. a: *S. palustris*, LM(s) ×1830; b: *T. maritima*, LM(s) ×1830.

Fig. 142. *Triglochin maritima* (sea arrowgrass) in a British Columbian coastal habitat (a) and with leaves and inflorescences (b, ×¾).

POTAMOGETONACEAE

Pondweed Family

Aquatic herbs, the pondweeds often have broad floating leaves and narrower, usually linear, submerged ones, the leaves sheathing basally and ligulate. Inflorescences are stalked spikes, the flowers bisexual or unisexual (usually monoecious), the perianth of four bractlike scales. There are four stamens, with primarily wind-pollinated pollen, and the ovary is superior and 4-carpellate with four sessile or short-styled stigmas. The fruit is a druplet or nutlet.

The pondweeds are a cosmopolitan family growing

in aquatic habitats with one major genus, *Potamogeton*. None is commonly found near major population centers.

Flowering of *Potamogeton* occurs usually from May (March in the southcentral region) to early autumn. Pollen shed is not great, and allergenicity is unknown.

Pollen Morphology (Fig. 144)

Grains are inaperturate, ellipsoidal to spheroidal, 20–30 μm in diameter; the exine only ca. 1 μm thick, reticulate, heterobrochate with winding muri and small lumina 0.3–0.4 μm wide; the sexine as thick as or thicker than the nexine; and the intine extremely thin, ca. <0.5 μm.

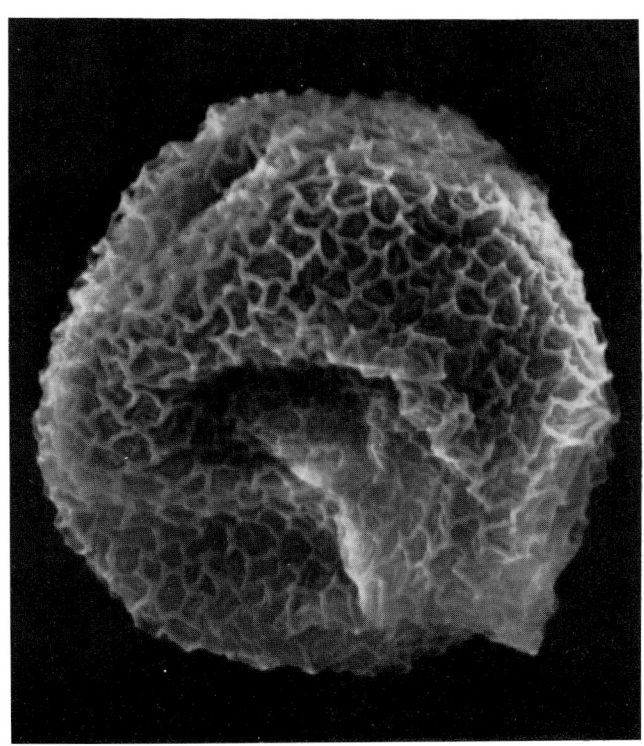

Fig. 144. Pollen of *Potamogeton pusillus* (small pondweed), SEM(a) ×4800.

TYPHACEAE
Cattail Family

Vegetatively, the cattails resemble tall grasses growing to 2 m high or more but having unisexual flowers borne cylindrically above the leaves, the staminate above the pistillate flowers in dense elongate inflorescences. There are two to five stamens with wind-dispersed pollen. The ovary has a single ribbon-shaped stigma, and both flowers have a perianth of bristles. The fruit is a minute achene dispersed by the wind.

This small family of one genus, *Typha* (cattail) (Fig.

Fig. 145. *Typha latifolia* (common cattail) with leaves and staminate (top) and pistillate inflorescences, ×¾.

145), occurs in shallow freshwater common in many temperate and tropical localities.

Species of cattails shed abundant pollen chiefly from June (as early as March in southern latitudes) to July (November in Florida). They have been minimally implicated in pollinosis in Virginia (Vaughan 1931) and *Typha latifolia* has been implicated in Minnesota (Ellis & Rosendahl 1933).

Pollen Morphology (Fig. 146)

Grains are either single, ovoidal to spheroidal with the longest axis 20–26 μm, or united in tetrads with a diameter of 40–50 μm, and 1-porate, the pores usually circular, 2–7 μm in diameter, often irregularly margined. The exine is reticulate with polygonal lumina, often elongate, 0.6–4.0 μm wide and with duplibaculate muri, and in places rugulate; the sexine thicker than the nexine; and the intine markedly thickened below the aperture to form an oncus ca. 6 × 15 μm.

Typha angustifolia pollen is single, whereas that of *T. latifolia* is united into tetrads in one plane that are either square, rhomboidal, linear, or irregular, with

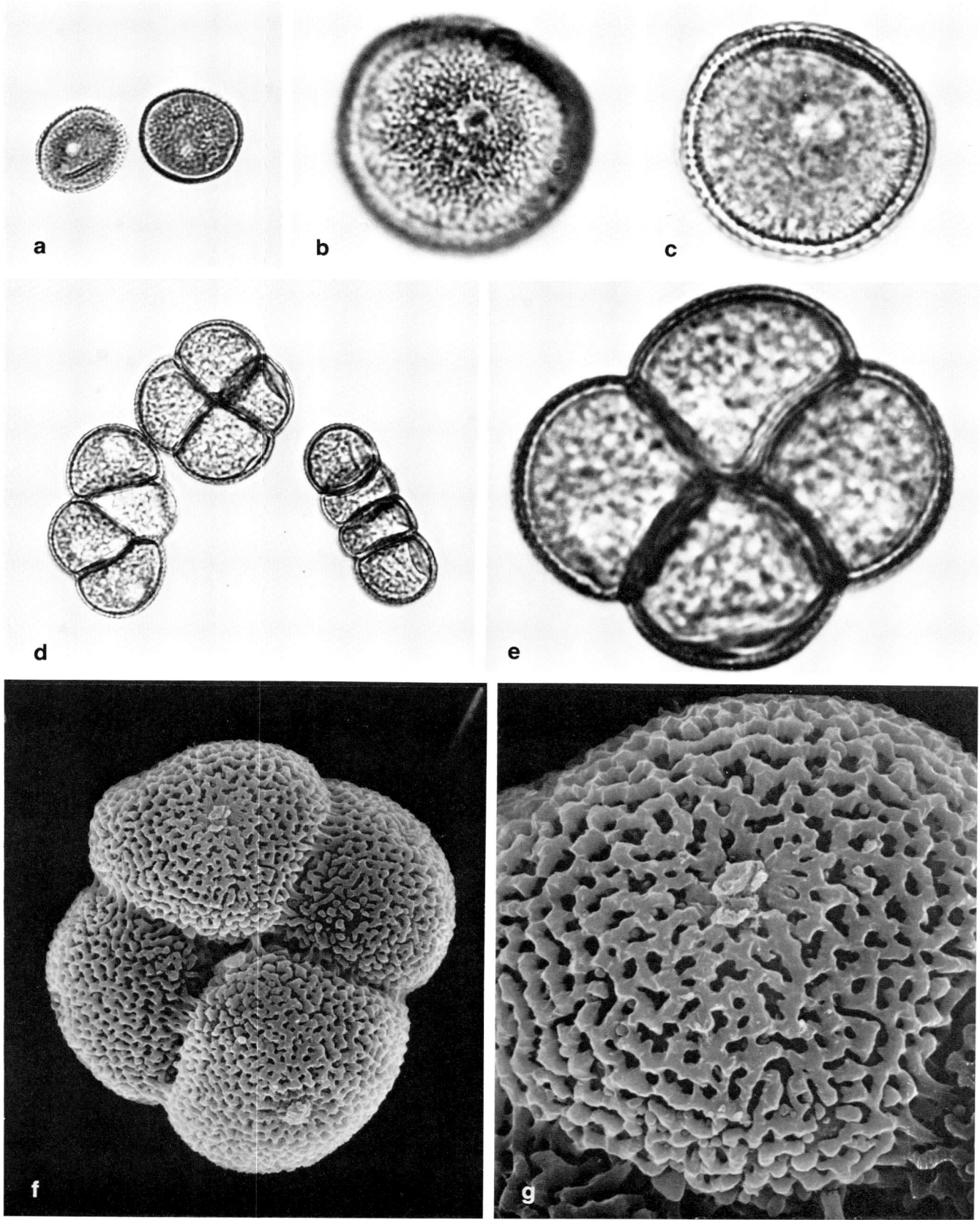

Fig. 146. Pollen of *Typha*. a–c: *T. angustifolia* (narrow-leaved cattail) having monads, a = LM(s) ×720, b = LM(s) surface view ×1830, c = LM(s) optical view ×1830; d–f: *T. latifolia* (common cattail) having tetrads, d = LM(s) ×720, e = LM(s) ×1830, f = SEM(a) ×2300, g = SEM(a) showing pore and sexine surface ×5600.

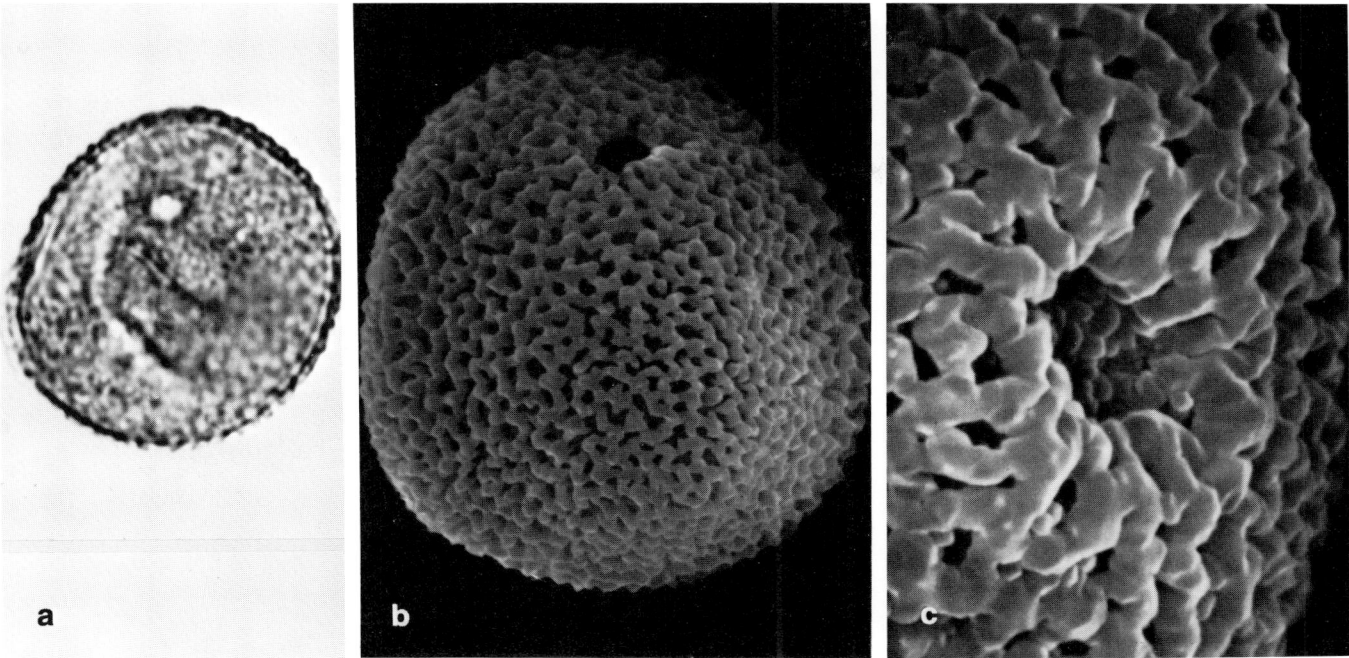

Fig. 147. Pollen of *Sparganium fluctuans* (bur-reed). a = LM(s) ×1830, b = SEM(a) ×3900, c = SEM(a) showing pore morphology and sexine surface ×8600.

the monads heteropolar, ±spheroidal, and 24–30 μm in diameter.

SPARGANIACEAE
Bur-reed Family

Related to the cattails, but separated by having unisexual flowers crowded in globose clusters along the stem axis, the single genus of bur-reeds, *Sparganium*, grows in marshy habitats mainly of temperate and arctic zones. It releases wind-dispersed pollen of unknown allergenicity from April or May through August or September.

Pollen Morphology (Fig. 147)

Grains are spheroidal to somewhat ellipsoidal, 22–28 × 24–29 μm and 1-porate; the pore ±circular but often irregularly margined, 3–6 μm in diameter, and operculate. The exine is ca. 1 μm thick, finely reticulate; the small lumina polygonal and smooth or with sparsely spaced granules; the sexine as thick as or thicker than the nexine; and the intine ca. 1 μm thick, but thickened markedly to form an oncus below the pore 4 × 10 μm. They resemble the single pollen grains of *Typha* (cattail).

Chapter 3

Weeds and Herbs

PTERIDOPHYTES

Ferns and Fern-allies

The Pteridophytes are ancient, green, spore-producing, vascular plants that are represented today by a very diverse assemblage of organisms. This diversity is undoubtedly reflected by disparate antigens borne by their spores. Allergenicity is known in the group, but research on its allergens is wanting.

Three genera, representing three very different classes of Pteridophytes, have been selected, because each produces spores captured by aerosamplers and two have been implicated as offenders in inhalant allergy. They are *Pteridium* (bracken fern) and the fern-allies *Equisetum* (horsetail) and *Lycopodium*) (clubmoss) (Fig. 148).

Sporulation

In eastern North America *Equisetum* spores are released from March to September, and those of *Lycopodium* and *Pteridium* may be found profusely on exposed slides in the northwestern region (Perlman 1955), in New Zealand (Licitis 1953), and elsewhere. *Equisetum* and *Lycopodium* spores have also been captured, but not in so high frequencies as *Pteridium* (Salén 1951; Bassett et al. 1978). Little attention has been given these spores in aerosampling and data are meager.

Spore Morphology (Fig. 149)

Spores of *Equisetum* are spheroidal, 38–56 μm in diameter (including perine), and 1-aperturate; the pore faintly demarked, circular (4.0–5.6 μm in diameter), and with a thickened rim ca. 4 μm wide, often appearing inaperturate. The perine (perispore) is 0.5 μm thick, faintly granular, wrinkled, forming broadly conical crestate folds 2–5 μm high; the exine (exospore) 1.0–1.5 μm thick and smooth; the intine (endospore) ca. 0.8 μm thick. In stained spores, elaters (four long bands with spatulate ends) can be observed—when dry, as when spores land on ground too dry for germination, elaters expand, giving boyancy to the spore and enabling it to be wind-dispersed, but when wet, elaters coil around the spore and decrease the opportunity of being carried off by wind.

Spores of *Lycopodium* and *Pteridium* are heteropolar; the apertures trilete; and the sculpturing either reticulate distally in *Lycopodium* or smooth, scabrous, or fimbriate in *Pteridium*.

Allergenicity

Allergenicity of *Lycopodium* spores is well documented. Perennial hay fever from inhaling dusting powders made of *L. clavatum* or *L. complanatum* spores was reported by Lambright & Albaugh (1934) and more recently by Salén (1951), who found 61 cases of immediate hypersensitivity due to a similar powder. Although less frequently used today, these spore powders are still found in fireworks (highly flammable) and cosmetics. Salén also reported one case of sporosis in the late summer from ambient exposure to sporulating plants of *Lycopodium*.

Spores of *Pteridium aquitinum* have also been implicated in inhalant allergies (Licitis 1953, Hyde 1972); no data exist for *Equisetum* spores (Bassett et al. 1978).

AMARANTHACEAE*

Amaranth Family

The amaranths are a family of about 65 genera and 900 species of herbs and some shrubs, a few of which are widely cultivated (cockscomb, globe amaranth) and others of which are known as noxious weeds (some species of *Amaranthus* and *Iresine*).

Leaves are entire, opposite, or alternate and without stipules. Flowers are usually in dense or congested spikelike or headlike inflorescences, bisexual or less commonly unisexual, with 4–5 perianth segments that are usually dry, membranaceous, and colorless, often united by their filaments to form a tube, and a superior ovary of 2–3 fused carpels. Flowers are subtended by dry, chaffy bracts that when massed are often very showy. The fruit is frequently a nut.

The most important allergenic plants are the predominantly anemophilous *Amaranthus* (amaranth, pigweed, waterhemp), and possibly *Dicraurus*, *Iresine* (bloodleaf), and *Tidestromia* (Fig. 150). Entomophily is widespread in the family, however, in such genera as *Achyranthes* and *Alternanthera* (chaff flowers), *Froelichia* (cottonweed), and *Gomphrena* (globe amaranth), and consequently their role in pollinosis is limited to incidental, local occurrences.

A common weed throughout the United States and

Fig. 148. Morphology of representative ferns and fern-allies. a: *Equisetum hyemale* (winter scouring rush) with vegetative and two fertile stems, ×⅔; b: *Pteridium aquilinum* (bracken fern) frond, ×⅘; c: Lower surface of a frond showing sori (c, ×2); d: *Lycopodium complanatum* (club-moss) stems, ×½.

WEEDS AND HERBS

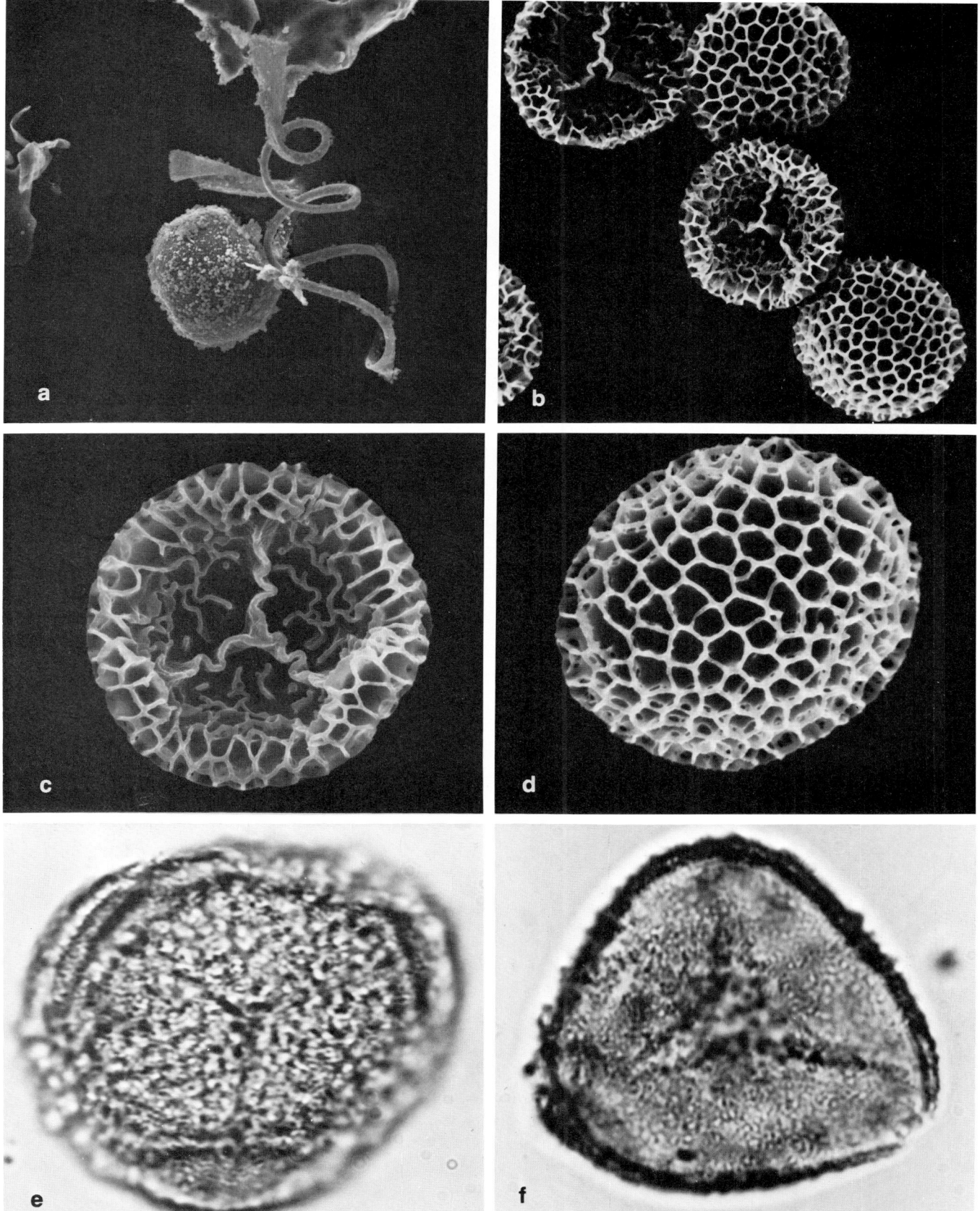

Fig. 149. Spores of *Equisetum*, *Lycopodium*, and *Pteridium*. a: *E. hyemale* (winter scouring rush), SEM(ad) with elators ×470; b–d: *L. clavatum* (running club-moss), b = SEM(a) ×950, c = SEM(a) proximal view showing trilete scar ×2200, d = SEM(a) distal view ×2200; e–f: *Pteridium aquilinum* (bracken fern), LM(a) proximal views ×1380 at surface (e) and optically (f).

Fig. 150. Morphology of representative Amaranthaceae. a: *Amaranthus retroflexus* (rough pigweed) stem with leaves and inflorescence, ×1; b–c: *Froelichia gracilis* (slender cottonweed) plant (b) and inflorescence (c, ×1); d–e: *Tidestromia lanuginosa* plant (d) and stem with leaves and flowers (e, ×1).

southern Canada, *Amaranthus* is particularly profuse along the Atlantic seaboard, the central United States, and the western Rocky Mountain region to California (Fig. 151).

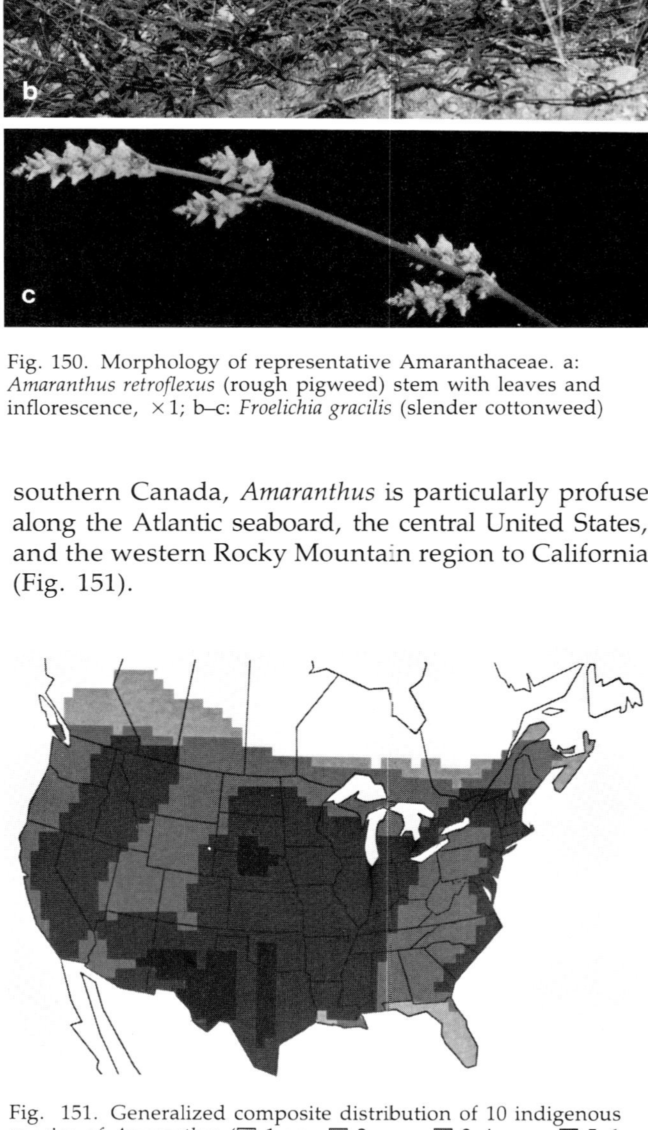

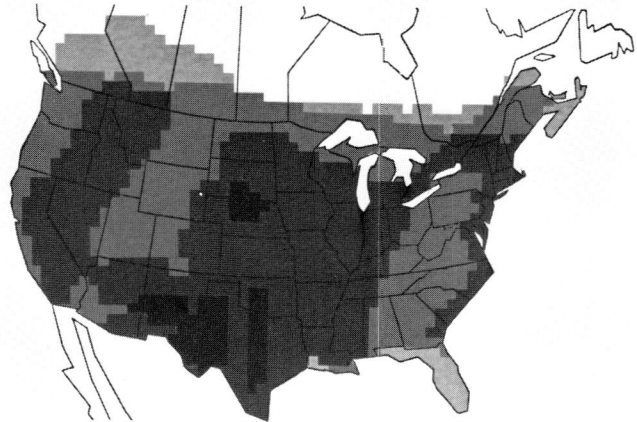

Fig. 151. Generalized composite distribution of 10 indigenous species of *Amaranthus* (☐ 1 sp., ▢ 2 spp., ▓ 3–4 spp., ■ 5–6 spp.).

Flowering and Pollen Aerobiology

Flowering peaks in midsummer and autumn, from July to October (Table 15). There is a wide range of flowering, however, that includes the spring months in southern latitudes and all year in southern Florida.

Pollen of the Amaranthaceae, together with that of the Chenopodiaceae, is airborne in the northeastern and northcentral regions from May to October, though principally from July to October (Table 16). Flowering data do not reflect the earlier minority times of anthesis during May and June, which are generally insignificant periods allergenically. Data for the southeastern region are not available, but aeropollen should closely approximate that of the southern part of the northeastern region. Airborne pollen has been retrieved in the southcentral region from April (rare) to November, with July and August predominant, but depending on the year representatives could be obtained year-round. The same pattern exists for the southwestern region, including southern California (see Table 16), where pollen shed is rare from January to April but is sufficiently common from

TABLE 15. Flowering by month of the Amaranthaceae.

GENUS	REGION							
	nNE	sNE	SE	sFL	NC	SC	NW	SW
Alternanthera			4–11	1–12		1–12		9–11
Amaranthus	(6–)7–10	6–10	5–11	1–12	7–10	5–11	6–11	6–11
Celosia		7–10	7–11	6–8		3–11		
Dicraurus						7–12		
Froelichia		5–9	5–10	1–12	7–8	3–11		
Gomphrena						4–9		
Iresine		8–10	8–10	1–12		(5–)7–10		
Tidestromia		7–10			7–9	3–10		4–12

June to December to be of major concern in pollinosis. In parts of the northwestern region pollen has been recovered with frequency from July through December.

Pollen Morphology (Fig. 152)

Grains are spheroidal, 14–50 μm in diameter, and pantoporate; the pores 20–65 per grain, usually circular (1.5–2.5 μm), either globally distributed or in luminoid areas separated by muroid ridges; and the opercular granular or with various processes. The sexine is often tegillate, undulating with a granular surface that is spinulose and tuberculate/punctate in SEM, or reticuloid with muroid ridges having granules or small spinulose processes; the nexine as thick as or often thinner than the sexine; and the intine generally thick or indistinct. (For pollen morphology, see Nowicke 1975.)

Amaranthus grains are 18–31 μm in diameter, pantoporate with 30–65 pores, and have a thin sexine (0.6–1.0 μm). In contrast, the grains of *Iresine* are smaller (14–17 μm in diameter) and polyporate, the ±20 pores in luminoid areas being separated by densely granular luminoid ridges; those of *Froelichia* are larger (43–50 μm in diameter) and polyporate (like *Iresine*) but with a thicker sexine (3–4 μm).

Pollen of the Amaranthaceae and Chenopodiaceae are similar, particularly the common genus in each family, *Amaranthus* and *Chenopodium*, respectively.

Allergenicity

By far the most frequent cause of pollinosis in the Amaranthaceae is *Amaranthus* (including *Acnida*). There are about 65 species in the genus, of which 25 occur in Texas alone (Correll & Johnston 1970). Many are widespread annual weeds in North America, some shedding limited pollen and perhaps of less significance allergenically (e.g., *A. retroflexus, A. torreyi*) (Solomon & Durham 1967) than those that shed abundant pollen (e.g., *A. palmeri, A. tamariscinus, A. tuberculatus*) (Wodehouse 1935). Species most often reported as important in pollinosis wherever common are: *A. australis* (southern water-hemp), *A. blitoides* (prostrate pigweed or tumbleweed, syn. *A. graecizans*), *A. hybridus* (green or spleen amaranth), *A. palmeri* (careless weed), *A. spinosus* (spiny amaranth), and *A. tamariscinus* (western water-hemp) (Balyeat & Stemen 1927; Heinberg 1930; Sellers 1935; Wodehouse 1935, 1971; Marchand 1948; Fein & Kamin 1962; Lindenbaum 1966b; Solomon & Durham 1967; Levetin & Buck 1980). Less significant, but nonetheless known to be local offenders, are *A. albus*

TABLE 16. Amaranthaceae-Chenopodiaceae aeropollen frequency in regions of North America based on percentage of pollen captured, by month (1974–78).[a]

REGION	MONTH											
	1	2	3	4	5	6	7	8	9	10	11	12
nNE					0.1	0.2	3.0	2.8	5.7	9.1		
sNE					0.1	1.1	10.1	3.2	2.7	0.5		
NC					0.7	1.1	9.3	66.4	21.8	9.4		
SC				0.1	4.2	1.0	26.4	27.8	7.0	2.8	12.0	
NW					1.5	1.1	17.0	36.5	42.5	60.5	26.5	20.0
SW	0.1			0.2	1.4	3.4	38.2	54.5	54.5	21.5	21.4	11.1
sCA	0.2	0.2	0.8	1.4	5.5	12.5	16.4	16.4	10.8	12.4	14.1	8.2

[a] No data available for SE.

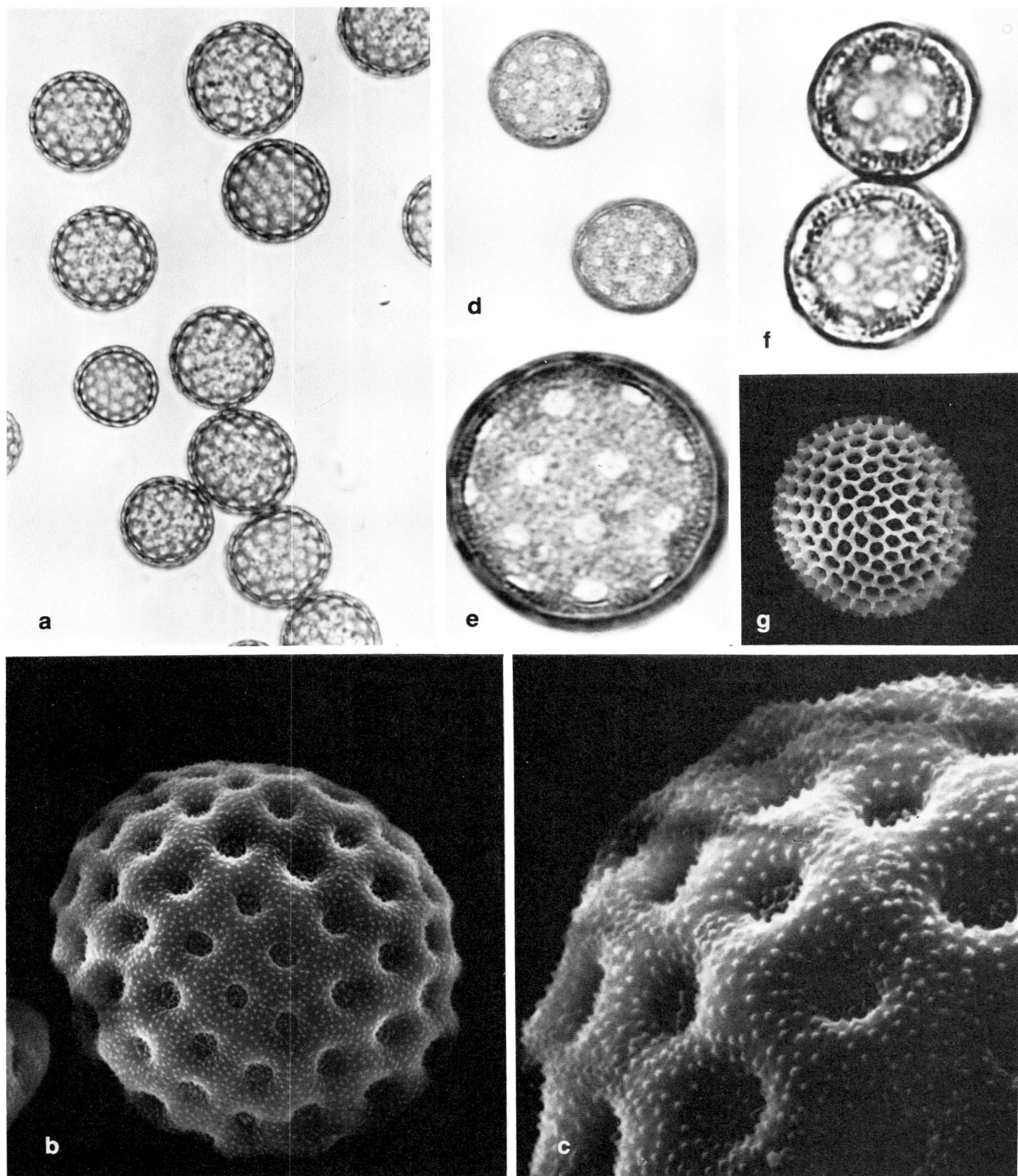

Fig. 152. Pollen of representative Amaranthaceae. a–c: *Amaranthus tamariscinus* (western water hemp), a = ×720, b = SEM(ad) ×4200, c = SEM(ad) showing pore morphology and sexine surface ×11000; d–e: *Amaranthus retroflexus* (rough pigweed), d = LM(s) ×720, e = LM(s) ×1830; f: *Iresine paniculata*, LM(a) ×1830; g: *Froelichia gracilis* (slender cottonweed), SEM(ad) ×1100.

(tumbleweed), *A. californicus*, *A. cannabinus* (waterhemp), *A. retroflexus* (redroot or rough pigweed), *A. torreyi*, and *A. tuberculatus*. *Amaranthus* species are less significant in eastern North America than in the central and western regions of the continent, where numerous reports of severe pollinosis from midsummer through September in Texas, California, and elsewhere correlate with plant abundance, anthesis, and strongly positive (3+–4+) skin test reactions.

Allergenicity of pollen from other genera of the family is unknown. Wodehouse (1971) suggested that the wind-pollinated *Tidestromia lanuginosa* is probably an offender in hay fever, and we would add *Iresine* and other ± anemophilous genera such as *Celosia* (cockscomb), which may release large amounts of aeropollen (e.g., *C. trigyna* in West Africa, Ybert 1980). Information on occurrence, frequency, and flowering of these genera is particularly relevant because pollen antigenic similarity is known in the family, suggesting that some common allergens may exist widely (Wodehouse 1957). Indeed, allergenic similarity extends to the related family Chenopodiaceae (Lamson 1931) to form an Amaranth-Chenopod allergenic group.

APIACEAE (*Umbelliferae*)
Carrot Family

Recognized by their umbellate inflorescences (Fig. 153), these herbs are well known as important foods and spices (carrot, celery, dill, fennel, parsley, parsnip), and poisons (hemlock). They are primarily insect-pollinated. Flowering begins as early as March in southern latitudes and terminates in June, but elsewhere flowering is from late spring through much of the summer.

Pollen Morphology (Fig. 154)

Grains of *Daucus* are perprolate to prolate, 27–30 × 13–14 μm, cylindrical or slightly constricted at the equator; the amb rounded-triangular and 3-colporate; the colpi very narrow, 14–20 μm long; the ora lalongate; and the opercula granular. The sexine is tegillate, smooth or undulating, thicker equatorially (to 2.2 μm) than at the poles (0.9–1.3 μm); the nexine thinner than the sexine, but thickened at the apertures; and the intine ca. 0.6 μm thick, somewhat thicker below the apertures.

Allergenicity

Daucus (carrot, Queen Anne's lace) pollen has been identified on atmospheric slides from Michigan, Missouri, and Virginia; unidentified umbelliferous pollen has also been retrieved from the ambient air in other states as well as in Europe, South Africa, and India (Lewis & Vinay 1979). Limited exposure to pollen

Fig. 153. *Daucus carota* (wild carrot, Queen Anne's lace) stem with inflorescence (a, ×⅔) and flowers (b, ×1¼).

can therefore be anticipated, resulting in local cases of pollinosis, as reported from New Jersey (Lewis & Vinay 1979).

ASTERACEAE (*Compositae*)
Aster or Composite Family

The Asteraceae are one of the largest families of flowering plants, with about 25,000 species. They are cosmopolitan in distribution, being partial to open habitats rather than to deep woods. In eastern North America they are by far the largest family of plants, both as native vegetation and garden ornamentals.

Herbs and subshrubs predominate. They possess simple, lobed, or toothed leaves that are alternate or opposite and often expand or sheath at the base. The family is easily recognized by its colorful headlike inflorescence (capitulum), made up of numerous small individual flowers (florets) and surrounded by an involucre of protective bracts. The florets are of several types: bisexual with the corolla tubular, having five short terminal lobes (disc flower); pistillate or sterile with a corolla tubular at the base, but above that ribbonlike and usually bent backward away from the center of the head and terminating in two or

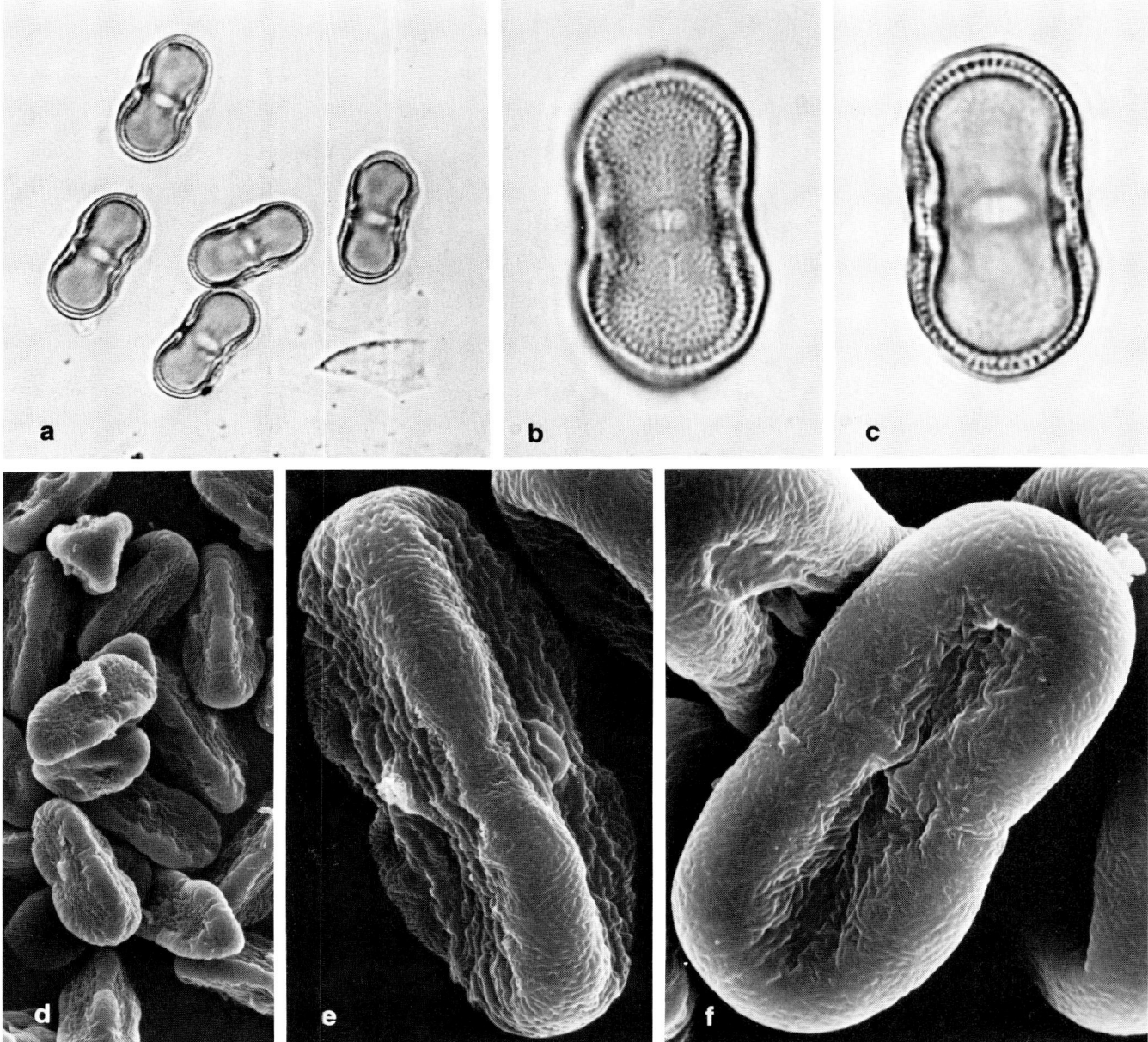

Fig. 154. Pollen of *Daucus carota*. a = LM(a) ×720, b–c = LM(a) equatorial views at surface (b) and optically (c) ×1830, d = SEM(ad) ×1000, e–f = SEM(ad) equatorial views ×3600.

three teeth (ray flower); or bisexual with a corolla, as for the ray flowers, but the ribbonlike portion having five terminal teeth (ligule flower). Various combinations of these florets occur, for example, a head with ray flowers found at the margin and disc flowers in the center is known as radiate, but if only disc flowers develop the head is discoid. Ligule flowers, however, make up the whole head of plants found in the Lactuceae (chicory tribe).

The arrangement of anthers and styles is associated with pollination and fertilization. The anthers mature before the stigmas and discharge their pollen into the tube formed by the cylinder of fused anthers (in the

Fig. 155. Morphology of representative Asteraceae. a: *Ambrosia artemisiifolia* (short ragweed) stem with staminate inflorescences, ×½; b: *Ambrosia bidentata* (southern ragweed) stem with staminate inflorescences, ×½; c: *Ambrosia trifida* (giant ragweed) stem with staminate inflorescences, the abundant pollen accumulating on leaves, ×½; d: *Iva annua* (rough marsh-elder) stems with staminate inflorescences predominant, ×½; e: *Iva axillaris* (poverty weed) stems with staminate and pistillate flowers, ×½; f: *Iva xanthifolia* stems with staminate inflorescences; ×½; g: *Xanthium strumarium* (cocklebur) stem with inflorescences, ×½; h: *Baccharis wrightii* stems and flowers, ×½; i: *Solidago canadensis* (Canada goldenrod) stem with inflorescence, ×½.

WEEDS AND HERBS

Fig. 156. Morphology of representative Asteraceae. a–b: *Artemisia ludoviciana* (western mugwort) plant (a) and inflorescence (b, ×⅕); c: *Cichorium intybus* (chicory) stem with flower head, ×⅘; d: *Bellis perennis* (English daisy) flower head, ×3.

Ambrosia subtribe the anthers merely converge). At this stage the style is short and the arms appressed. The style then elongates up the anther tube pushing the pollen ahead by hairs found along the style arms, eventually presenting the pollen at the top of the tube for dispersal by vectors or occasionally by wind. Only later do the style arms separate to expose the stigmatic surfaces in order to affect cross-pollination, but should this fail the style arms may recurve sufficiently for the stigmas to contact pollen from the same floret and in this way self-pollinate when compatible.

The most important composite genera allergenically are *Ambrosia, Iva,* and *Artemisia* (Figs. 155–56). *Ambrosia* is particularly common in North America, not only in population frequency but also in number of species. About 17 are recognized and many are widely distributed, with the majority occurring in the central part of the continent and extending to the northeastern and southwestern regions. Between 5 and 8 species of *Ambrosia* occur from eastern Wyoming south to western Texas and west to southern California (Fig. 157). *Iva* is much less common and is absent in the western part of the northwestern region and in areas around the Great Lakes and the Appalachian Mountains. The genus is most frequent along the Atlantic and Gulf coasts and from Oklahoma to Utah (Fig. 158). *Artemisia* is most frequent in the western half of the continent; as many as 11 species are found in some localities in the Rocky Mountains and surrounding areas (Fig. 159).

Classification

Twelve major tribes are recognized in the family (Cronquist 1955, 1977) and separate into three groups: the radiate tribes (having ray and disc flowers), the discoid tribes (have disc flowers only), and the ligulate tribe Lactuceae. This separation is sometimes formalized by recognizing three subfamilies, with the tribe Arctotideae (number 7) forming a link between the radiate and discoid groups. The Lactuceae, on the other hand, stand clearly apart from all other tribes (Fig. 160).

The few but important wind-pollinated genera in North America are found in only three tribes. *Ambrosia, Dicoria, Hymenoclea, Iva, Xanthium* (all subtribe Ambrosiinae), and the more distantly related *Parthenice* belong to Heliantheae (number 1). The subtribe Ambrosiinae is considered a specialized group within this large and most primitive of composite tribes. Sometimes it is recognized as a distinct tribe, but the evidence is not overwhelming, for the Ambrosiinae are strongly linked with other members of the sunflower tribe through both *Parthenice* and the insect-pollinated *Parthenium*. *Artemisia* is found in the more evolved but related radiate tribe Anthemideae (number 6), and *Baccharis,* some species of which are clearly wind-pollinated (Quintero 1955; Wodehouse 1971), is classified in the radiate tribe Astereae (number 2).

The many primarily insect-pollinated genera that have been implicated in localized pollinosis are found widely scattered throughout this classification. These genera (see Table 22 for flowering data), together with the wind-pollinated genera (in boldface), are listed according to their current affiliation by tribe.

Group 1: Radiate tribes

Heliantheae (sunflower tribe): subtribe Ambrosiinae with **Ambrosia** (ragweed), **Dicoria, Hymenoclea** (burro-brush), **Iva**, (marsh-elder), **Xanthium** (cocklebur); and *Encelia* (brittle bush), *Helenium* (sneezeweed), *Helianthus* (sunflower), **Parthenice**, *Parthenium* (feverfew, guayule), *Rudbeckia* (coneflower), *Tagetes* (marigold), and *Viguiera*.

Astereae (aster or daisy tribe): *Aster* (daisy), **Baccharis**, *Bellis* (English daisy), *Erigeron* (fleabane), *Grindelia* (rosinweed), and *Solidago* (goldenrod).

Senecioneae (groundsel tribe): *Senecio* (ragwort) and *Tussilago* (coltsfoot).

Anthemideae (mayweed tribe): *Achillea* (yarrow), *Anthemis* (chamomile, dog fennel), **Artemisia** (sage, sagebrush, wormwood), *Chrysanthemum* (syn. *Leucanthemum*) (chrysanthemum, pyrethrum), and *Tanacetum* (tansy).

Group 2: Discoid tribes

Eupatorieae (boneset tribe): *Ageratum* (floss flower), *Eupatorium* (boneset), and *Piqueria* (florists' stevia).

Vernonieae (ironweed tribe): *Vernonia* (ironweed).

Cynareae (thistle tribe): *Carduus* (plumeless thistle), *Centaurea* (knapweed), and *Cirsium* (thistle).

Group 3: Ligulate tribe:

Lactuceae (chicory tribe): *Cichorium* (chicory), *Sonchus* (sow thistle), and *Taraxacum* (dandelion).

These infrafamilial groupings and tribes may epitomize major allergenic similarities and/or differences within this vast family. Future research should consider this arrangement as fundamental to chemical characteristics as a whole.

The confusing generic names used for species of the three most important wind-pollinated composite genera (Wodehouse 1935, 1971) need clarification and updating. *Ambrosia* includes *Acanthambrosia, Franseria* (false ragweed), and *Gaertneria* (Payne 1964); *Iva* includes *Chorisiva, Cyclachaena, Leuciva,* and *Oxytenia* (Jackson 1960); and *Artemisia* includes *Artemisiastrum, Crossostephium, Picrothamnus,* and *Vesicarpa* (Airy Shaw 1973). Two additional genera, *Chamartemisia* and *Sphaeromeria,* supposedly related to *Artemisia,* are in fact *Tanacetum*.

Of the 17 to 18 species of *Ambrosia* indigenous to North America, 15 of the most important have been grouped by possible evolutionary affinities (Mabry 1970). Characteristics of their pollen allergens may also reflect this classification, which in outline is as follows:

Group 1—*A. ambrosioides* (canyon ragweed), *A. cordifolia, A. deltoidea* (Arizona bur-sage), and *A. dumosa* (bur-sage); Group 2—*A. acanthicarpa* (bur ragweed), *A. chamissonis, A. grayi, A. tomentosa,* and *A. trifida* (giant ragweed); and Group 3—*A. artemisiifolia* (short ragweed), *A. bidentata* (southern rag-

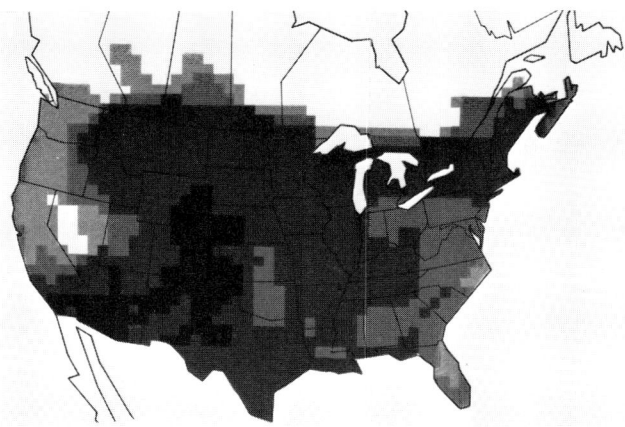

Fig. 157. Generalized composite distribution of 17 indigenous species of *Ambrosia* (□ 1 sp., ▨ 2 spp., ▧ 3–4 spp., ■ 5–8 spp.).

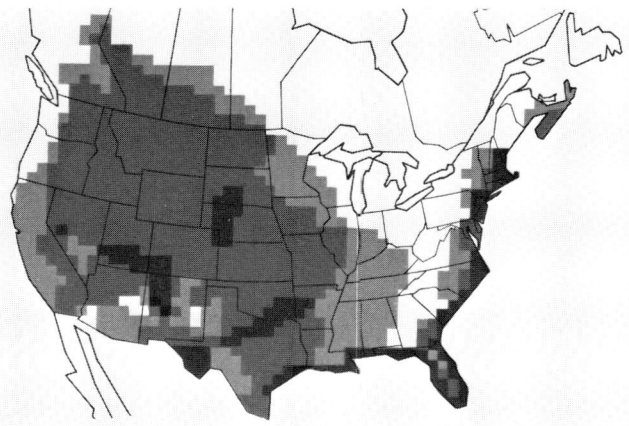

Fig. 158. Generalized composite distribution of 11 indigenous species of *Iva* (□ 1 sp., ▨ 2 spp., ▧ 3 spp., ■ 4 spp.).

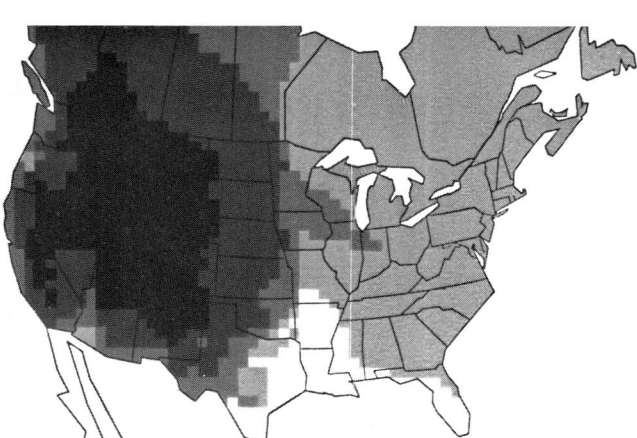

Fig. 159. Generalized composite distribution of 14 indigenous species of *Artemisia* (□ 1 sp., ▨ 2–3 spp., ▧ 4–6 spp., ■ 7–11 spp.).

weed), *A. confertiflora*, *A. hispida* (coastal ragweed), *A. psilostachya* (perennial or western ragweed), and *A. pumila*. Unfortunately, only a few of these species have been examined allergenically, none from Group 1.

Flowering and Pollen Aerobiology

Throughout most of temperate North America composites flower during midsummer or late summer to autumn as exemplified by *Ambrosia* (Table 17), *Iva* (Table 18), *Artemisia* (Table 19), *Xanthium*, and other anemophilous genera (Table 20). In Florida and the Southwest flowering may be much earlier or later for *Ambrosia* and may be year-round in some species (western *A. chamissonis*, Floridian *A. hispida*). Flowering in the winter months in parts of Arizona and California is also known for *A. acanthicarpa*, *A. ambrosioides*, *A. chenopodiifolia*, *A. deltoidea*, *A. dumosa*, and *A. ilicifolia*. Species of *Baccharis* and *Dicoria* also have prolonged flowering in the Southwest, the former often shedding abundant, long-spined pollen unrelated to the *Ambrosia* group or to *Artemisia*.

Short-spined (1.0–2.5 μm long) pollen is characteristic of wind-pollinated *Ambrosia* (except *A. grayi*, *A. tomentosa*), *Iva* (except *I. imbricata*, *I. microcephala*), *Dicoria*, and *Parthenice*. Only the long-spined (3.2+ μm) pollen of *Baccharis* would commonly become airborne in aerosamples of the western regions. Pollen of many other genera are long-spined, however, but these are largely insect-pollinated and would only occasionally be found in aerosamples, as reported by O'Rourke (1980) from Tuscon, Arizona (probably *Encelia farinosa*, O'Rourke, personal communication). A third group of composites includes those having pollen with very reduced spinules (0.5–0.6 μm long), which, using light microscopy, are difficult to observe as spiny, the pollen surface appearing only granular

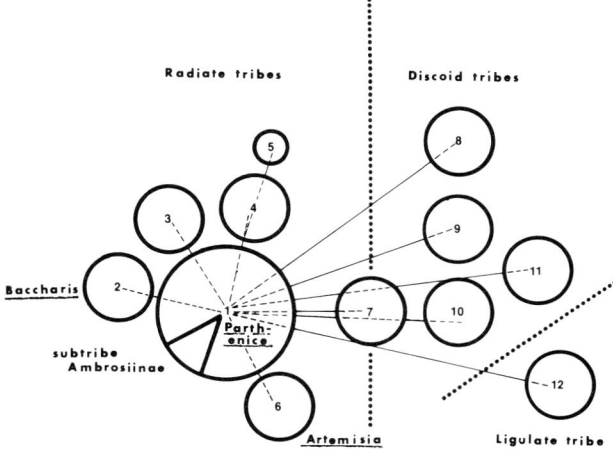

Fig. 160. Schematic representation of probable relationships of the tribes of Asteraceae. 1: Heliantheae, 2: Astereae, 3: Inuleae, 4: Senecioneae, 5: Calenduleae, 6: Anthemideae, 7: Arctotideae, 8: Eupatorieae, 9: Vernonieae, 10: Cynareae, 11: Mutisieae, 12: Lactuceae. The subtribe Ambrosiinae, including *Ambrosia* and *Iva*, is grouped in the tribe Heliantheae.

TABLE 17. Flowering by month of *Ambrosia*.

SPECIES	REGION							
	nNE	sNE	SE	sFL	NC	SC	NW	SW
A. acanthicarpa		8–10			8–9	6–9		8–12
A. ambrosioides								(2–)3–5
A. artemisiifolia	7–10	7–10	8–11	3–11	8–9	8–11	8–10	8–10
A. bidentata		7–10	8–11			8–9		
A. chamissonis							6–9	1–12
A. chenopodiifolia								1–6
A. confertiflora						6–11		5–11
A. deltoidea								1–4, 10–12
A. dumosa								2–6, 9–12
A. eriocentra								4–7
A. grayi						6–9		
A. hispida				1–12				
A. ilicifolia								2–6
A. psilostachya	7–9	8–10	7–11		8–9	8–11	7–10	7–11
A. pumila								6–9
A. tomentosa		7–9			7–8	5–8		5–8
A. trifida	6–9	7–9	7–11		8–9	8–11	7–10	6–9

TABLE 18. Flowering by month of *Iva*.

SPECIES	REGION							
	nNE	sNE	SE	sFL	NC	SC	NW	SW
I. acerosa					7–9			(7–)8–12
I. ambrosiaefolia						8–11		5–10
I. angustifolia			9–10			8–11		
I. annua	7–10	7–10	8–11		8–9	8–11		
I. axillaris					6–7	4–7	5–9	5–9
I. dealbata						8–11		
I. frutescens	8–10	8–10	8–11	6–11		6–10		
I. imbricata			7–11	6–11				
I. microcephala			9–10	6–10				
I. nevadensis								6–10
I. xanthifolia	7–9	6–10	8–10		7–8	8–11	8–10	7–10

or undulating. These grains are often mistakenly excluded from the Asteraceae, even though typical of all species of the wind-pollinated *Artemisia, Xanthium, Hymenoclea*, and of the exceptional species of *Ambrosia* and *Iva* noted earlier.

Abundant pollen is shed from all species found in these genera, usually en masse in August and September in the eastern half of the continent but extending well beyond this period in the southcentral and southwestern regions (Table 21). In fact, *Ambrosia*-like pollen is found throughout the year in California, though very limited from January to March.

Pollen is occasionally released incidentally by plants known to be principally pollinated by insects. Wodehouse (1971), for example, implied that most members of the tribes Heliantheae and Astereae release some pollen that becomes atmospheric (e.g., *Erigeron, Helianthus, Rudbeckia*, and *Solidago*) so that their contribution to pollinosis must not be ignored. Lewis & Vinay (1979) reported pollen of *Anthemis, Chrysothamnus, Eupatorium, Helianthus, Solidago*, and *Taraxacum* becoming airborne in limited quantities in the United States and abroad. These and other genera implicated in pollinosis have a wide range of flowering times according to region of distribution (Table 22); all have been implicated in endemic outbreaks of inhalant allergy.

Pollen Morphology

Grains are mostly oblate-spheroidal to prolate-spheroidal, 15–48 × 16–40 μm; the amb triangular, 3–4 lobate, pentagonal or hexagonal, and 3–4 colporate; the colpi short (*Ambrosia*) or long (*Artemisia*); and the ora various. The sexine is generally thick, tectate, and provided with long or short spines, or spinules, and often permeated with micropores; the nexine

TABLE 19. Flowering by month of *Artemisia*.

SPECIES	REGION						
	nNE	sNE	SE	NC	SC	NW	SW
A. absinthium			6–10	8–9		7–9	
A. annua	8–10	8–10	8–10			8–10	
A. arbuscula						7–9	7–8
A. biennis	6–9	6–11		8–9		8–10	8–10
A. californica							8–12
A. campestris	7–10	7–10	9–10	7–8	9–10	7–9	
A. cana				8–9		8–9	7–9
A. carruthii		8–10			7–9		8–11
A. douglasiana						6–10	6–10
A. dracunculus		6–10		8–9		7–10	8–10
A. frigida	8	7–9		7–8	8	7–9	
A. ludoviciana	7–9	6–9	7–10	7–9	6–11	6–10	7–9
A. michauxiana				7–8		6–8	5–8
A. norvegica						7–9	7–9
A. spinescens						4–6	4–5
A. stelleriana			5–8				
A. tridentata				7–9		7–9	8–10
A. vulgaris	7–10	7–10	7–10	7–9		8–10	5–10

TABLE 20. Flowering by month of allergenically minor, but primarily anemophilous, genera of the Asteraceae.

GENUS	REGION							
	nNE	sNE	SE	sFL	NC	SC	NW	SW
Baccharis	9–10		9–11	9–12		4–11	8–9	(1–)3–12
Dicoria						6–1		
Hymenoclea						9–10		2–4,9
Parthenice								7–12
Xanthium	8–10	7–12	7–11		7–9	6–11	5–10	6–10

usually thinner than the sexine; and the intine thin, but slightly thickened below the apertures.

Long-spined (3.2–5.6 μm long) grains occur in *Antennaria, Aster, Baccharis, Carduus, Chrysanthemum, Cichorium, Eupatorium, Helenium, Matricaria, Parthenium, Rudbeckia, Senecio, Solidago, Tagetes, Tanacetum,* and *Taraxacum.* Grains with shorter spines (1–3 μm long) are found in *Ambrosia, Centaurea, Dicoria, Iva,* and *Parthenice.* In grains of *Ambrosia grayi, A. tomentosa, Arctium, Artemisia, Euphrosyne, Hymenoclea, Iva imbricata, I. microcephala,* and *Xanthium,* the spinules are further reduced to small pointed or blunt projections <0.6 μm long.

Ambrosia pollen grains are oblate-spheroidal (22–29 x 22–32 μm) to prolate-spheroidal (ca. 21 x 19 μm) and brevicolpate (4–8 x 1.6–2 μm); the opercula slightly granular; and the ora lolongate (2.4–4 x 1.5–2 μm) to subcircular. The sexine is tectate, 2.3–3.2 μm thick; the largest spines typically 2–3 μm long with pointed

TABLE 21. *Ambrosia*-like aeropollen frequency in regions of North America based on percentage of pollen captured, by month (1974–78).

REGION	MONTH											
	1	2	3	4	5	6	7	8	9	10	11	12
nNE							1.5	91.5	85.6	[a]		
sNE							25.6	76.9	86.5	[a]		
SE					5.8	26.8	56.6	68.4	[a]			
NC							1.1	20.0	17.9	12.0		
SC					0.5	0.5	2.4	20.0	71.5	90.3	30.6	
SW	0.4	0.1	0.8	3.7	1.4	0.5	0.3	2.6	7.6	17.0	9.3	2.6
sCA				1.1	3.7	12.1	7.1	17.5	16.1	21.7	38.4	5.4

[a] Few *Ambrosia* pollen observed, though frequency high.

apices and broad bases, sometimes with intermixed spinules, or occasionally with very short spinules (<0.6 μm long) interspersed with small piloid elements (*A. grayi, A. tomentosa*). The nexine is 0.8–1 μm thick; and the intine 0.6–1.2 μm thick, somewhat thicker below the apertures, forming lens-shaped thickenings (Fig. 161).

TABLE 22. Flowering by month of entomophilous genera of the Asteraceae that may be incidentally anemophilous.

GENUS	VERNACULAR	REGION							
		nNE	sNE	SE	sFL	NC	SC	NW	SW
Achillea	yarrow	5–9	5–11	4–11		6–7	3–8	6–8	3–8
Ageratum	floss flower		8–10	7–12	1–12				1–9
Anthemis	chamomile, dog fennel	6–9	5–10	3–8		6–7	3–6	5–10	3–8
Aster	daisy	6–10	7–11	4–11	3–12	7–10	3–4, 6–11	5–10	1–12
Centaurea	knapweed	6–10	5–10	3–11		7–9	3–8	5–10	5–11
Chrysanthemum	pyrethrum	6–10	5–10	4–8		5–7		5–10	2–9
Cichorium	chicory	7–9	5–10			7–9	3–11	7–10	6–10
Cirsium	thistle	6–10	5–11	3–11	1–12	6–9	3–12	5–10	3–10
Encelia	brittle bush						3–4		2–7
Erigeron	fleabane	5–10	4–11	3–11	3–8	5–8	5–11	4–8	3–11
Eupatorium	boneset	7–9	7–11	6–11	1–12	7–9	2–11	6–9	1–12
Grindelia	rosinweed	7–9	7–10	7–10		8–9	5–12	6–10	3–10
Helenium	sneezeweed	7–11	6–11	4–11	3–5, 9–11	8–9	3–11	7–9	6–9
Helianthus	sunflower	7–10	(5–) 7–10	6–11	3–11	8–9	6–11	(4–) 7–10	1–12
Parthenium	feverfew, guayule	6–9	5–10	5–8	6–8		3–11		
Rudbeckia	coneflower	6–10	5–11	7–11	3–8	7–9	3–11	6–9	6–9
Senecio	ragwort	5–10	4–10	3–8	1–12	5–9	3–11	5–10	12–10
Solidago	goldenrod	7–10	5–12	4–11	1–12	7–10	7–1	6–10	5–11
Sonchus	sow thistle	6–10	5–10	5–11	3–8	7–9	3–5	7–10	1–12
Tagetes	marigold	8–10	7–10	7–11			8		
Tanacetum	tansy	7–9	6–9	7–10		7–9		5–10	5–10
Taraxacum	dandelion	4–9	2–11	2–11		4–7	3–8	4–9	1–12
Tussilago	coltsfoot	4–6		3–5					3–4
Vernonia	ironweed	8–9	5–10	6–9	1–12	8–9	6–8		

Grains of *Ambrosia trifida*, *A. deltoidea*, and *A. psilostachya* are small (15–26 x 16–28 μm); 4-colporate apertures are found for *A. cheiranthifolia*, *A. confertifolia*, *A. hispida*, and *A. tomentosa*, often an indication of polyploidy.

Pollen of *Iva* is similar to that of *Ambrosia*, except that the sexine is conspicuously thickened equatorially (to 3.2 μm), and the spines tend to be shorter (0.3–1.6 μm long). *Iva xanthifolia* and *I. dealbata* differ from the other species by having relatively long colpi (to 14 μm) (Fig. 162).

Artemisia grains are prolate-spheroidal (21–26 x 19–23 μm) to subspheroidal (ca. 24 x 23 μm) and 3-colporate; the colpi long (18–21 x 1.2–1.6 μm); and the ora lalongate (2.4–3.2 x 1.6–3.2 μm). The sexine is thickened equatorially (ca. 2.4 μm) and thinner at the poles (ca. 1.6 μm), tectate, undulating, and with short spinules only to 0.6 μm long interspersed with granules or microverrucae. The nexine is ca. 1.6 μm thick; and the intine is considerably thickened to ca. 3 μm below the apertures (Fig. 163). Grains of *Artemisia* are easily distinguished from those of *Ambrosia* by their long colpi and absence of significant spines.

Long-spined pollen in the Asteraceae is typified by species of *Baccharis*, *Carduus*, *Eupatorium*, *Solidago*, *Tagetes*, and *Tanacetum* (Fig. 164).

One of the most distinct types of pollen is found in the tribe Lactuceae (Fig. 165). As exemplified by *Taraxacum* and *Cichorium*, grains are subprolate to prolate-spheroidal (28–48 x 26–40 μm) and 3-colporate in lacunae; the colpi short (10–12 x ca. 4.8 μm); the ora lolongate (4.8–6.4 x 4.4–4.8 μm); and the opercula granular. The sexine is echinolophate, ca. 10 μm thick, consisting of ridged tecta having large spines (3.2–4.2 μm long) and micropores surrounding the lacunae that number 12–14. The aperturate lacunae are ± elliptical, ca. 13 μm wide, and the nonaperturate ones angular. The nexine is thinner than the sexine; and the intine is ca. 1 μm thick, slightly thicker to 2 μm below the apertures.

Allergenicity

It is probably correct to generalize that sufficient exposure to pollen of any member of the Asteraceae will cause inhalant allergic reactions among a fraction of atopic individuals. Frequency and level of these reactions is often high; in fact, higher than for plants of any other family involving comparable exposures. Since sufficient repeated exposures are essential, wind-pollinated plants that include members of the subtribe Ambrosiinae (*Ambrosia*, *Dicoria*, *Hymenoclea*, *Iva*, *Xanthium*), *Parthenice*, some *Baccharis* species, and *Artemisia* are the major offenders. Exposure to other genera in the family, however, particularly if sensitivity to these wind-pollinated plants already exists, could trigger allergic reactions, for some major allergens appear widespread in the family and unexpected cross-reactions undoubtedly occur.

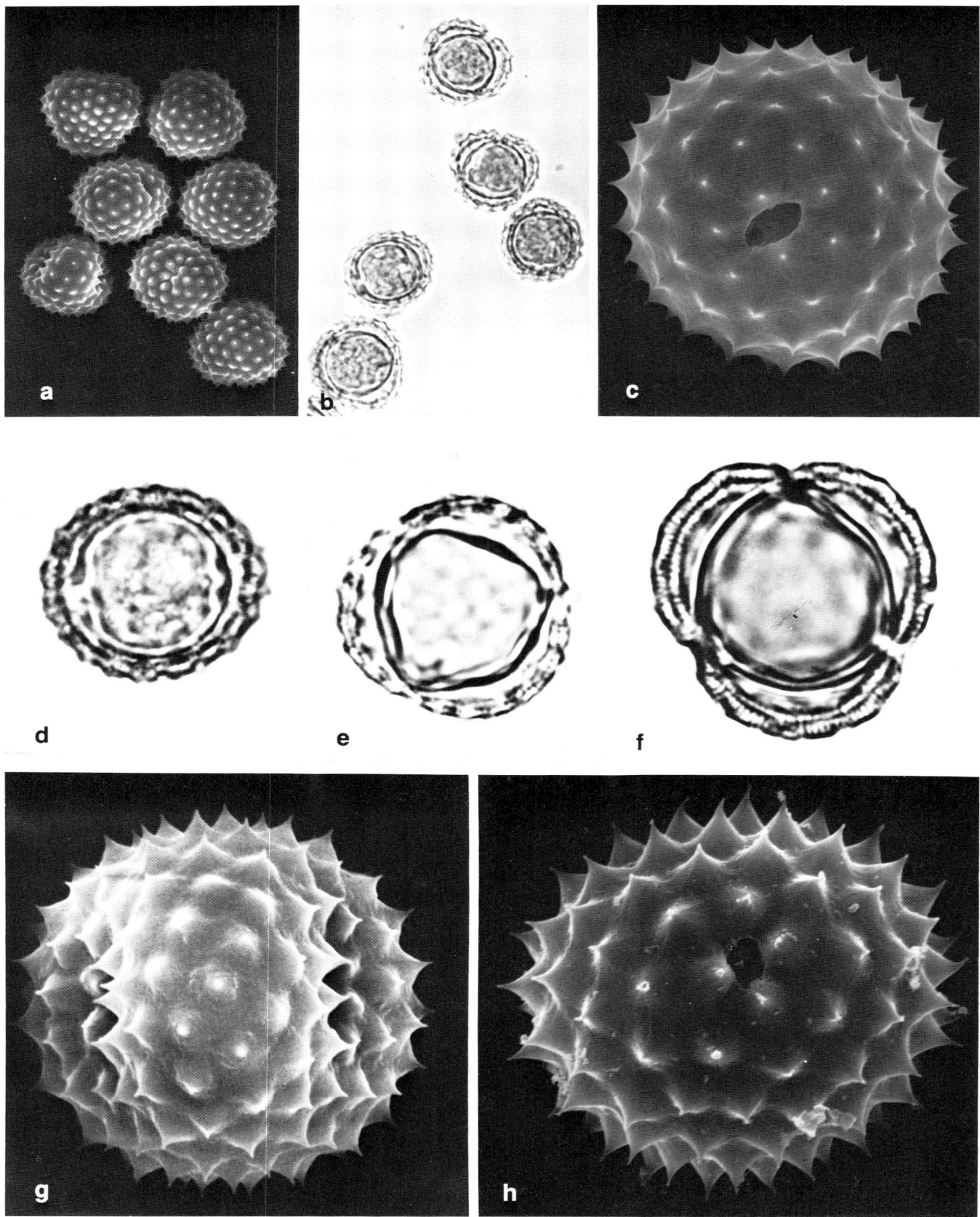

Fig. 161. Pollen of *Ambrosia*. a: *A. artemisiifolia* (short ragweed), SEM(ad) ×1000; b–d: *A. bidentata* (southern ragweed), b = LM(s) ×720, c = SEM(ad) equatorial view showing 1 aperture ×2800; d = LM(s) equatorial view ×1830; f: *A. hispida* (coastal ragweed), LM(a) polar view ×1830; g: *A. artemisiifolia*, SEM(ad) equatorial view showing 2 apertures ×4000; h: *A. trifida* (giant ragweed), SEM(ad) equatorial view showing 1 aperture ×4400.

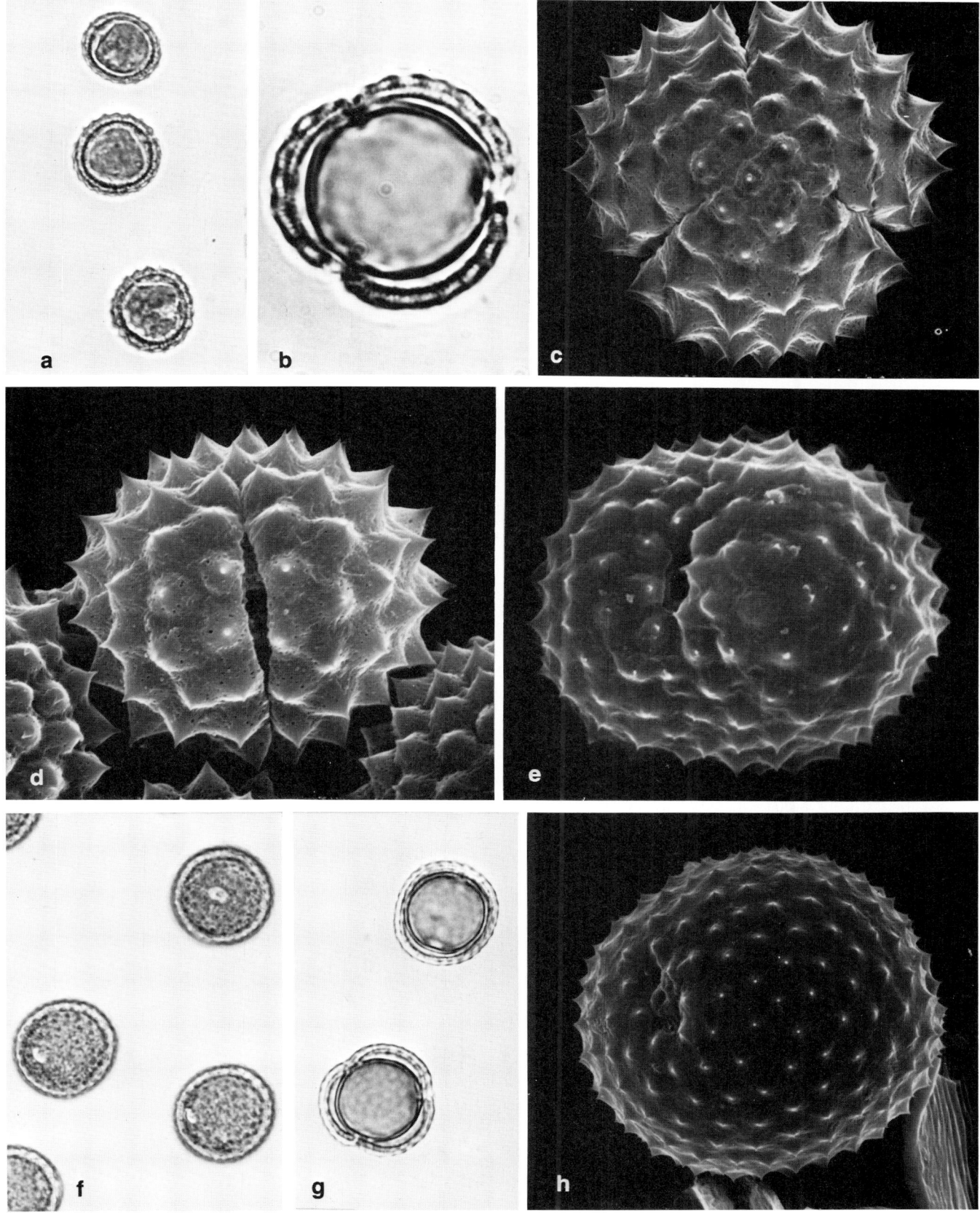

Fig. 162. Pollen of *Iva*, *Hymenoclea*, and *Xanthium*. a: *Iva annua* (rough marsh-elder), LM(s) ×720; b,e: *Hymenoclea salsola*, b = LM(a) polar view ×1830, e = SEM(a) equatorial view showing 1 aperture ×5000; c–d: *Iva xanthifolia*, c = SEM(ad) polar view ×4200, d = SEM(ad) equatorial view showing 1 aperture ×4400; f–h: *Xanthium strumarium* (cocklebur), f = LM(s) ×720, g = LM(a) ×720, h = SEM(ad) equatorial view showing 1 aperture ×3000.

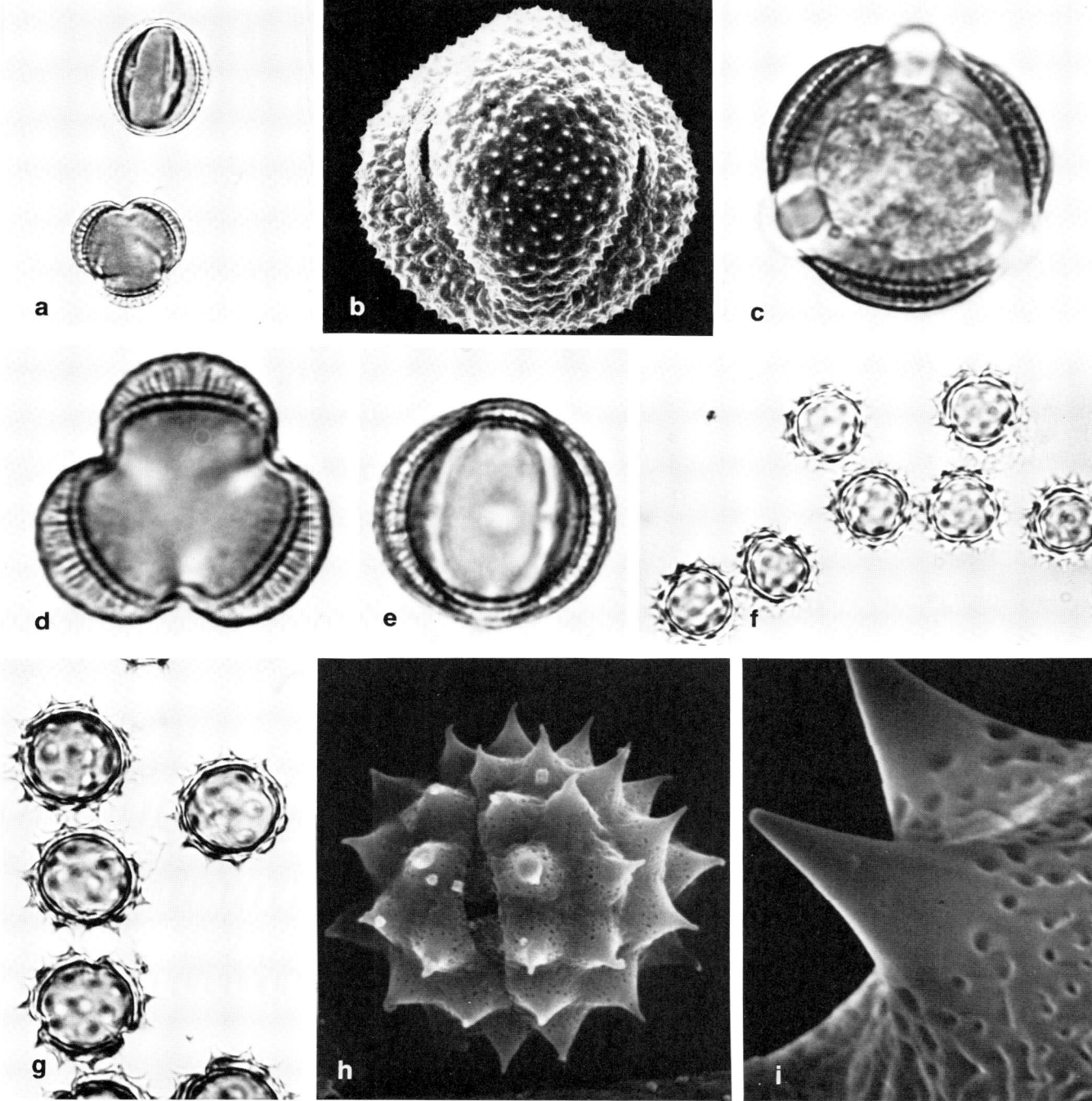

Fig. 163. Pollen of *Artemisia*, *Parthenice*, and *Parthenium*. a,c: *Artemisia ludoviciana* (western mugwort), a = LM(a) ×720, c = LM(a) ×1830; b: *A. tridentata* (common sagebrush), SEM(a) equatorial view showing 2 apertures, ×2400; d–e: *A. absinthium* (common wormwood), d = LM(s) polar view ×1830, e = LM(a) equatorial view ×1830; f: *Parthenice mollis*, LM(a) ×720; g–i: *Parthenium integrifolium* (American feverfew), g = LM(a) ×720, h = SEM(a) equatorial view showing 1 aperture ×4300, i = SEM(a) showing spines and micropores ×14000.

Species of *Ambrosia* are by far the most important cause of pollinosis in North America, with *A. artemisiifolia* and *A. trifida* alone accounting for more hay fever than all other plants together (Wodehouse 1971). Pollen sensitivity to one species implies some degree of sensitization to the pollen of all others (Prince and Secrest 1939; Lewis and Imber 1975c), although al-

Fig. 164. Pollen of representative Asteraceae. a–b: *Baccharis pilularis* (chaparral broom), a = SEM(a) polar view ×4000, b = LM(a) polar view ×1830; c: *Carduus nutans* (musk-thistle), LM(s) polar (upper) and near equatorial views ×720; d–e: *Eupatorium serotinum*, d = LM(s) ×720, e = SEM(a) ×1700; f–g: *Solidago canadensis* (Canada goldenrod), f = LM(a) ×1830, g = SEM(a) equatorial view showing 2 apertures ×4200; h: *Tagetes patula* (French marigold), SEM(a) ×850; i: *Tanacetum vulgare* (tansy), SEM(a) equatorial view showing 2 apertures ×3200.

WEEDS AND HERBS

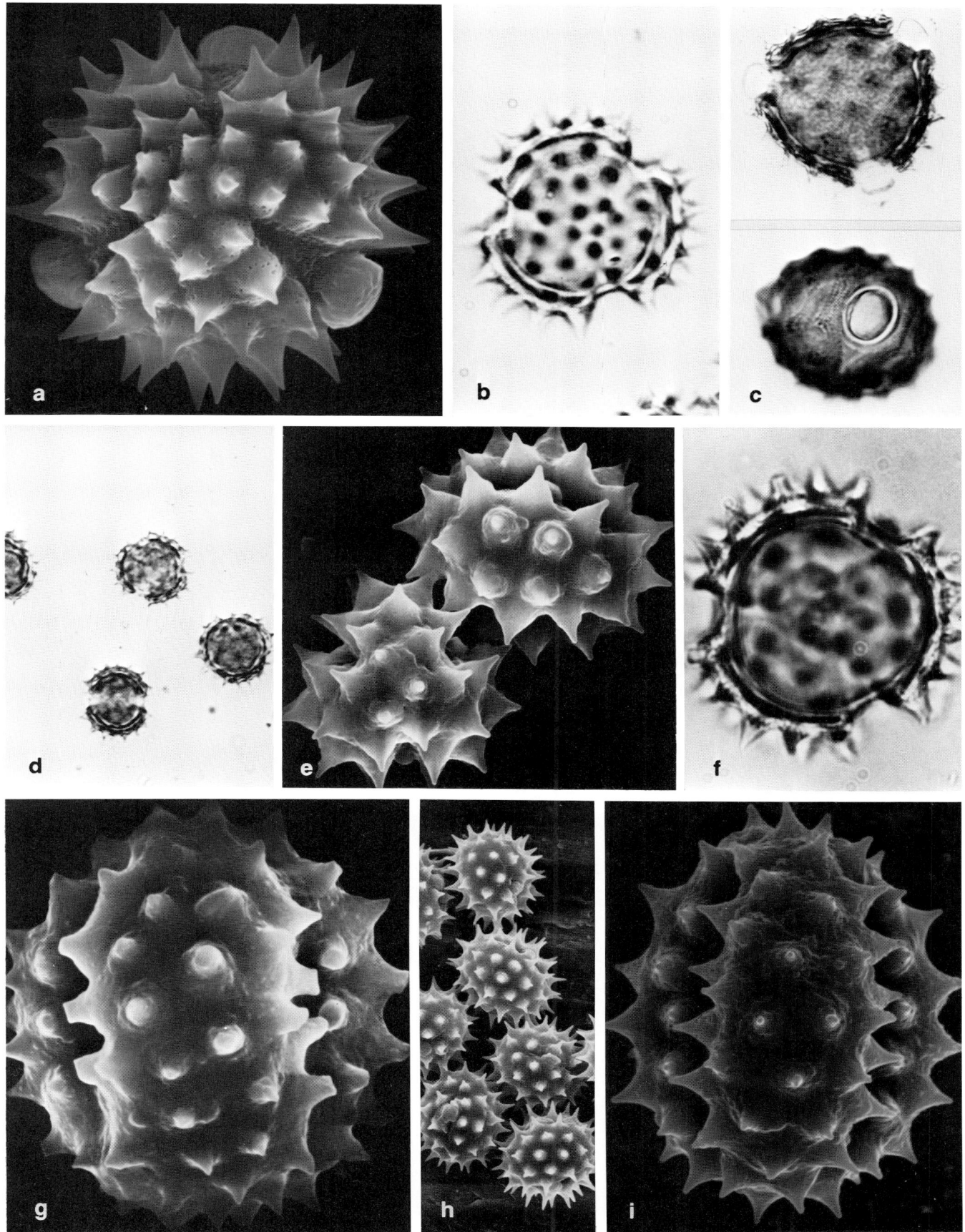

147

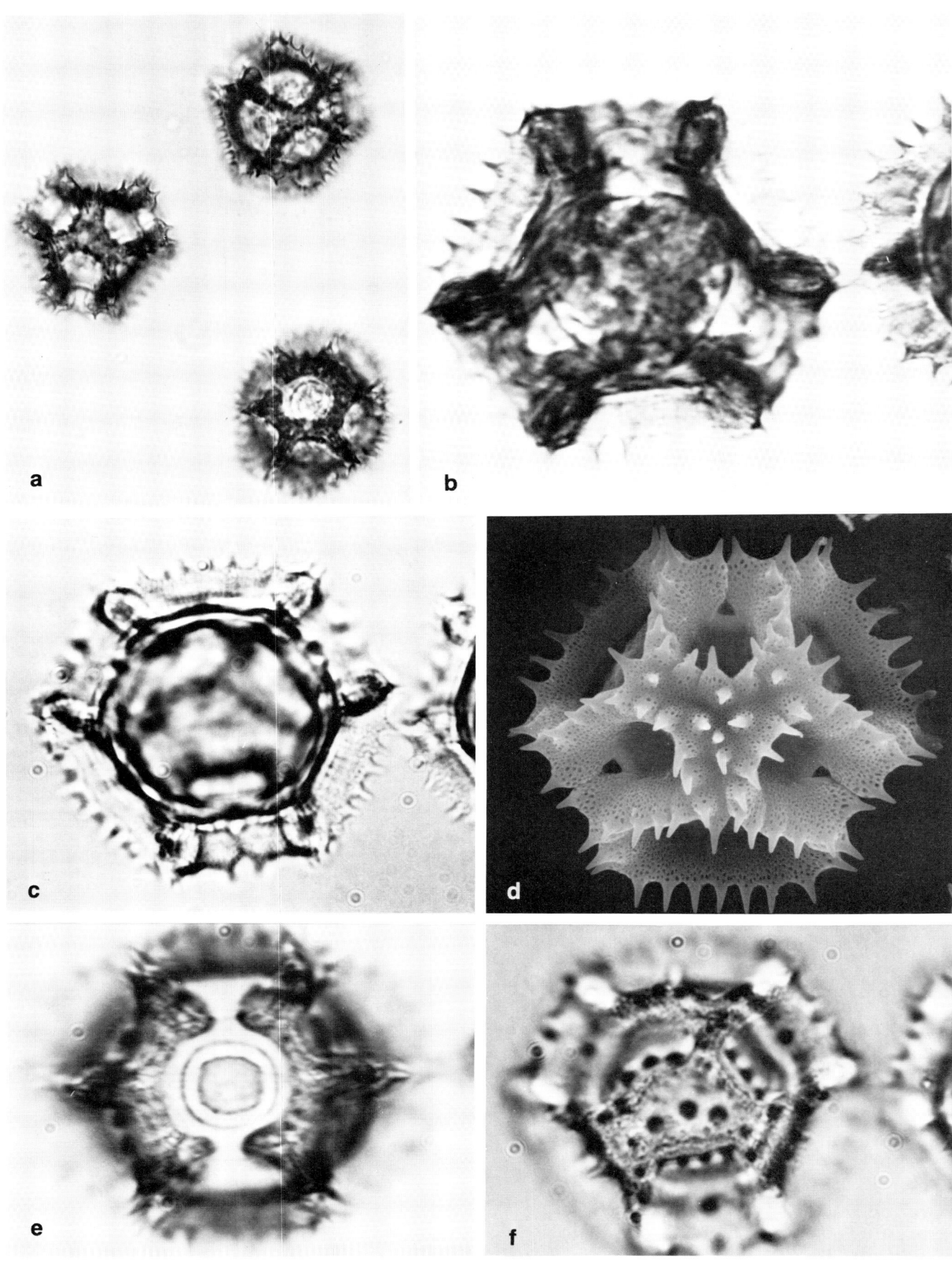

lergens and their frequency are not identical from species to species. Care must be exercised when traveling to or living in southern latitudes of the continent so as to anticipate exposure to ragweed pollen "out of season," during the winter and spring months (see Table 17). A relevant example is the Caribbean species *A. hispida*, native to southern Florida: it flowers year-round, though commonly from January to March, and is particularly abundant along coastal regions.

Because of high levels of patient cross-reactivity to pollen extracts, some major allergens may also be shared between ragweed and other members of the ragweed subtribe, viz., with *Iva* and even with the tribally distinct but related *Artemisia* (Lewis & Imber 1975c). Indeed, all members of the allied Radiate group (see Fig. 160) may share some important allergens while possessing others that are distinctive. Data are conflicting, however, and further research with patients is needed (Prince & Secrest 1939; Leiferman et al. 1976).

A number of specific examples of pollinosis elicited by asteraceous pollen in North America and abroad are arranged by tribe:

Tribe 1—Heliantheae

Hymenoclea: This pollen is regarded as an important offender in inhalant allergy in the Southwest (Wodehouse 1971). In our survey four allergists from California reported 345 patients examined annually who suffered from pollinosis and had an average skin reaction of 3.1+ on testing with *Hymenoclea* pollen.

Iva: *I. annua* (syn. *I. ciliata*) pollen is a significant allergen needing further botanical surveys in the south, where frequency of plants is apparently extensive (Tocker 1956) and perhaps increasing.

Parthenium: A case of inhalant allergy as well as several of dermatitis and edema were attributed to *Parthenium* pollen in Texas (Kahn & Grothaus 1936). Its pollen may occasionally become airborne for short distances.

Viguiera: *V. helianthoides* flowers in December and January in Cuba. Sufficient pollen is released by this insect-pollinated species (140 pollen/m³) to elicit pollinosis among those living near the flowering populations (Quintero 1955). Quintero also observed that those who were sensitive to ragweed also reacted markedly to *Viguiera* pollen (both are members of the tribe Heliantheae and may share some major allergens).

Tribe 2—Astereae

Baccharis: Some species of this genus are clearly wind-pollinated, either primarily or facultatively. *B. halimifolia* sheds wind-dispersed pollen as observed in Cuba (Quintero 1955), and in Arizona Hafford (in Wodehouse 1971) reported that *B. sarothroides* sheds so much pollen that it can be seen in the air as the wind blows it off the bushes. Allergenicity has been reported for pollen of *B. halimifolia* from Florida and *B. sarothroides* from Arizona (Wodehouse 1971).

Solidago: During a particularly profuse growing season for goldenrod in Virginia Vaughan & Crockett (1932) found that for every 100 persons sensitive to ragweed, 30 were found sensitive to goldenrod. Most goldenrod-sensitive individuals were also sensitive to ragweed, suggesting the existence of some common pollen allergens between these genera of different tribes but of the same Radiate group of tribes.

Tribe 6—Anthemideae

Anthemis: In Oregon south of Portland Perlman (1955) found that sufficiently abundant pollen of dog fennel was released by large populations to produce symptoms of inhalant allergy during April and May.

Artemisia: In addition to the many species already summarized as allergenic by Wodehouse (1971), we add the offending species *A. montana* from Japan (Wagatsuma & Shida 1944) and *A. verlotorum* from Argentina (Dumm & Zarate 1944).

Chrysanthemum: Ox-eye daisy (*C. leucanthemum*) pollen, as well as that of dandelion (*Taraxacum officinale*, tribe Lactuceae), was so frequent in the air close to the ground when plants were disturbed that Kupias et al. (1981) suggested it could cause allergic symptoms.

Tribe 8—Eupatorieae

Piqueria: Langley (1937) reported a case of occupational pollinosis in which this winter-flowering greenhouse plant (florists' stevia) caused parallel symptoms to those following exposure to ragweed to which the patient was sensitive.

Allergens

Naturally occurring atopic allergens are all proteins or glycoproteins, the major ones weighing between 20,000 and 40,000 daltons (King 1979). Among the best known are two allergens from *Ambrosia*, antigen E (AgE) and antigen K (AgK). AgE is a protein of 343 amino acids, having a molecular weight of 37,800 daltons, and accounts for about 6% of the total protein of whole ragweed extract. It comprises 90% of the allergenic activity of ragweed pollen. In contrast, AgK (MW 38,200 daltons) accounts for about 3% of extractable ragweed protein and is less active than AgE in patients allergic to ragweed.

Three additional antigens, AgRa3, AgRa4, and AgRa5, and cytochrome c are smaller than AgE and

Fig. 165. Pollen of *Cichorium* and *Taraxacum*. a–d: *Cichorium intybus* (chicory), a = LM(s) ×720, b–c = LM(s) polar views ×1830, d = SEM(a) polar view ×2000; e–f: *Taraxacum officinale* (dandelion), e = LM(a) equatorial view ×1830, f = LM(a) polar view ×1380.

BRASSICACEAE (*Cruciferae*)
Mustard Family

Mostly temperate herbs centered in the Mediterranean basin and western Asia, the mustards are known best for vegetable and oilseed crops (e.g., broccoli, cabbage, mustard, turnip, rape). Insect-pollination predominates during anthesis of *Brassica* from January to October (southwest) and from spring to autumn, depending on locality, in other areas of the continent (Fig. 166).

Pollen Morphology (Fig. 167)

Grains of *Brassica* are prolate to subspheroidal, 24–27 x 18–19 μm; the amb rounded-triangular or with convex sides and 3(–4)-colpate; the colpi long (17–18 x 2.0–3.5 μm); and the opercula granular. The sexine is 1.0–1.2 μm thick, and reticulate with small, polygonal lumina (0.6–2.0 μm in diameter) and duplibaculate muri; the nexine 0.6–0.8 μm thick; and the intine ca. 0.8 μm thick, thickened to 1.8 μm below the apertures.

Allergenicity

Most incidentally dispersed pollen of *Brassica* is from weedy or cultivated species. In California and Oregon strongly positive skin test reactions among sensitized individuals are relatively common (Lewis & Vinay 1979). Hypersensitivity to rape (*B. napus*) was documented in Sweden in a patient suffering from conjunctivitis, rhinitis, and asthma who responded favorably to therapy using rape pollen extract (Coldahl 1954).

CANNABACEAE
Hemp Family

A small family consisting of hemp or marijuana (*Cannabis*) and hops (*Humulus*), these weedy herbs have become widespread in North America. The greatest concentration of naturalized *Cannabis* is in the central continental area. Leaves are palmately lobed or divided; flowers are dioecious; the staminate has five stamens and five calyx segments; the pollen is wind-pollinated; and the pistillate is densely clustered, the entire calyx enclosing the ovary, the style with two

Fig. 166. *Brassica kaber* (charlock) stems with flowers, ×¾.

arms. The fruit is usually a glandular achene (Fig. 168).

Flowering and Pollen Aerobiology

Cannabis flowers from June to August (northcentral region) or from September to October (northeast region). *Humulus* begins flowering somewhat later in July and also extends until August (northcentral and northwest regions) or October (eastern region).

Pollen of both genera is shed in large quantities. It is extremely buoyant (Wodehouse 1971). Air contamination from marijuana pollen, for example, has been as high as 15% of the total ragweed figure in Omaha, Nebraska (MacQuiddy 1955). Large amounts were caught on exposed slides in Ottawa (Bassett et al. 1978).

Pollen Morphology (Fig. 169)

Grains are suboblate to oblate, 14–23 x 20–28 μm; the amb rounded-triangular to ± circular and 3(–4)-porate; the pores circular or somewhat oval, 1.5–1.6 x

WEEDS AND HERBS

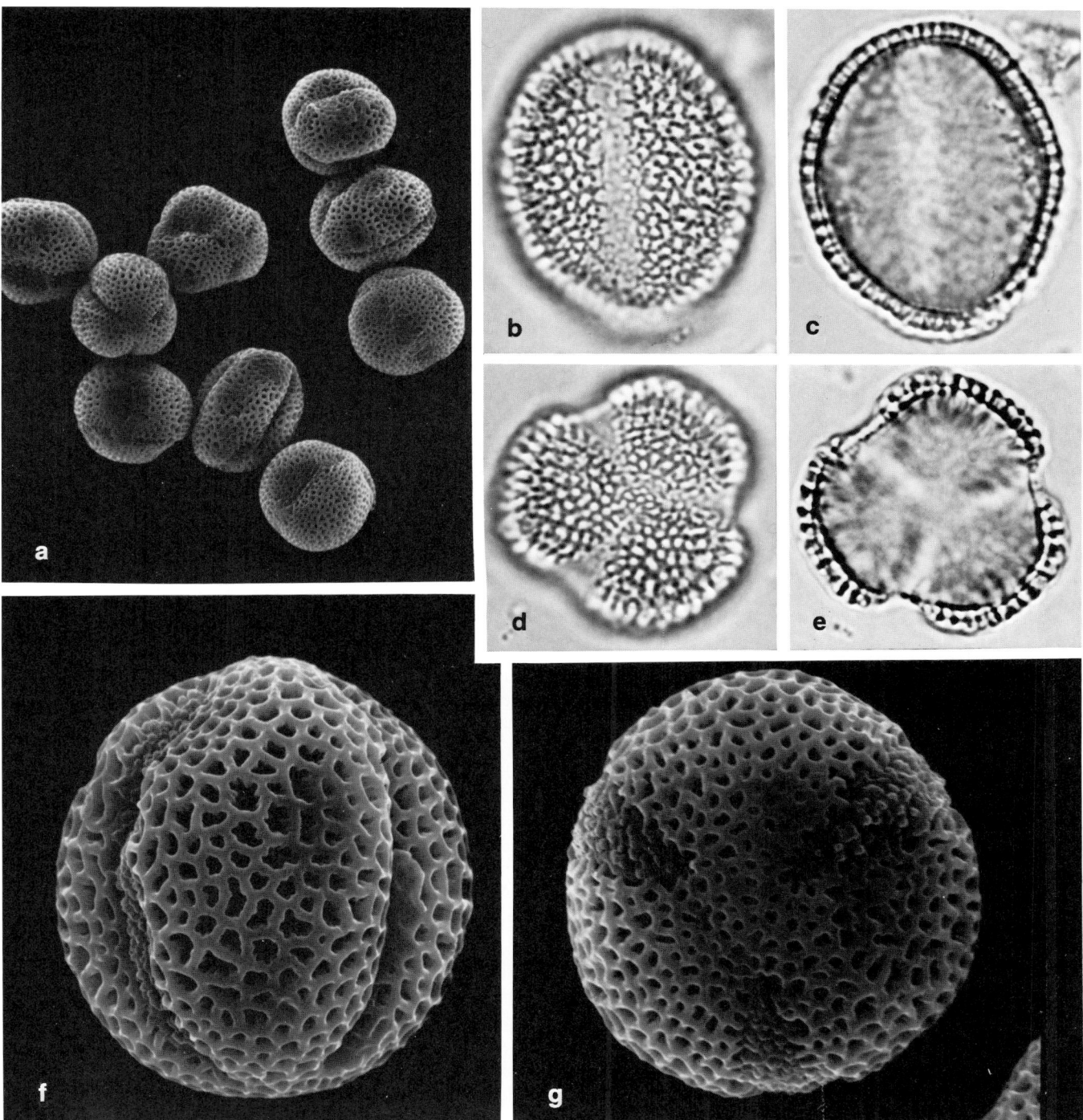

Fig. 167. Pollen of *Brassica rapa* (field mustard). a = SEM(ad) ×1000, b–c = LM(a) equatorial views at surface (b) and optically (c) ×1830, d–e = LM(a) polar views at surface (d) and optically (e) ×1830, f–g = SEM(ad) equatorial view showing 2 apertures (f) and polar view (g) ×3800.

2–3 μm in diameter, slightly aspidate; and the margins thickened. The sexine is granular, thin (0.5–0.6 μm thick), somewhat thickened at the pores; the nexine thinner than the sexine; and the intine considerably thickened below the pores, forming an oncus 5–6 x 8–10 μm in size. Grains of *Cannabis* and *Humulus* are very similar and they resemble pollen of the Moraceae.

Allergenicity

Where *Cannabis sativa* is common, as in the north-central region from the Dakotas to Nebraska, mid-summer and late-summer inhalant allergies are a well-known consequence of exposure to its pollen (Maloney & Brodkey 1940; MacQuiddy 1955). The allergic response is specific since the pollen is entirely unlike

151

Fig. 168. Morphology of representative Cannabaceae. a: *Cannabis sativa* (marijuana) stems showing leaves and inflorescences ×⅛; b–c: *Humulus scandens* (Japanese hops) stem with leaves and inflorescence (b, ×½) and staminate flowers (c, ×2½).

that of any other common offender (hops are not frequent), the symptoms of rhinitis being followed by a severe type of bronchial asthma. Allergenicity of *Humulus* pollen is not known, but exposure ought to produce reactions similar to those of *Cannabis* among sensitized individuals.

CHENOPODIACEAE*

Goosefoot Family

The family consists of about 100 genera and 1,500 species, predominantly of perennial herbs adapted to soils containing high percentages of inorganic salts (halophytes). Of 22 genera found in North America, 18 are represented in western regions in salt-rich soils, such as salt marshes and flats, and since saline soils are often associated with arid conditions, many species exhibit xerophytic adaptations (Fig. 170). Some important species are introduced from Europe and Asia. The most significant plants economically are sugar beet (*Beta vulgaris*) and spinach (*Spinacea oleracea*).

The herbs have deep, penetrating roots, and small, often hairy, lobed or spiny alternate leaves without stipules, but sometimes the leaves are fleshy or absent, with the jointed stems giving the plants a cactuslike appearance. Flowers are inconspicuous, greenish, usually arranged in spikelike inflorescences, bisexual or unisexual, and mostly wind-pollinated. They consist typically of five (varying from two to five) united sepals with as many stamens as sepal lobes, the filaments usually distinct, and one pistil consisting of three fused carpels and usually three to four stigmas. The fruit is a small round nut or achene.

Atriplex (Fig. 171) and *Chenopodium* (Fig. 172) are predominant in the western half of the continent, particularly in the Rocky Mountains and adjacent regions.

The genera divide into two subfamilies, of which the following are considered in this treatment: subfamily Chenopodiodeae—*Allenrolfea, Atriplex* (orach, saltbush), *Axyris* (Russian pigweed), *Bassia* (smother weed), *Beta* (sugar beet), *Chenopodium* (goosefoot, wormseed), *Corispermum* (bugseed), *Cycloloma* (winged pigweed), *Eurotia* (winter fat), *Grayia* (hop-sage), *Kochia* (burning bush, summer-cypress), *Monolepis* (poverty weed), and *Salicornia* (glasswort, saltwort); and subfamily Salsoloideae—*Halogeton, Salsola* (Russian thistle), *Sarcobatus* (greasewood), and *Suaeda* (syn. *Dondia*) (sea-blite, seepweed).

Flowering and Pollen Aerobiology

As a whole, the Chenopodiaceae flower typically in the summer and autumn, from June to November (Table 23). In the southcentral and western regions flowering can begin much earlier (March), but peak pollen shed is not until the summer months (see Table 16).

Although most members of the family shed abundant pollen and are wind-pollinated, others are self-pollinating and some are at least facultatively or incidentally insect-pollinated and may shed little pollen. *Suaeda* flowers, for example, contain nectar and are pollinated by ants, butterflies, bees, and thrips (Blackwell & Powell 1982). Wind-pollination cannot be excluded under certain circumstances, for *S. fruticosa* can release huge quantities of buoyant pollen (Wodehouse 1935). Blackwell & Powell also reported *Halogeton, Kochia*, and *Salsola* as being sometimes entomophilous, but the primary mode of pollination may still be anemophilous. *Chenopodium album* (lamb's quarters) produces meager amounts of pollen (Solomon & Durham 1967) that is of low allergenicity (Wodehouse 1935), although other species (e.g., *C.*

WEEDS AND HERBS

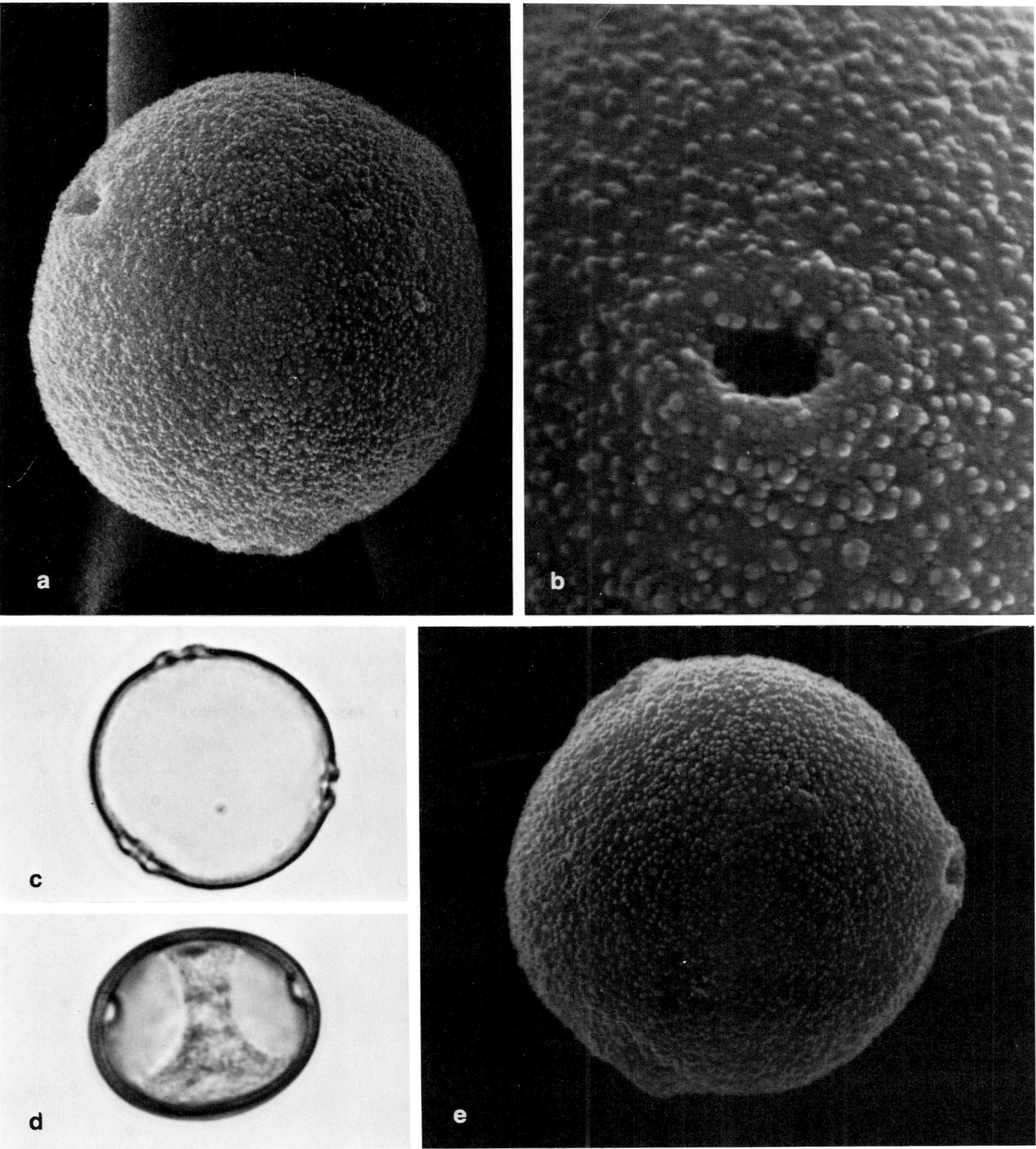

ambrosioides) release abundant pollen. Thus, even within genera the mode of pollination and potential role of pollen in inhalant allergy may vary markedly.

Pollen Morphology (Figs. 173–74)

Grains are basically similar to those of the Amaranthaceae. Within the fundamental spheroidal structure with pantoporate apertures there is some pollen var-

Fig. 169. Pollen of *Cannabis* and *Humulus*. a–b: *Cannabis sativa* (marijuana), a = SEM(a) oblique polar view ×1900, b = SEM(a) showing pore and sexine surface ×6000; c–f: *Humulus scandens* (Japanese hops), c = LM(a) polar view ×1830, d = LM(s) equatorial view showing greatly thickened intine below pores ×1830, e = SEM(a) near polar view ×1850.

AIRBORNE AND ALLERGENIC POLLEN OF NORTH AMERICA

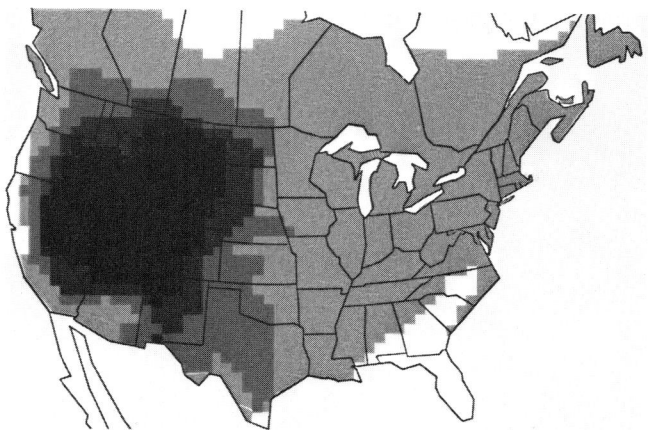

Fig. 171. Generalized composite distribution of 9 indigenous species of *Atriplex* (☐ 1 sp., ▨ 2–3 spp., ▩ 4–5 spp., ■ 6–8 spp.).

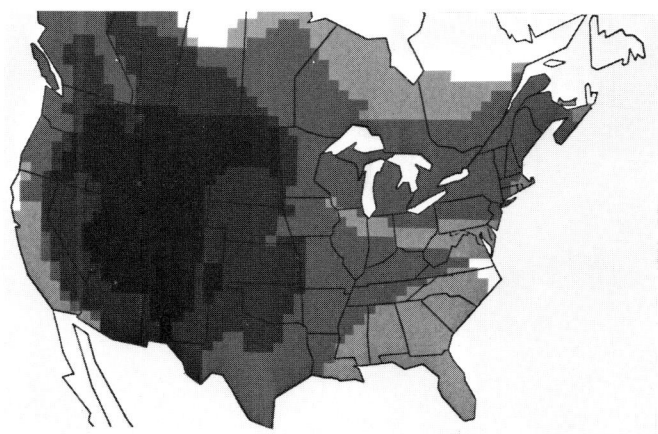

Fig. 172. Generalized composite distribution of 8 indigenous species of *Chenopodium* (☐ 1 sp., ▨ 2–3 spp., ▩ 4–5 spp., ■ 6–7 spp.).

TABLE 23. Flowering by month of the Chenopodiaceae.

GENUS	REGION						
	nNE	sNE	SE	NC	SC	NW	SW
Allenrolfea					4–9		6–8
Atriplex	8–10	6–11	7–11	7–8	4–11	6–9	4–9
Axyris		6–10		7–8			
Bassia					6–9	7–9	7–10
Beta							7–10
Chenopodium	(5–)7–11	5–11	5–11	6–9	(3–)4–10	5–10	(3–)6–10
Corispermum		8–10				6–11	7–10
Cycloloma		6–10	5–11	7–8			5–9
Eurotia					4–9	5–7	3–6
Grayia						4–6	3–6
Halogeton						7–9	6–8
Kochia	8–10	7–10		7–8	6–8	7–9	5–10
Monolepis	6–10	5–10		5–6		5–7	4–9
Salicornia	8–10	7–10	7–10		8–10	6–9	7–11
Salsola	7–10	7–10	6–11	7–8	7–10	6–8	7–10
Sarcobatus				6–7		5–7	5–8
Suaeda	8–10	7–10	8–11	7–9	4–10	6–9	(5–)7–10

Fig. 170. Morphology of representative Chenopodiaceae. a–b: *Atriplex truncata* (wedgescale) plant (a, ×¼) and stem with flowers (b, ×2½); c: *Bassia hyssopifolia* (smother weed) stems with inflorescences, ×1; d: *Chenopodium album* (lamb's quarters) stem with leaves and inflorescences, ×¼; e: *Kochia scoparia* (burning bush) stem with leaves and flowers, ×1¼; f: *Salicornia rubra* (red glasswort) stems with inflorescences, ×1½; g: *Salsola kali* (Russian thistle) stem with flowers in leaf axes, ×2; h: *Sarcobatus vermiculatus* (greasewood) stems with inflorescences, ×⅘; i: *Suaeda americana* (American seepweed) stems with flowers, ×½.

OVERLEAF

Fig. 173. Pollen of representative Chenopodiaceae (*Atriplex-Beta*). a–b: *Atriplex patula* (orach), a = LM(a) ×720, b = LM(s) ×1830; c: *Atriplex argentea* (saltbush), SEM(a) ×2300; d: *Atriplex truncata* (wedgescale), LM(s) ×1830; e: *Axyris amaranthoides* (Russian pigweed), LM(s) ×720; f–g: *Bassia hyssopifolia* (smother weed), f = LM(a) ×1830, g = SEM(ad) ×2600; h–i: *Beta vulgaris* (sugar beet), h = LM(a) ×1830, i = SEM(ad) ×3700.

Fig. 174. Pollen of representative Chenopodiaceae (*Chenopodium-Suaeda*). a–b: *Eurotia lanata* (winter fat), a = SEM(a) ×3800, b = LM(a) ×1830; c–d: *Chenopodium album* (lamb's quarters), LM(a) at surface (c) and optically (d) ×1830, e–f: *Kochia scoparia* (burning bush), e = LM(a) ×720, f = SEM(a) ×3100; g: *Salicornia rubra* (red glasswort), SEM(ad) ×3400; h–i: *Salsola kali* (Russian thistle), LM(a) at surface (h) and optically (i) ×720; j–k: *Suaeda americana* (American seepweed), LM(a) at surface (j) and optically (k) ×1830.

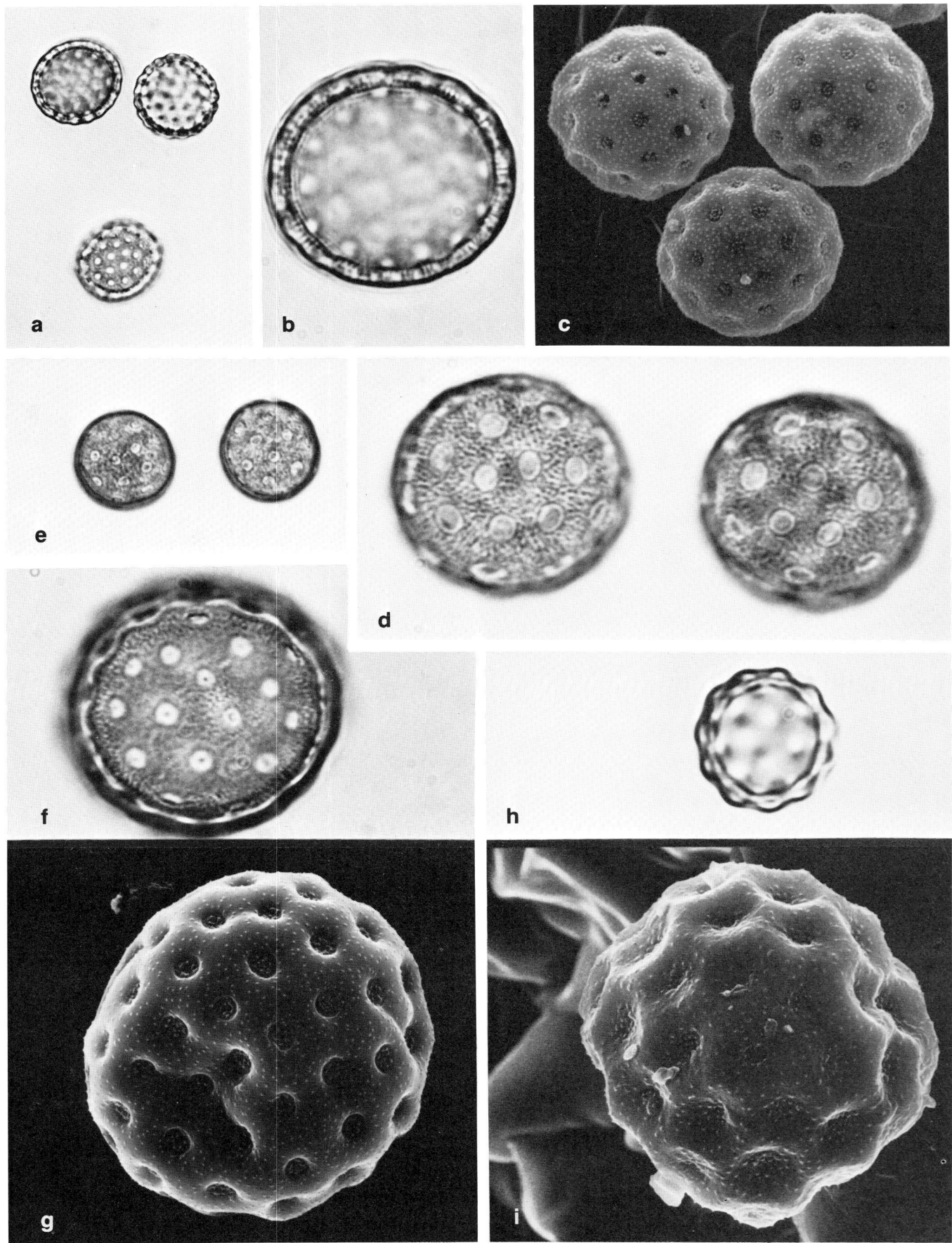

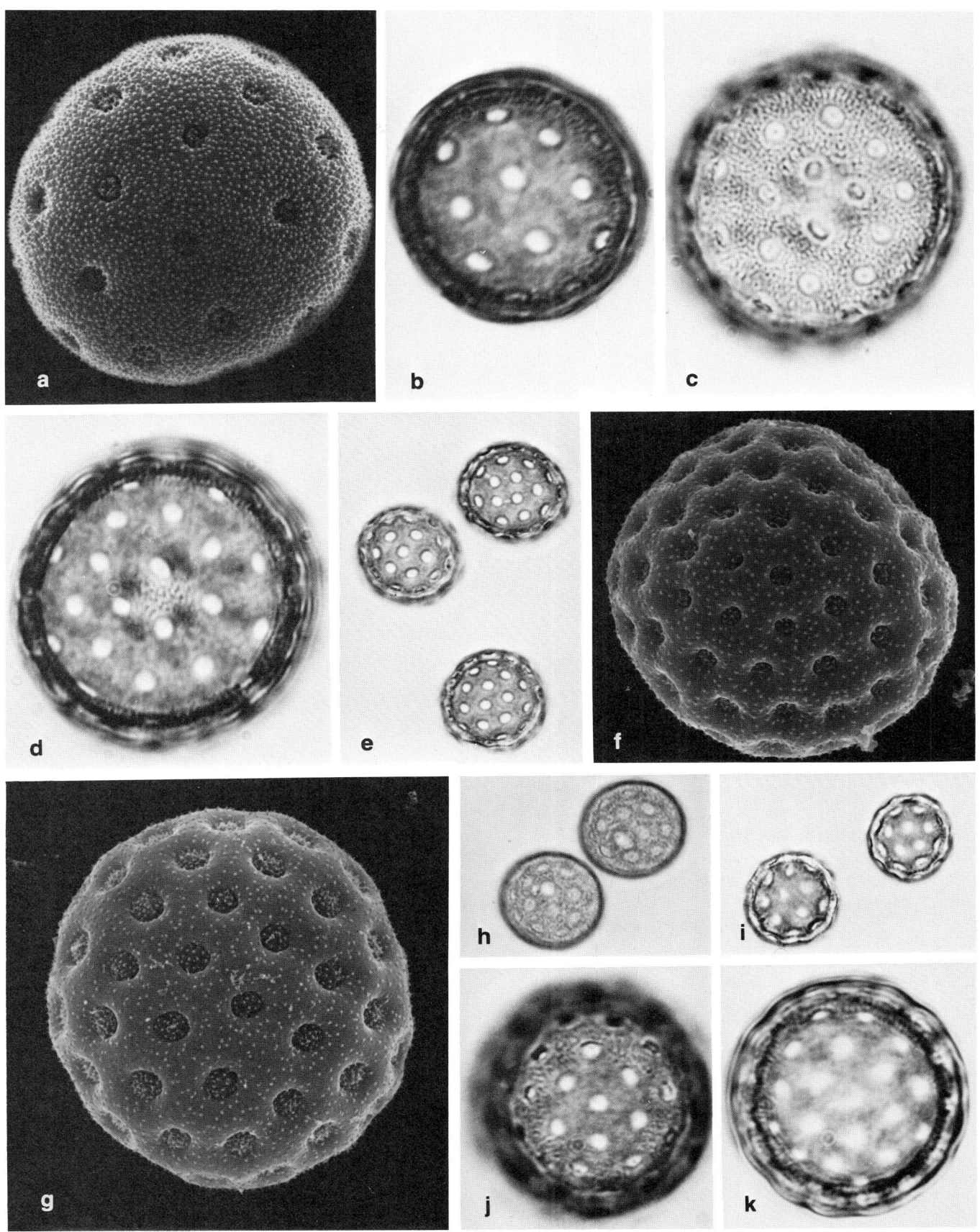

iation between chenopodiaceous genera. For example, *Kochia* pollen has many pores (65–80), whereas pollen of *Sarcobatus* has only 14–20; pollen of *Beta* is very small, only 14–15 μm in diameter, the majority 20–30 μm in diameter; pores of *Axyris* and *Eurotia* are spaced widely apart; and pores of *Salsola* are large (2–2.4 μm) and surrounded by muroid ridges formed by a considerably thickened sexine interporally. However, pollen of *Chenopodium* closely resembles that of *Atriplex, Bassia, Grayia, Salicornia, Suaeda,* and *Amaranthus* in the allied Amaranthaceae (Nowicke 1975).

Allergenicity

One of the most highly allergenic pollen is produced by *Salsola kali,* the introduced Russian thistle (Solomon & Durham 1967). Important also are species of *Atriplex, Bassia, Beta,* some species of *Chenopodium, Kochia,* and undoubtedly species of other genera that release sufficient wind-dispersed pollen for sensitization. This occurs predominantly in the western half of the continent where allergic responses are common and often acute and where cross-reactions are sufficiently great to suggest major common pollen allergens as well as minor ones among members of the family (Lamson 1931; Sellers & Adamson 1932; Phillips 1932, 1939; Ellis & Rosendahl 1933; Lamson & Watry 1933; Sellers 1935; Wodehouse 1935, 1971; Durham 1943; Ballestero & Monticelli 1944; Small & Small 1946; Blue 1955b; Ursing 1968; Levetin & Buck 1980; Weber 1981). In the eastern part of the continent late-summer and early-autumn pollinosis due to chenopod pollen may be masked by the effects of *Ambrosia* (ragweed) pollen, so individuals not responding to desensitization using ragweed extracts ought to be tested with *Chenopodium* pollen (Homan 1963) and perhaps others. Nonetheless, there are fewer acute sensitivities in the East, undoubtedly because pollen production levels are often below the effective level needed for widespread sensitization (Lewis & Imber 1975c).

EUPHORBIACEAE

Spurge Family

The spurges are a large and diverse family of herbaceous and woody plants. The woody and shrublike species have been discussed in Chapter 1.

Only pollen from the herbaceous, wind-pollinated *Acalypha* (three-seeded mercury) (Fig. 175) is expected with some frequency east of the Rocky Mountains. Unfortunately, because of small size and rare illustration, its pollen is not well known, even though several species are common weedy components of urban areas. Flowering is year-round in southern Florida, from June to autumn as far north as the southern part of the Northeast, and from July or August to autumn in the northern part of the northeastern and northcentral regions.

Airborne pollen of *Mercurialis* (mercury) has been captured in Britain (Hyde 1950), and where plants have become established in the northeast and southwest regions ambient pollen can be expected when mercury flowers from July to November and from March to September, respectively.

Fig. 175. *Acalypha rhomboidea* (three-seeded mercury) stems with leaves and inflorescences (a, ×1) and staminate spike with bracts (b, ×5).

Pollen Morphology (Fig. 176)

Grains of *Acalypha* are oblate-spheroidal, ca. 10–11 × 12 μm; the amb triangular to ±circular, rarely tetragonal, and 3(–4)-porate; the pores avoid, ca. 2.4 × 1.8 μm in diameter, with thickened margins; and the opercula granular. The sexine is thin (ca. 0.5 μm), tegillate, undulating, and scabrate; the nexine as thick as the sexine and slightly thickened at the pores; and the intine thin (ca. 0.5 μm). Grains of *Mercurialis* are larger (23–24 × 20–21 μm) and 3-colporate with long colpi and lalongate ora. Those of *Croton* are spheroidal, large (68–70 μm), and inaperturate.

Allergenicity

Although pollen of *Mercurialis* is known to be allergenic in Europe (Alemany-Vall 1955), there is no report of allergenicity in North America either of *Mercurialis* or *Acalypha* pollen. Their pollen probably contributes somewhat to pollinosis but is unrecognized as such.

FABACEAE (*Leguminosae*)
Legume Family

The woody plants of the Fabaceae have already been considered in Chapter 1. Because of their enclosed stamens and animal-pollination, members of the large herbaceous subfamily Faboideae (syn. Papilionoideae) (Fig. 177), like the woody members of the subfamily, are not prime candidates for releasing aeroallergens. Yet, marked levels of positive skin tests correlated with pollinosis were reported among agricultural workers in response to pollen extracts from *Medicago* (alfalfa) and *Trifolium* (clover) from nine states (Lewis & Vinay 1979). These unlikely events occur because farmers use dried, friable hay, thereby contacting pollen from easily broken flowers when the hay is used. Similarly, *Melilotus* (sweet clover) and *Vicia* (vetch) are suspected offenders whenever used as dried fodder. In South Africa dried alfalfa (*Medicago sativa*) incited allergic rhinitis and respiratory allergy among six patients, all being successfully desensitized using alfalfa pollen extract (Ordman 1958).

Pollen Morphology (Fig. 178)

Grains are prolate to subprolate, 25–50 × 18–42 μm; and the amb triangular, rounded-triangular, or with convex sides and 3-colporoidate or 3-colporate. The sexine is punctitegillate or reticulate, only ca. 0.6 μm thick; nexine as thick as or slightly thicker than the sexine; and the intine about 1 μm thick, thinner in *Melilotus*.

Grains of *Medicago* are punctitegillate, whereas those of *Trifolium* and *Melilotus* are reticulate. The latter are also small (25–28 × 18–20 μm).

HALORAGACEAE
Milfoil Family

The Haloragaceae are a family of aquatic and moist terrestrial herbs commonly represented in North America by *Myriophyllum* (water-milfoil). The plants are spreading as troublesome weeds of waterways. Pollen is wind-dispersed from spring to autumn depending on locality. Allergenicity is not known.

Pollen Morphology (Fig. 179)

Grains are suboblate, ca. 20 × 24 μm; the amb ±rhomboidal or infrequently triangular, and (3–)4–5-porate, aspidate; the pores lolongate, ca. 3.2 × 1.7 μm; and the annulus 3.5 μm in diameter. The sexine is thin (ca. 0.5 μm), scabrate; the nexine as thick as the sexine, but both thickened adjacent the pores; and the intine thin.

Myriophyllum pollen resembles that of the Betulaceae, for which it is undoubtedly confused during the spring.

ONAGRACEAE
Evening-primrose Family

Mostly an herbaceous family well-recognized for its vector pollination, *Epilobium* (fireweed) is, however, often found in such large populations that small amounts of pollen may become airborne for short distances (Bassett & Crompton 1967). This incidental pollen release is sufficient to sensitize some individuals and to trigger allergic rhinitis among loggers in the Northwest who are repeatedly exposed to fireweed pollen in recently cleared forest areas (Fig. 180).

Pollen Morphology (Fig. 181)

Grains of *Epilobium* are subisopolar, large (80–92 μm, E), peroblate, and 3-porate; the pores markedly aspidate, lolongate, ca. 8 × 15 μm in diameter; the annuli thick; and the opercula granular and exserted. The sexine is tectate and densely granular, ca. 0.8 μm thick, particularly thickened toward the pores (to 3 μm); the nexine about twice as thick as the sexine, crassinexinous adjacent the pores; and the intine ca. 0.8 μm thick.

PAPAVERACEAE
Poppy Family

The poppies are a family of mainly herbaceous plants found principally in the north temperate zone. All produce a latex, the most important being opium from *Papaver somniferum*. Pollination is largely by insects, but the numerous exposed anthers suggest that some pollen may incidentally become airborne. The

AIRBORNE AND ALLERGENIC POLLEN OF NORTH AMERICA

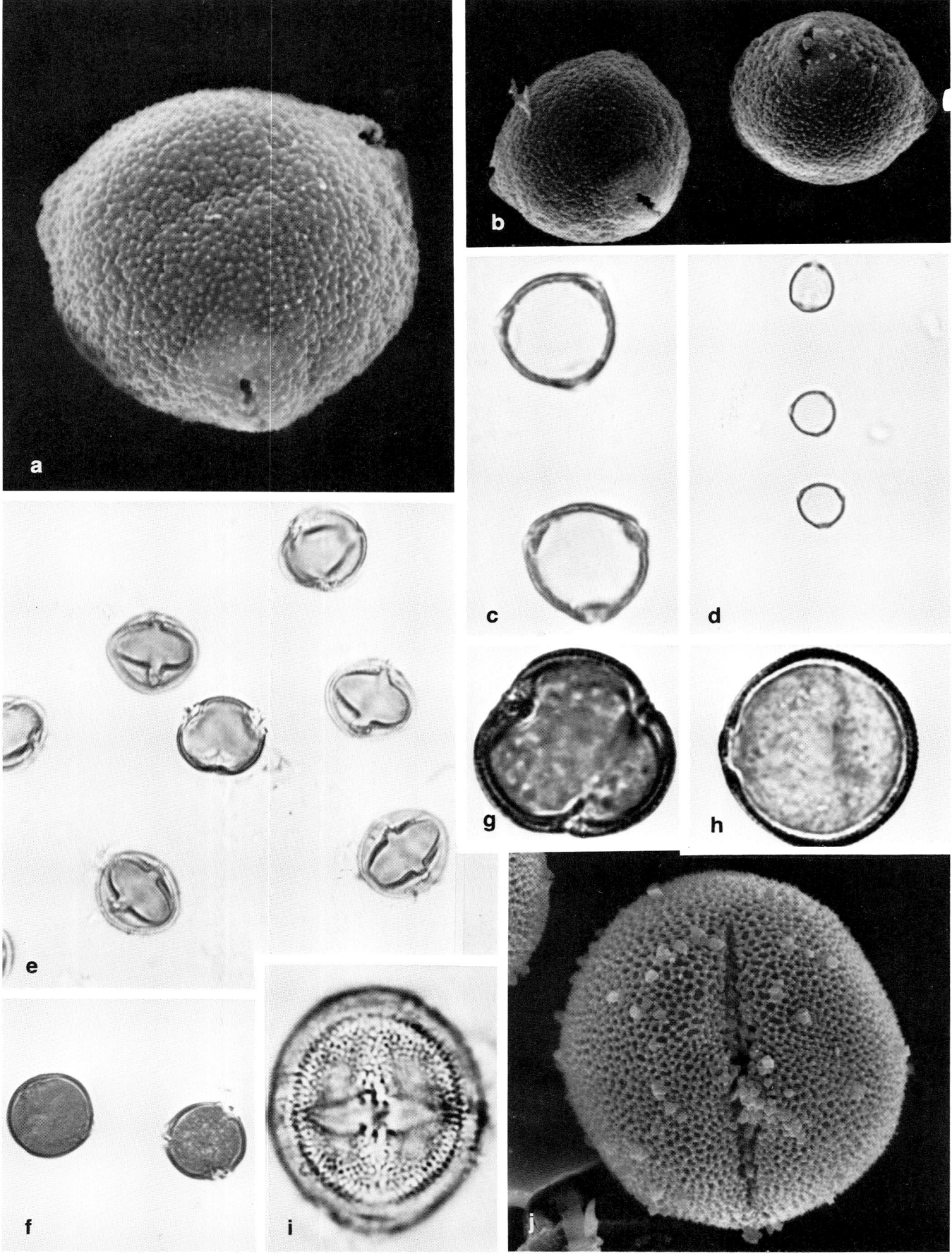

WEEDS AND HERBS

Fig. 176. Pollen of *Acalypha* and *Mercurialis*. a–d: *A. rhomboidea*, a = SEM(a) near polar view ×6100, b = SEM(a) ×3600, c = LM(a) ×1830, d = LM(a) ×720; e–j: *Mercurialis annua* (mercury), e = LM(a) ×720, f = LM(s) ×720, g = LM(s) near polar view ×1830, h = LM(s) optical equatorial view ×1830, i = LM(s) surface equatorial view, j = SEM(a) equatorial view showing 1 aperture ×3800.

Fig. 177. Morphology of representative herbaceous Fabaceae. a: *Medicago sativa* (alfalfa) stem with inflorescences, ×½; b: *Melilotus alba* (white sweet clover) stem with inflorescences, ×⅘; c: *Trifolium pratense* (red clover) stem with inflorescence, ×1; d: *Vicia sativa* (vetch) flowers, ×1½.

161

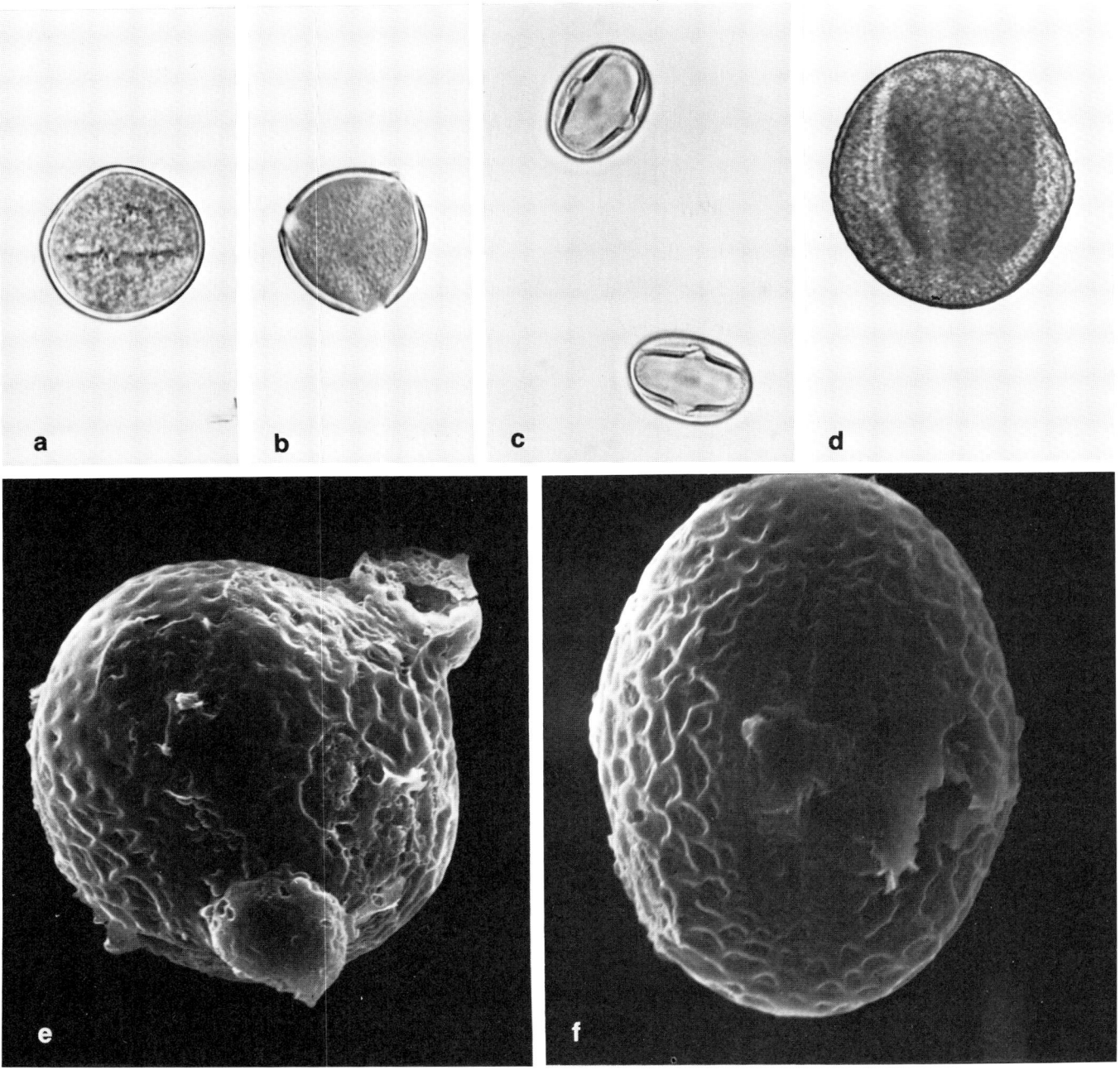

Fig. 178. Pollen of representative herbaceous Fabaceae. a–b: *Medicago sativa* (alfalfa), LM(s) equatorial (a) and polar (b) views ×720; c: *Melilotus officianalis* (yellow sweet clover), LM(a) equatorial views ×720; d–f: *Trifolium pratense* (red clover), d = LM(s) ×720, e = SEM(a) oblique-polar view ×2500, f = SEM(a) equatorial view showing 1 aperture ×2500.

woody *Bocconia* grown in the southern United States is wind-pollinated.

Eschscholzia californica (California poppy) (Fig. 182) has been implicated as a rare offender in pollinosis (Lewis & Vinay 1979).

Pollen Morphology (Fig. 183)

Grains of *Eschscholzia* are subspheroidal (19–24 × 23–24 μm) to subprolate (32 × 27 μm); the amb circular and 4–5(–6)-colpate; the colpi long (15–27 × 0.6–1.0 μm) and narrow; the opercula granular. The sexine is ca. 0.9 μm thick, reticulate; the lumina polygonal (1–2 μm in diameter) with free piloid elements and the muri simplibaculate to duplibaculate; and the nexine and intine slightly thinner than the sexine.

Grains of *Bocconia* are ±spheroidal (22–28 μm in diameter) and 4–6-porate.

WEEDS AND HERBS

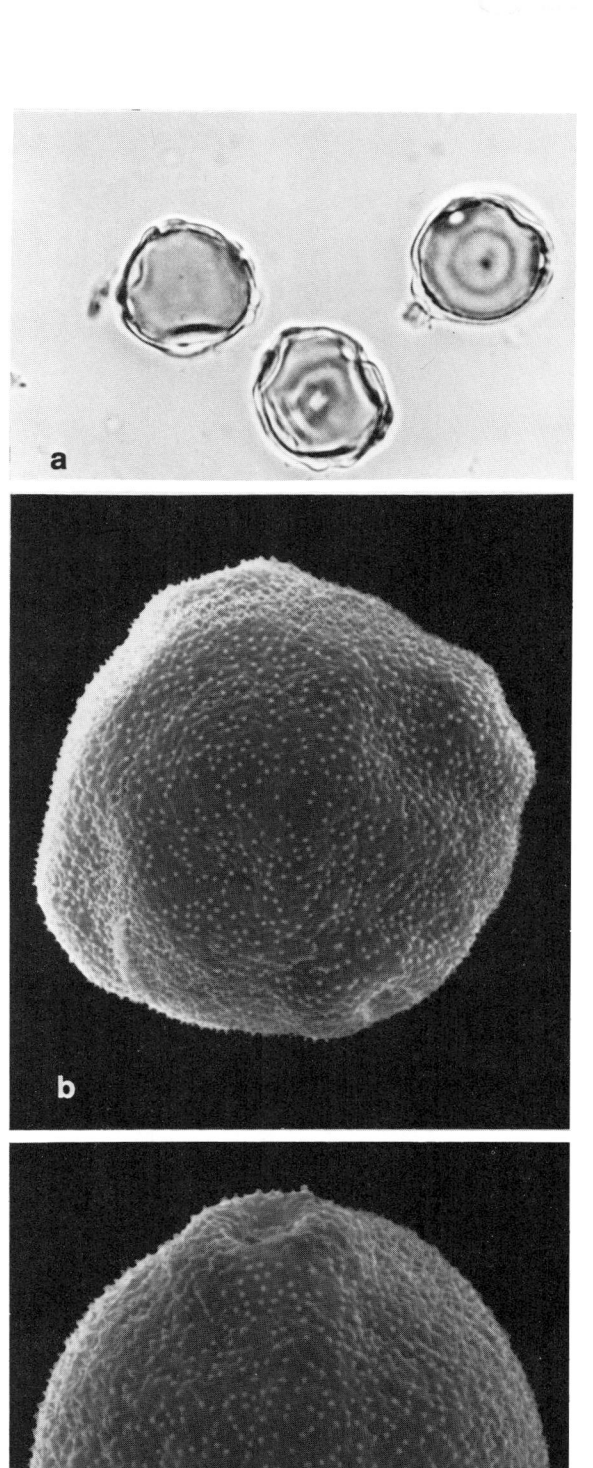

Fig. 180. *Epilobium angustifolium* (fireweed) stem with inflorescence (a, ×¾) and showing flower detail (b, ×2).

Fig. 179. Pollen of *Myriophyllum exalbescens*. a = LM(a) ×720, b = SEM(a) near polar view ×3000, c = SEM(a) equatorial view showing aspidate pores ×3400.

163

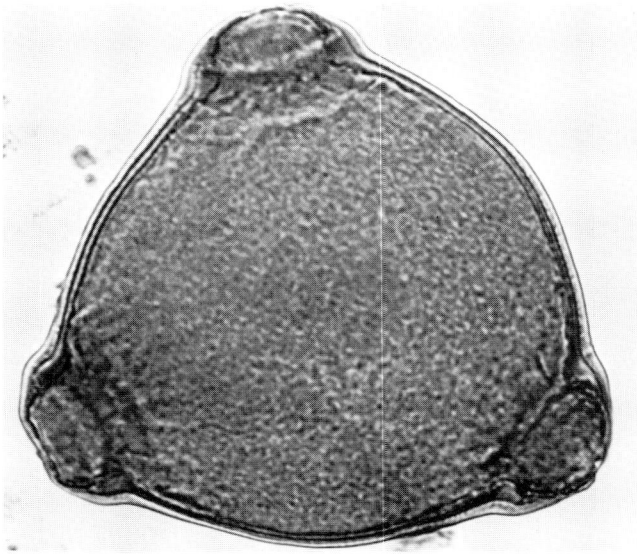

Fig. 181. Pollen of *Epilobium angustifolium*, LM(s) polar view showing 3 aspidate pores ×720.

Fig. 182. *Eschscholzia californica* (California poppy) plant with flowers, ×½.

PLANTAGINACEAE*
Plantain Family

The family consists of three herbaceous genera of which about 250 species belong to *Plantago* (plantain) (Fig. 184) and two to *Littorella*, plus the monotypic South American *Bougueria nubigena*. All are primarily wind-pollinated. Economically, the family is important in a negative sense, since several species are troublesome weeds and offenders in inhalant allergy.

Leaves are often in basal rosettes. Flowers are small, usually bisexual, and numerous in capitate spikes; the floral parts are in fours, excepting the superior ovary of two fused carpels with a simple style and stigma. The fruit is a capsule.

Native plantains are found throughout the United States and southern Canada, but four or more species form a broad distribution from the eastern Rocky Mountains to the Appalachian Mountains. The greatest concentration of species is within this range, from Nebraska and Missouri to Texas and Tennessee (Fig. 185).

Flowering and Pollen Aerobiology

Peak flowering in the family occurs from May to July, but some species begin anthesis much earlier or continue later (Table 24). The introduced *Plantago lanceolata* (English plantain), reputedly responsible for more pollinosis than any other *Plantago* species (Solomon & Durham 1967), flowers and may continue to flower until October (northeast region) or November (southeast and southcentral regions).

Ambient pollen of *Plantago* has been recovered from throughout the continent (see Table 25). In the Northeast, significant frequencies are atmospheric

TABLE 24. Flowering by month of the Plantaginaceae.

SPECIES	REGION						
	nNE	sNE	SE	NC	SC	NW	SW
Littorella americana	7–9						
Plantago aristata	6–11	5–11	4–7		4–7	6–8	6–7
P. cordata		5–7	3–5				
P. elongata		4–6		6	3–5	4–6	
P. heterophylla		4–5	3–5			3–5	
P. lanceolata	5–10	4–10	4–11	5–7	3–11	4–8	(3–)4–8
P. major	6–10	5–10	6–11	6–7	3–11	5–8	4–9
P. maritima	6–9					6–8	5–9
P. patagonica				6–7	3–6	5–6	
P. purshii	5–8	5–8	5–6				7–8
P. pusilla	4–6	4–6					
P. rhodosperma		5–6			3–5		5
P. rugelii	7–10	5–10	6–11	7–8			
P. virginica	4–6	4–6	3–6	5–6	3–6		5

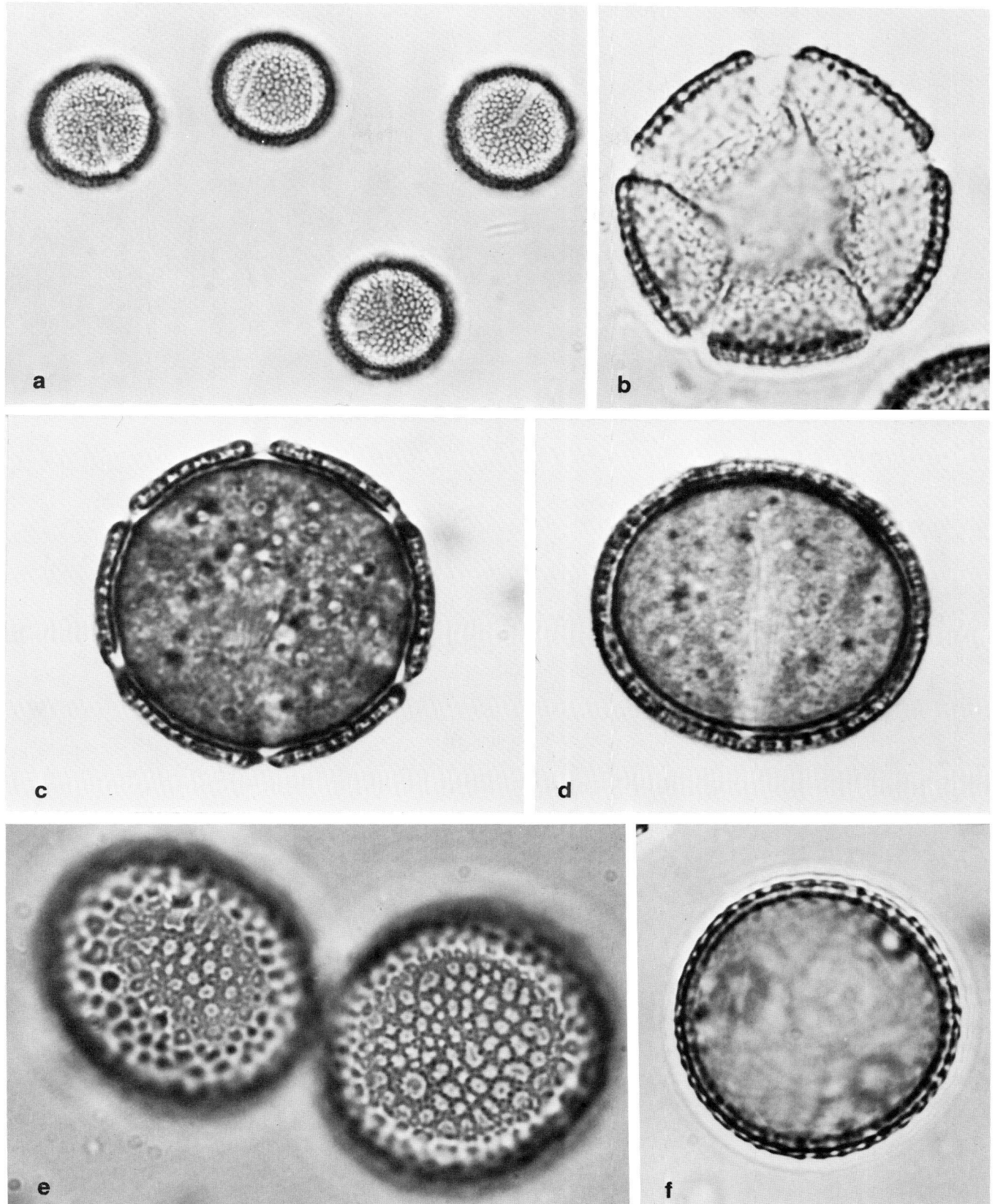

Fig. 183. Pollen of *Eschscholzia* and *Bocconia*. a–d: *E. californica* (California poppy), a = LM(s) ×720, b = LM(a) ×1830, c–d = LM(s) polar (c) and equatorial (d) views ×1830; e–f: *B. frutescens*, LM(a) surface (e) and optical (f) views ×1830.

Fig. 184. *Plantago lanceolata* (English plantain) plant (a, ×⅓) and inflorescences showing the exposed anthers (b ×1½).

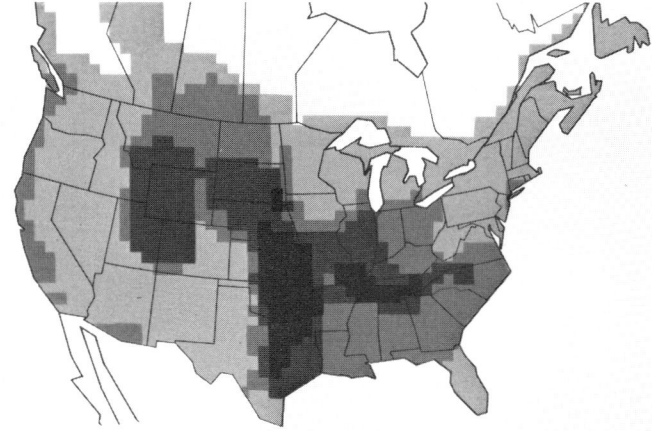

Fig. 185. Generalized composite distribution of 12 indigenous species of *Plantago* (☐ 1–2 spp., ▨ 3 spp., ▦ 4 spp., ■ 5–7 spp.).

TABLE 25. *Plantago* (plantain) aeropollen frequency in regions of North America based on percentage of pollen captured, by month (1974–78).

REGION	MONTH								
	2	3	4	5	6	7	8	9	10
nNE				0.4	5.0	18.9	1.9	0.5	0.6
sNE			0.3	0.5	13.6	18.2	2.3	0.4	0.1
SE					?	?	6.6	4.6	12.8
NC				0.2	0.5	1.4	0.4		
SC				0.9	2.0	2.6	0.1	0.1	
NW				0.3	8.0	23.9	23.7	16.4	2.9
SW	0.9	0.3	1.0	1.0	3.3	2.6	2.7	0.2	
sCA				2.5	1.1	4.3			

during June and July, in the Southeast data are incomplete but substantial quantities are found from August to October, and in the Northwest plantain pollen is common from June to September. Lesser frequencies are found in the northcentral and southwestern regions.

Pollen Morphology (Fig. 186)

Grains are spheroidal, 26–31 μm in diameter, and pantoporate; the (3–)6–10 pores circular to slightly lalongate, 1.6–3 μm in diameter, each with an annulus 1.2–2.0 μm wide; and the pluglike opercula granular, sometimes indistinct. The sexine is ca. 0.5 μm thick, scabrate with an undulating and densely granular surface; the nexine as thick as the sexine; and the intine ca. 0.8 μm thick.

Plantago aristata pollen is 4-porate, whereas that of *P. lanceolata* is 6–10-porate. Sculpturing of *P. maritima* pollen is characterized by granular "islands." Pollen of *Littorella americana* is 12–14-porate without annuli.

WEEDS AND HERBS

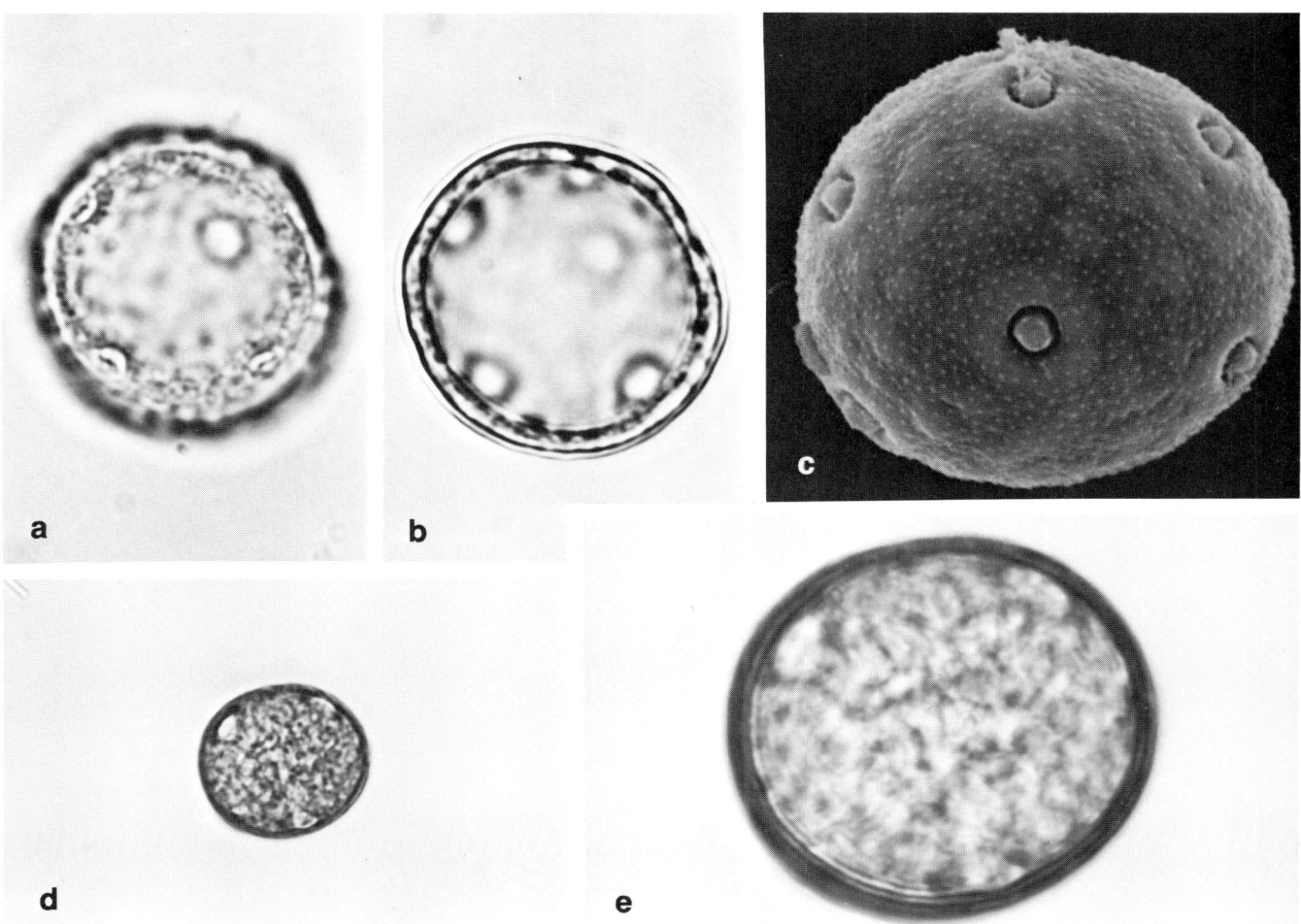

Fig. 186. Pollen of *Plantago*. a–d: *P. lanceolata* (English plantain), a–b = LM(a) ×1830, c = SEM(a) ×3000; d–e: *P. maritima* (seaside plantain), d = LM(s) ×720, e = LM(s) ×1830.

Allergenicity

Bernton (1925) found that 4.3% of hay fever patients in Washington, D.C., reacted to *Plantago* pollen, and Blumstein & Tuft (1937) reported from Philadelphia even greater numbers (17.5%) of early summer patients suffering from inhalant allergies sensitive to its pollen. Rogers (1944) also found limited sensitivity there. Solomon & Durham (1967) observed that English plantain is only a minor contributing factor in pollinosis after the grass pollen season in the East, but that in the Northwest, where higher atmospheric concentrations occur (Table 25), it may be of more specific relevance.

In our survey we found that plantain is considered an important offender by some allergists in several eastern coastal states as well as in the Northwest and California. In Idaho, for example, along with sagebrush and Russian thistle, English plantain is of major importance in pollinosis among weedy plants (W.L. Venning, personal communication), and in central and western Oregon over 2,000 atopic patients have responded positively (2+–4+) to skin tests using plantain pollen extracts (R.L. Cutter and N.M. Kudelko, personal communications). It should be pointed out, however, that in Britain plantains and grasses often flower simultaneously and that some positive skin test responses may be due to contamination of plantain pollen extracts with grass pollen (A.W. Frankland, personal communication). These false positive reactions can also occur elsewhere whenever extracts are contaminated with other allergens.

POLYGONACEAE

Buckwheat Family

The Polygonaceae are a very large, mostly herbaceous family of northern temperate regions. Economic significance of the family is limited, the better known species being common sorrel (*Rumex acetosa*),

rhubarb (*Rheum rhabarbarum*), and buckwheat (*Fagopyrum esculentum*).

In North America only *Oxyria* (mountain sorrel) and *Rumex* (dock or sorrel) (Fig. 187) are characteristically wind-pollinated and often disperse abundant pollen; others may shed some pollen incidentally, but it will not be dispersed far from the plants. *Muehlenbeckia* (wire plant), introduced into California gardens, may also release wind-dispersed pollen. *Rheum* is said to be wind-pollinated (Solomon & Durham 1967; Wodehouse 1971), but most botanists find it primarily entomophilous (Knuth 1909; J.L. Reveal, personal communication).

Species of *Rumex* are found essentially throughout the continent. In our sample they are most common in the western half of the United States, particularly from southern Wyoming to eastern Arizona and New Mexico (Fig. 188).

Flowering

Members of the large genus *Rumex* with, for example, 20 species in the northeastern region and 24 species in California alone, flower largely in the spring and summer, but in California flowering is year-round (Table 26). The single species of *Oxyria*, *O. digyna*, flowers during the summer. Of the entomophilous genera, species of *Eriogonum* and *Polygonum* have long flowering periods, particularly in California and Florida, respectively.

Pollen Morphology (Fig. 189)

Grains are oblate-spheroidal to prolate, 19–56 × 18–48 μm; the amb rounded-triangular to circular, and 3(–6)-colpate, 3-colporoidate, or polyporate; the colpi long and narrow; and the pores/ora circular (*Polygonum*), lolongate (*Eriogonum, Rumex*), lalongate to synclinorate (*Polygonum*), or only faintly demarked (*Fagopyrum, Oxyria*). The sexine is tegillate or reticulate, thin (ca. 0.6 μm, *Oxyria, Rumex*), or thick (ca. 2.2–3.2 μm, *Eriogonum, Fagopyrum*), or thin to thick (0.6–4.5 μm, *Polygonum*); the nexine thinner than or as thick as the sexine; and the intine ca. 0.5 μm thick, thicker below the apertures.

Allergenicity

Rumex acetosella, *R. hymenosepalus*, and, by implication, *R. crispus* and *R. obtusifolius* may be repsonsible for pollinosis about the time grass allergenicity is

Fig. 187. Morphology of representative Polygonaceae. a–b: *Rumex crispus* (curley dock) upper stem with inflorescences (a, ×⅔), and flowers (b, ×3); c–d: *Polygonum bicorne* (pink knotweed) stem with inflorescences (c, ×⅘) and two inflorescences (d, ×3).

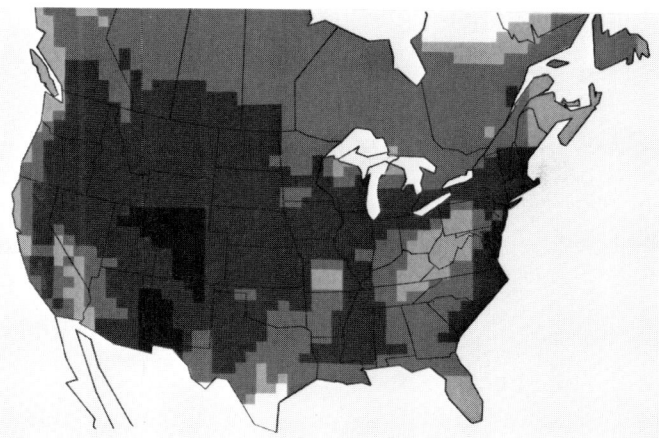

Fig. 188. Generalized composite distribution of 9 of the most important indigenous species of *Rumex* (☐ 1 sp., ▨ 2 spp., ▩ 3–4 spp., ■ 5–6 spp.).

TABLE 26. Flowering by month of the Polygonaceae.

GENUS	REGION							
	nNE	sNE	SE	sFL	NC	SC	NW	SW
Eriogonum		7–10	7–10		6–8	5–11	4–8	3–12
Fagopyrum	6–9	5–9	6–11		7–8	5–11		
Oxyria[a]	6–8						6–8	7–9
Oxytheca							6–8	4–10
Polygonella	8–10	7–10	6–10	6–11				
Polygonum	5–11	5–11	5–11	1–12	6–9	4–11	3–11	3–11
Rumex[a]	5–8	4–9	3–6	3–5, 9–11	5–9	3–5	3–9	1–12

[a]Typically wind-pollinated, the remainder principally insect-pollinated.

paramount (Wodehouse 1971). Most species shed abundant pollen, but some are largely self-pollinating and release little (e.g., *R. maritimus*, *R. pulcher*). Solomon (1969) found many patients experiencing pollinosis who responded positively to skin and nasal tests using *Rumex* pollen extracts; *R. acetosella* was implicated as the offending species. The allergenic activity of *Rumex* is probably greater than has been implicated because its pollen is often shed concomitant with grass pollen dispersal. Patients suffering from pollinosis at the same time of year who are unresponsive to grass pollen hyposensitization should be tested for *Rumex* sensitivity.

Oxyria, and possibly the cultivated *Muehlenbeckia*, may shed sufficient pollen to be offending agents in pollinosis, but data are wanting. Species of insect-pollinated genera could be responsible for localized cases of inhalant allergy wherever abundant populations exist.

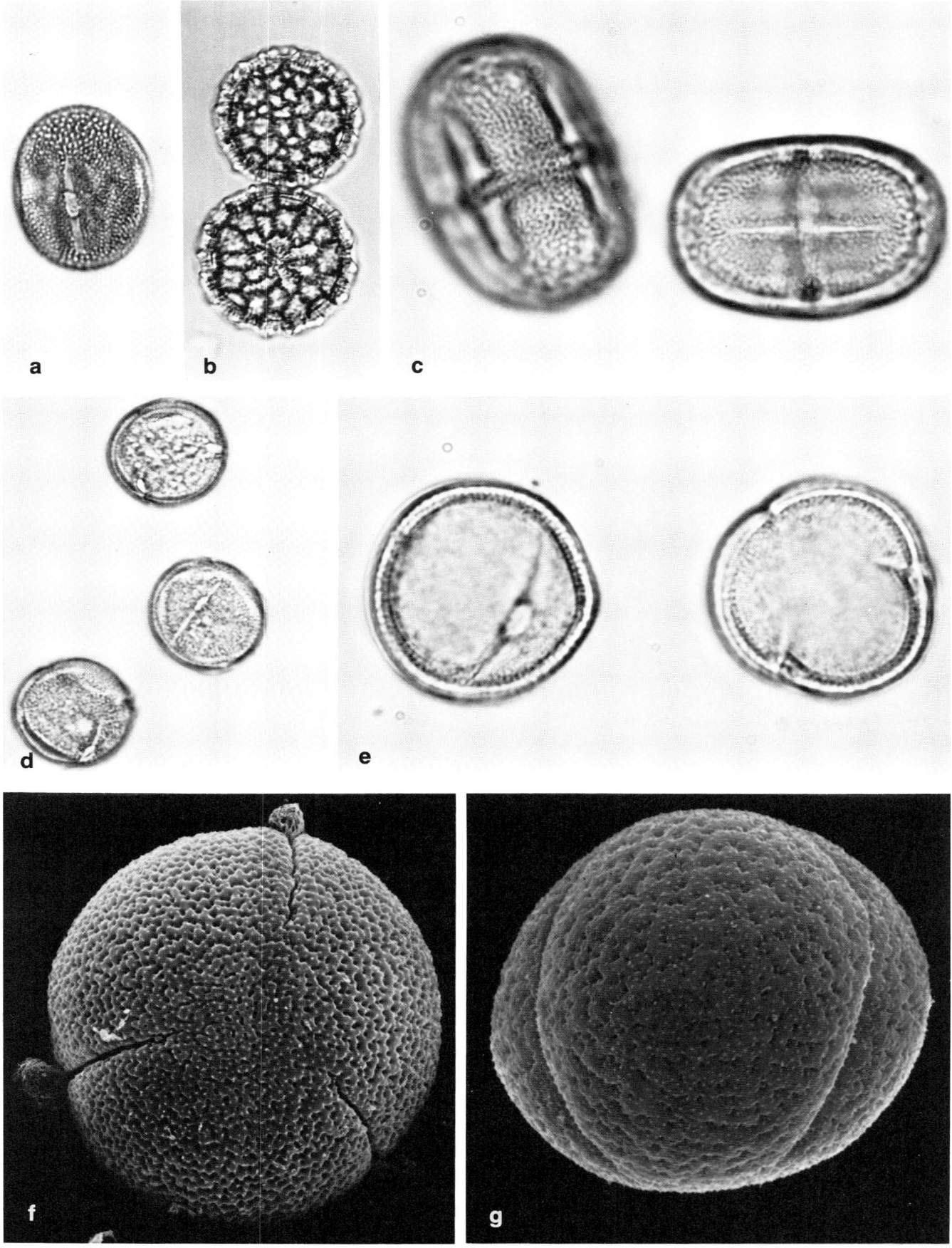

RANUNCULACEAE

Buttercup Family

The family contains a number of well-known wild flowers and ornamentals, such as anemones, buttercups, and larkspurs, and some poisonous plants, like monkshood (*Aconitum*). Many species are centered in the temperate and cold regions of both hemispheres.

Flowering and Pollen Aerobiology

The only wind-pollinated plants in the family are the meadow-rues (*Thalictrum*) (Fig. 190). Although the majority are primarily anemophilous, others are visited by pollen-seeking insects and may be at least facultatively entomophilous. The pollen retains traces of the entomophilous ancestry characteristic of the family by its slight cohesiveness and by the fact that the anthers dehisce successively (Airy Shaw 1973).

Thalictrum consists of 150 species, of which about 25 cultivated and indigenous species are found in North America. They flower mostly in the spring until July, but as late as August in the Northeast and as early as January in the Northwest. The common genera *Clematis* (Fig. 190) and *Ranunculus* (buttercup), both entomophilous, also flower by March or April and continue to July or later. In the Southwest they

Fig. 189. Pollen of representative Polygonaceae. a: *Eriogonum sphaerocephalum* (wild buckwheat), LM(s) near equatorial view ×720; b: *Polygonum lapathifolium* (dock-leaved smartweed), LM(a) ×720; c: *Polygonum scandens* (climbing false buckwheat), LM(a) equatorial views ×1830; d: *Rumex crispus* (yellow dock), LM(s) ×720; e–f: *Rumex verticillatus* (swamp dock), e = LM(a) ×1830, f = SEM(a) polar view ×4100; g: *Rumex acetosella* (common or sheep sorrel), SEM(a) equatorial view showing 2 apertures ×4400.

Fig. 190. *Thalictrum dioicum* (early meadow-rue) stem with leaves and staminate flowers (a, ×⅘) and *Clematis dioscoreifolia* vine showing leaves and flowers (b, ×1½).

begin flowering in January and February, respectively.

Airborne ranunculaceous pollen has been identified in Britain, including a few of *Thalictrum* (Hyde 1950) and *Ranunculus* (Tinsley & Smith 1974). The latter has been found in high frequencies in Finland (Kupias et al. 1981). Bassett et al. (1978) have also found *Thalictrum* on atmospheric slides in Canada. Undoubtedly, its pollen becomes airborne wherever the plant occurs, but the pollen has missed recognition in the United States.

Pollen Morphology (Fig. 191)

Grains of *Thalictrum* are spheroidal, small (ca. 16–17

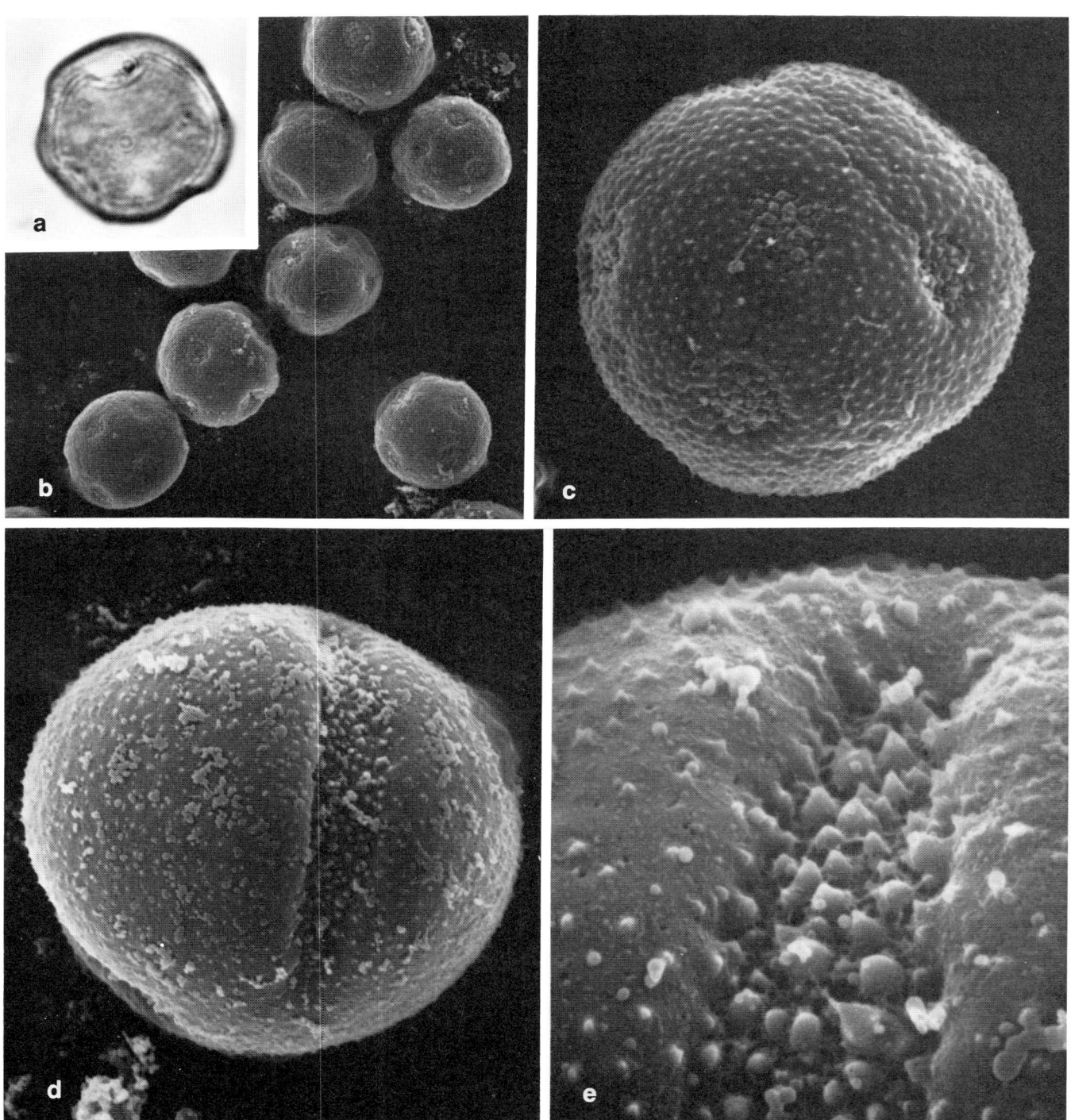

Fig. 191. Pollen of *Thalictrum* and *Clematis*. a–c: *T. dioicum*, a = LM(s) ×1830, b = SEM(a) ×1200, c = SEM(a) ×3200; d–e: *C. dioscoreifolia*, d = SEM(a) equatorial view showing 1 aperture ×3800, e = SEM(a) showing part of aperture and sexine surface ×12000.

μm in diameter), and pantoporate; the pores 6–8, large (ca. 4–5 μm in diameter), circular to elongate; the margins slightly irregular; and with densely granular opercula. The sexine is 0.8 μm thick, tectate with a granular surface; the nexine as thick as the sexine; and the intine only 0.5 μm thick.

Grains of *Ranunculus* differ by their larger size (to 42 μm wide), apertures (3–5-colpate), and sculpturing (tegillate, undulating, and with small excrescences).

Allergenicity

Offenders in inhalant allergy have occasionally been recognized in the family. Thiberge & Hauser (1931) described a case of hay fever due to *Clematis*, the pollen being found on slides exposed in the patient's bedroom. This typically entomophilous garden vine clearly may release wind-dispersed pollen that is allergenic. Pollen of *Thalictrum* is a more likely generalized candidate, however, among undiagnosed cases of pollinosis during late spring and early summer.

URTICACEAE
Nettle Family

A family of predominantly herbaceous plants widely dispersed in tropical and temperate regions, the nettles are best known for the stinging hairs typical of *Urtica* (stinging nettle) and *Laportea* (wood nettle) (Fig. 192). The hairs contain histamine, acetylcholine, and 5-hydroxytryptamine and cause an irritating dermatitis.

The five important genera indigenous to North America—*Boehmeria* (false nettle), *Parietaria* (pellitory), and *Pilea* (clearweed, richweed), and those just noted—are all wind-pollinated. In addition, several Eurasian species of *Urtica* are common weeds, and the allergenically significant Old World *Parietaria officinalis* (wall pellitory) is occasionally cultivated in California (Charpin & Charpin 1980).

Flowering and Pollen Aerobiology

Members of the family flower predominantly during the summer and early autumn months, although *Parietaria* and *Urtica* may flower as early as January or February in southern latitudes (Table 27).

Urtica, *Parietaria*, and *Pilea* disperse their pollen in puffs to the ambient air. Their stamens are bent down inward in the floral bud, and when mature they spring violently upward and bend out of the flower, the anthers turning inside out so that the loose, powdery pollen is ejected as a little cloud to be borne away by the wind (Airy Shaw 1973). In Britain, where urticaceous plants are common, Hyde (1959) estimated that their ambient pollen closely approximates the frequency of grass pollen. Dominance of *Boehmeria*

TABLE 27. Flowering by month of the Urticaceae.

GENUS	nNE	sNE	SE	sFL	NC	SC	NW	SW
Boehmeria	7–9	6–10	7–8	3–8	7–8	6–11		
Laportea	7–9	5–8	6–8		7–8			
Parietaria	6–10	5–10	3–11 (1–12)	3–5	7–8	2–6	5–7	2–7
Pilea	8–9	7–10	8–9		7–8	6–11		
Urtica	5–10	4–10	4–7		7–9	2–10	1–9	1–9

pollen is evident in the atmosphere of Taipei, Taiwan, where it sheds more pollen than any other agent except *Trema* (Ulmaceae) (Chen & Huang 1980). In North America little is known of airborne Urticaceae pollen even though the plants are frequent throughout most of the continent. They superficially resemble spores of fungi, for which they may be mistaken.

Pollen Morphology (Figs. 193–94)

Grains are oblate-spheroidal to spheroidal, small, the average diameters 12–15 μm; the amb rounded-triangular, and 2–4-porate; and the pores ± circular, 1.0–1.5 μm in diameter, annulate, but sometimes indistinctly so. The sexine is thin (0.4–0.5 μm thick), tectate, smooth, or with minute excrescences; the nexine indistinct; and the intine considerably thickened (to 3 × 4 μm) to form cup-shaped onci below the pores. Grains of *Parietaria* and *Urtica* are often 3–4-porate, while those of *Boehmeria*, *Pilea*, and *Laportea* are usually 2–3-porate. Otherwise, grains of the family are very similar.

Allergenicity

As incitants of inhalant allergy, pollen of the Urticaceae remains virtually unknown and unexplored in North America. In Europe, however, *Parietaria* pollen is well known to elicit severe pollinosis (Panzani 1956; Serafini 1957), and it is only reasonable to assume that where the plant is common in the United States and Canada, repeated exposure will result in parallel allergic symptoms among sensitized individuals.

No cross-reaction was found between pollen allergens of *Parietaria* and *Urtica* (Serafini 1957). The genera are not closely related (classified in different tribes, Parietarieae and Urticeae) and even though the grains of both genera are very similar, major allergens may differ. *Urtica* pollen hypersensitivity has been reported from southern California (Piness 1955) and Oklahoma (Levetin & Buck 1980), and J.T. Chiu (personal communication) from southern California has found moderately positive (average 2+) skin test reactions to *Urtica* pollen extracts among approximately 150 patients suffering from pollinosis.

Fig. 192. Morphology of representative Urticaceae. a: *Boehmeria cylindrica* (bog-hemp) stem with leaves and inflorescences, ×½; b: *Laportea canadensis* (wood nettle) stems with leaves and inflorescences, ×⅕; c: *Pilea pumila* (clearweed) stem with leaves and inflorescences, ×¾; d: *Urtica dioica* (stinging nettle) stem with leaves and inflorescences, ×¾.

Fig. 193. Pollen of representative Urticaceae (*Boehmeria-Pilea*). a–e: *Boehmeria cylindrica* (bog-hemp), a–b = LM(a) equatorial views ×1830, c = LM(s) equatorial view showing thickened intine below pores ×1830, d = SEM(a) ×8400, e = SEM(a) showing pore morphology and sexine surface ×17000; f–g: *Laportea canadensis* (wood nettle), f = SEM(a) × 8000, g = LM(a) ×1830; h–k: *Pilea pumila* (clearweed), h = LM(s) ×720, i–j = LM(a) surface (i) and optical (j) views ×1830, k = LM(s) equatorial views ×1830.

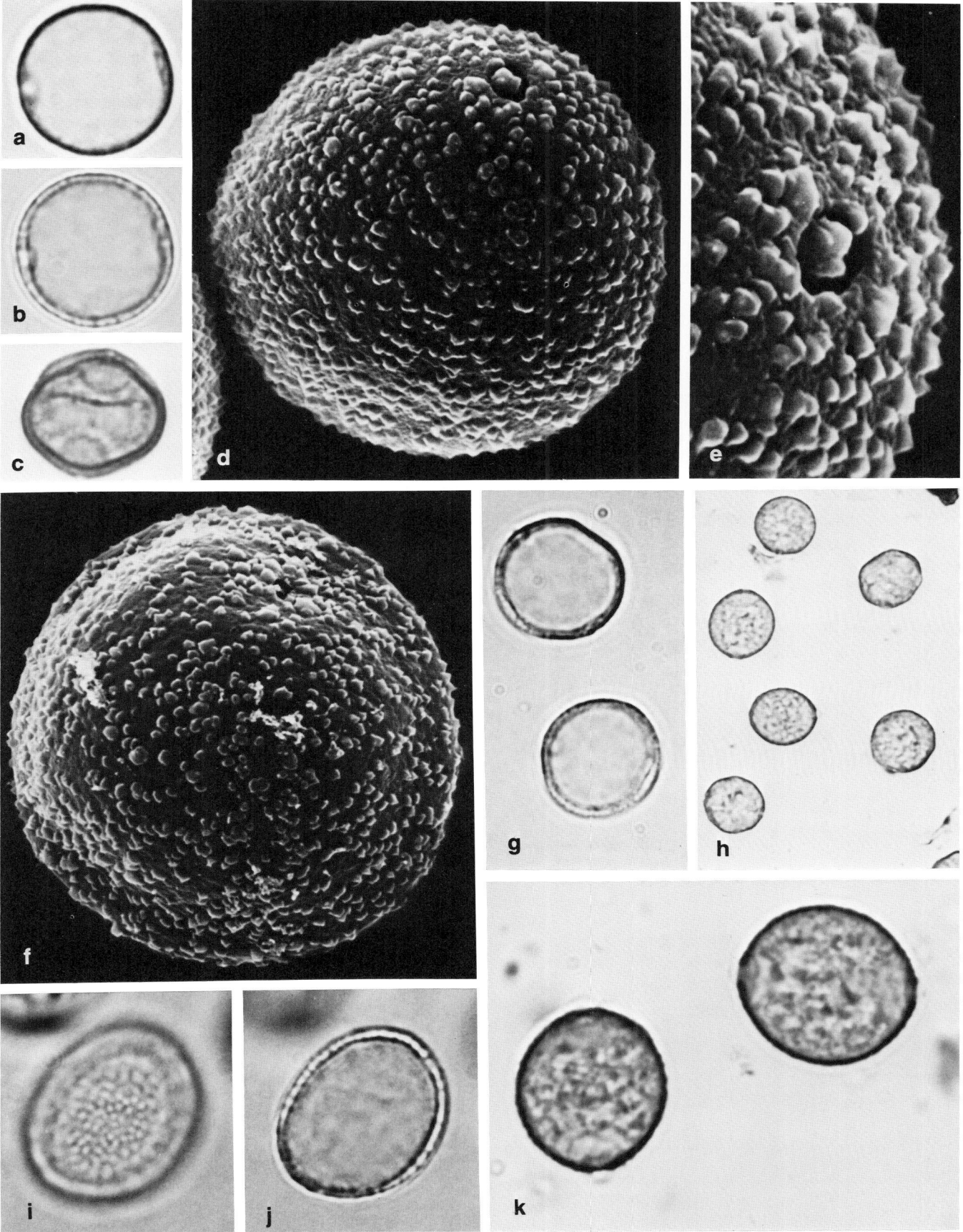

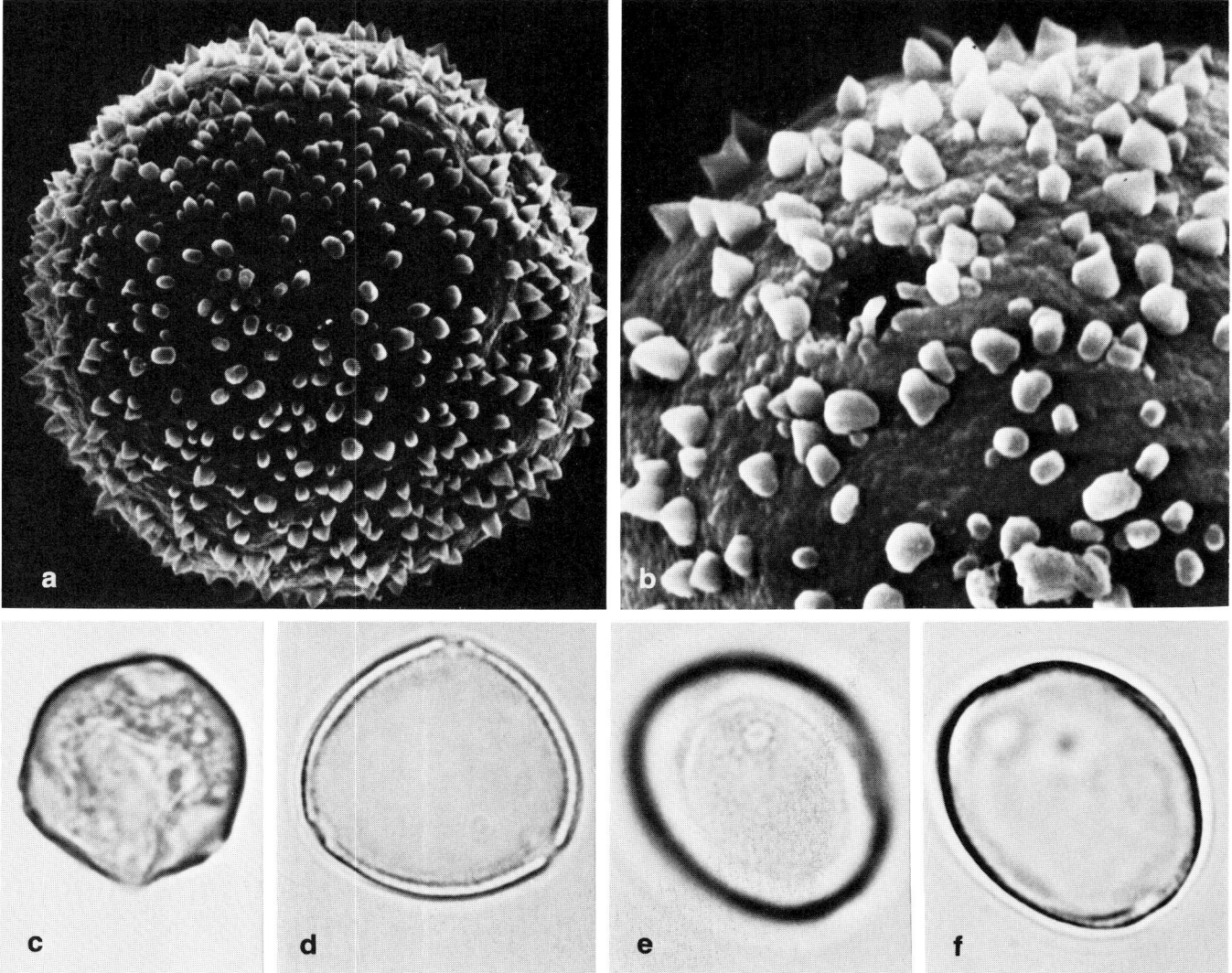

Fig. 194. Pollen of representative Urticaceae (*Pilea-Urtica*). a–b: *Pilea pumila*, a = SEM(a) ×7500, b = SEM(a) showing pore and sexine surface; c–f: *Urtica dioica* (stinging nettle), c = LM(s) polar view showing thickened intine below pores ×1830, d = LM(a) polar view ×1830, e–f = LM(a) surface (e) and optical (f) views ×1830.

Appendix 1

Distribution Maps of Species

Maps giving generalized distributions of important wind-dispersed species indigenous to North America are arranged following the text format, viz., plant-seasonal sections—(1) trees and shrubs, (2) grasses and grasslike plants, and (3) weeds and herbs—and within sections alphabetically by family. Also included are many maps showing distributions of species whose mode of pollination is secondarily wind-dispersed as well as a few examples of cases where incidental dispersal of pollen is known or suspected to occur by wind.

Introduced species, excepting grasses, are listed first with their zone of hardiness—they typically do not occur above the zone given (see Fig. 2), but may be widespread below it either as cultivated plants common in urban areas or as adventive (often weedy) elements of the flora. Maps of both indigenous (black) and introduced (stippled) species of grasses and grasslike plants are given.

Legends serve not only to identify the maps but to provide data on species origin (when introduced), vernacular names, important synonyms, and authorities of species. Most authors of species including parenthetical names are given in full, except a few commonly used or long names.* Only the author(s) publishing the species is given; species names ascribed to another person are unnecessary for the correct author citation and are omitted (e.g., "Parry ex" is deleted from the authority for *Picea engelmannii* Engelmann, the latter being the sole author of the published name).

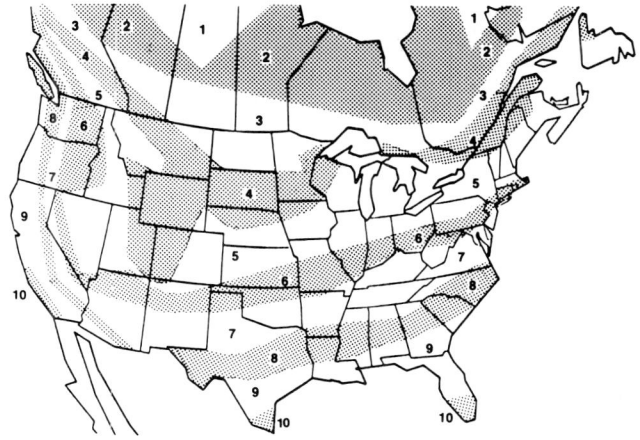

Fig. 2. Map showing hardiness zones from northern Canada (1) to subtropical Florida, Texas, and California (10). A plant recorded for Zone 6, for example, will survive in that zone and to south and west, but not north of it. Most data are for cultivated plants (Hortus Third 1976).

* BSP—Britton, Sterns & Poggenberg; DC.—A.P. de Candolle (father); A.DC.—Alphonse de Candolle (son); HBK—Humboldt, Bonpland & Knuth; L.—Linnaeus. The abbreviations f. = filius (son) and × = hybrid.

Trees and Shrubs

CUPRESSACEAE

INTRODUCED SPECIES

Cupressus sempervirens L.
  Italian cypress, Eurasian, Zone 8

Juniperus chinensis L.
  Chinese juniper, Asian, Zone 5

Platycladus orientalis (L). Franco (syn. *Thuja orientalis* L.)
  Oriental aborvitae, Asian, Zone 5

INDIGENOUS SPECIES

Cupressus arizonica Greene
Arizona cypress

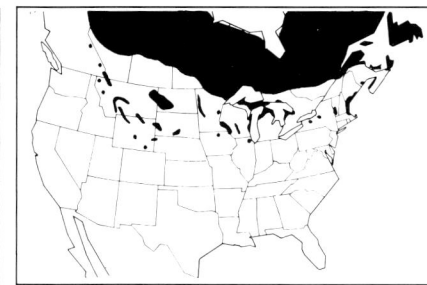
Juniperus horizontalis Moench
creeping juniper

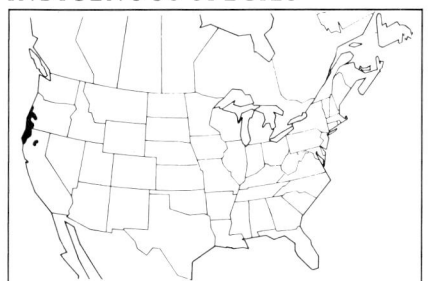
Chamaecyparis lawsoniana (A. Murray) Parlatore
Lawson's false cypress, Port Orford cedar

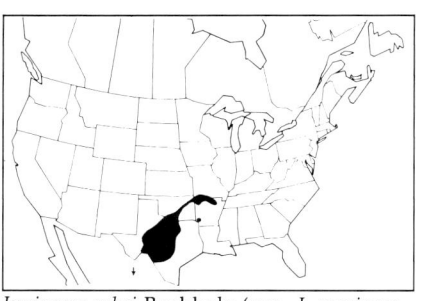
Juniperus ashei Buchholz (syn. *J. mexicana* of authors, *J. sabionoides* (HBK) Nees)
Ashe's juniper, mountain cedar

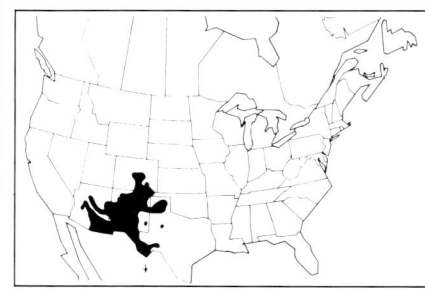
Juniperus monosperma (Englemann) Sargent
one-seeded or cherrystone juniper

Chamaecyparis nootkatensis (D. Don) Spach
Alaska cedar, yellow cypress

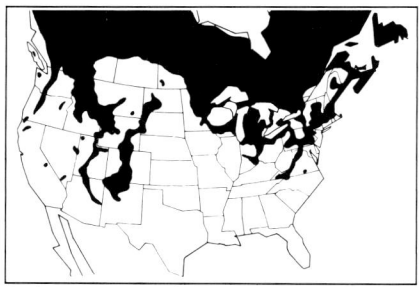
Juniperus communis L.
common juniper

Juniperus occidentalis Hooker
western juniper

Chamaecyparis thyoides (L.) BSP
Atlantic white cedar

Juniperus deppeana Steudel
alligator juniper

Juniperus pinchotii Sudworth
Pinchot's or red-berry juniper

DISTRIBUTION MAPS OF SPECIES

Juniperus scopulorum Sargent
Rocky Mountain juniper

Thuja plicata D. Don
giant arborvitae, western red cedar

Abies concolor (Gordon & Glendinning) Hildebrand
white fir

PINACEAE

INTRODUCED SPECIES

Abies alba Miller
silver fir, European, Zone 5

Cedrus deodara (D. Don) G. Don
deodar cedar, Asian, Zone 7

Cedrus libani A. Richard (syn. *C. libanotica* Link)
cedar of Lebanon, Mediterranean, Zone 6

Larix decidua Miller
European larch, European, Zone 3

Picea abies (L.) Karsten
Norway spruce, European, Zone 3

Pinus mugo Turra
mountain pine, European, Zone 3

Pinus nigra Arnold
Austrian or black pine, Eurasian, Zone 5

Pinus sylvestris L.
Scotch pine, Eurasian, Zone 3

Pseudolarix amabilis (Nelson) Rehder
golden larch, Asian, Zone 6

INDIGENOUS SPECIES

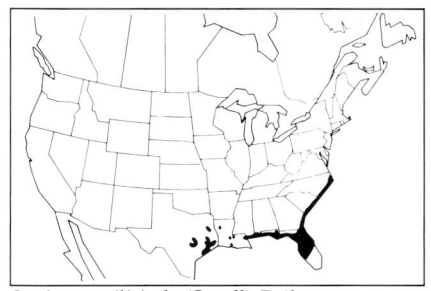

Juniperus silicicola (Small) Bailey
sand or southern red cedar, coast juniper

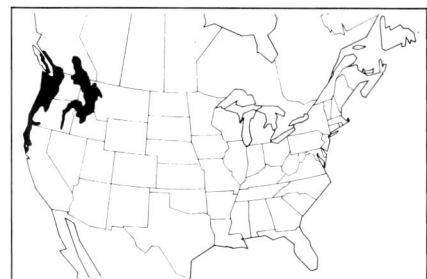

Abies grandis (Lambert) Lindley
giant fir

Juniperus virginiana L.
eastern juniper or red cedar

Abies lasiocarpa (Hooker) Nuttall
alpine fir

Abies amabilis J. Forbes
Cascade or silver fir

Larix laricina (Du Roi) C. Koch
American larch, tamarack

Thuja occidentalis L.
American arborvitae, eastern white cedar

Abies balsamea (L.) Miller
balsam fir

Larix occidentalis Nuttall
western larch

179

APPENDIX 1

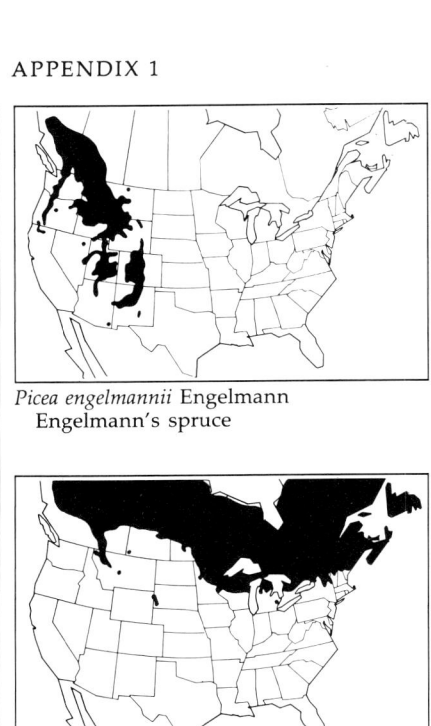
Picea engelmannii Engelmann
Engelmann's spruce

Pinus albicaulis Engelmann
whitebark pine

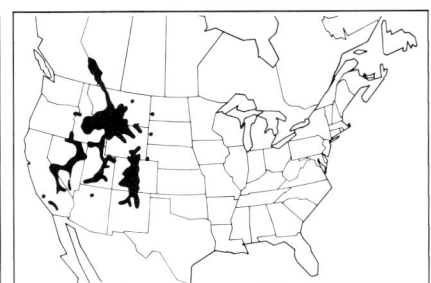
Pinus flexilis James
limber pine

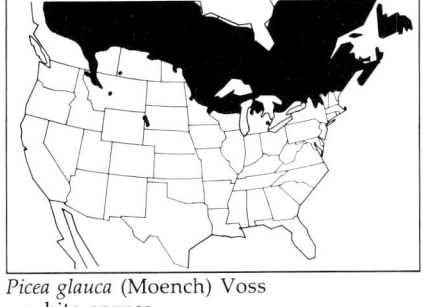
Picea glauca (Moench) Voss
white spruce

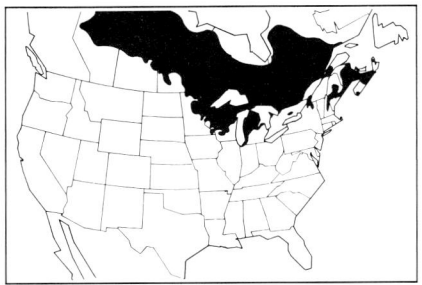
Pinus banksiana Lambert
jack pine

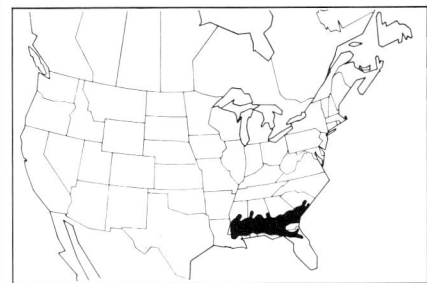
Pinus glabra Walter
cedar or spruce pine

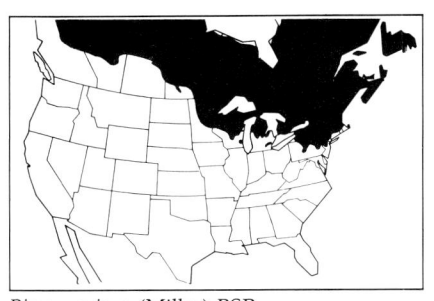
Picea mariana (Miller) BSP
black spruce

Pinus contorta Loudon
lodgepole pine

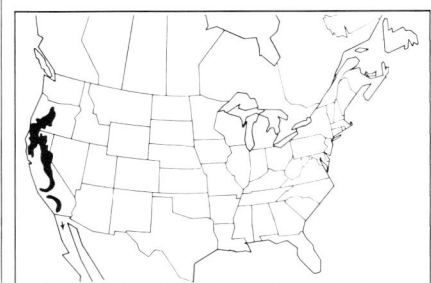
Pinus lambertiana Douglas
sugar pine

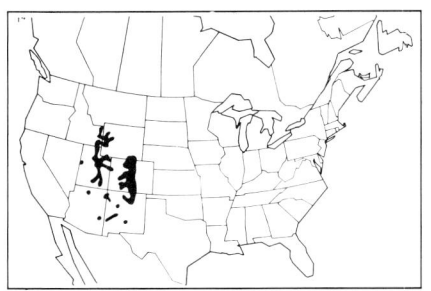
Picea pungens Engelmann
Colorado spruce

Pinus echinata Miller
short-leaved pine

Pinus monophylla Torrey & Fremont
single-leaved pine

Picea sitchensis (Bongard) Carrière
Sitka spruce

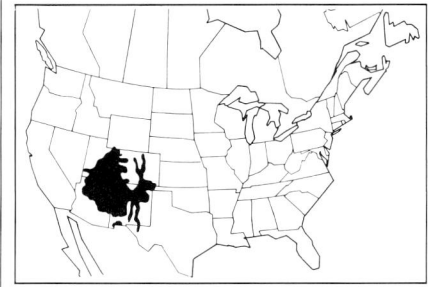
Pinus edulis Engelmann
pinyon

Pinus monticola D. Don
western white pine

DISTRIBUTION MAPS OF SPECIES

Pinus palustris Miller
long-leaved pine

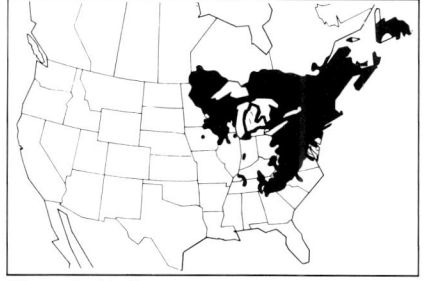
Pinus strobis L.
white pine

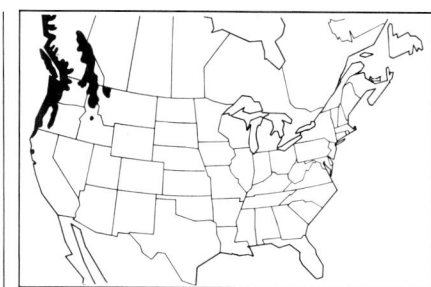
Tsuga heterophylla (Rafinesque) Sargent
western hemlock

TAXACEAE

INTRODUCED SPECIES

Taxus baccata L.
English yew, Eurasian, Zone 7

Taxus cuspidata Siebold & Zuccarini
Japanese yew, Asian, Zone 5

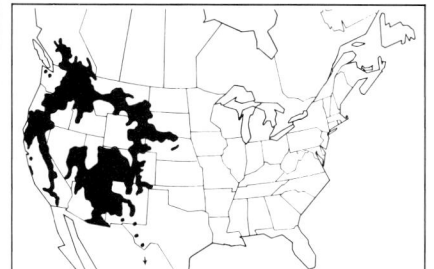
Pinus ponderosa Lawson & Lawson
ponderosa pine

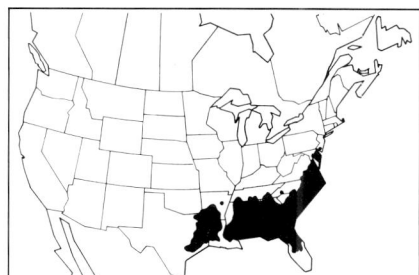
Pinus taeda L.
loblolly pine

INDIGENOUS SPECIES

Pinus resinosa Aiton
Norway or red pine

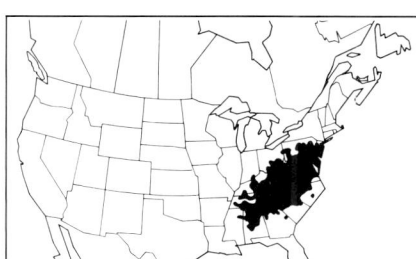
Pinus virginiana Miller
scrub or Virginia pine

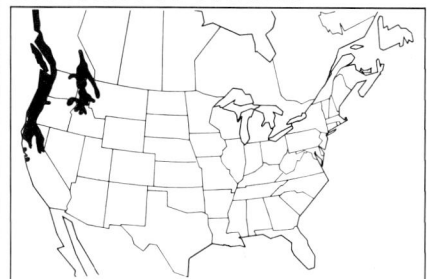
Taxus breviofolia Nuttall
western yew

Pinus rigida Miller
pitch pine

Pseudotsuga menziesii (Mirbel) Franco
Douglas fir

Taxus canadensis Marshall
Canada yew

Pinus serotina Michaux
pond pine

Tsuga canadensis (L.) Carrière
eastern hemlock

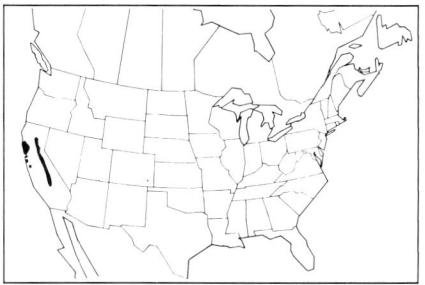
Torreya californica Torrey
California nutmeg

APPENDIX 1

TAXODIACEAE

INDIGENOUS SPECIES

Sequoia sempervirens (D. Don) Endlicher
coastal redwood, sequoia

Taxodium distichum (L.) Richard
bald cypress

GINKGOACEAE

INTRODUCED SPECIES

Ginkgo biloba L.
ginkgo, maidenhair tree, Asian, Zone 5

ACERACEAE

INTRODUCED SPECIES

Acer palmatum Thunberg
Japanese maple, Asian, Zone 6

Acer platanoides L.
Norway maple, European, Zone 4

Acer pseudoplatanus L.
sycamore-maple, Eurasian, Zone 5

Acer tataricum L.
Tatarian maple, Eurasian, Zone 5

INDIGENOUS SPECIES

Acer circinatum Pursh
vine maple

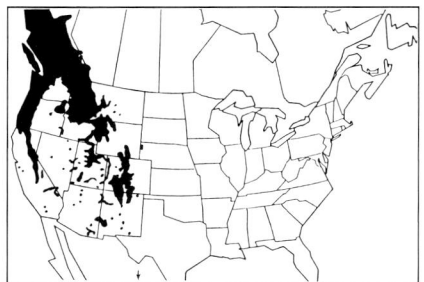

Acer glabrum Torrey
Rocky Mountain maple

Acer grandidentatum Nuttall
big-toothed maple

Acer macrophyllum Pursh
big-leaved maple

Acer negundo L.
box-elder

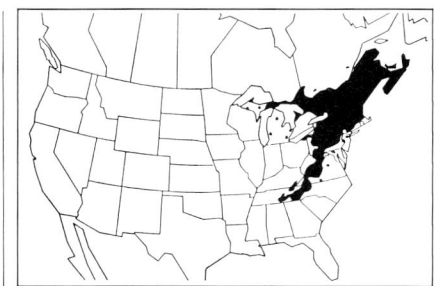

Acer pensylvanicum L.
striped maple

Acer rubrum L.
red maple

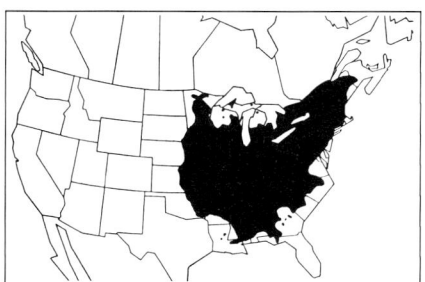

Acer saccharinum L.
silver maple

Acer saccharum Marshall (syn. *A. barbatum* Michaux, *A. nigrum* Michaux)
sugar maple (Forida and black maples)

Acer spicatum Lamarck
mountain maple

ANACARDIACEAE

INTRODUCED SPECIES

Anacardium occidentale L.
  cashew, tropical American, Zone 10

Cotinus coggygria Scopoli
  smokebush, smoketree, Eurasian, Zone 5

Mangifera indica L.
  mango, Asian, Zone 10

Pistacia chinensis Bunge
  Chinese pistachio, Asian, Zone 6 (east)

Pistacia lentiscus L.
  mastic tree, Mediterranean, Zone 9 (west)

Pistacia vera L.
  pistachio, Asian, Zone 9 (cultivated California)

Rhus chinensis Miller
  nutgall tree, Asian, Zone 8

Rhus verniciflua J. Stokes
  lacquer or varnish tree, Asian, Zone 9

Schinus molle L.
  Peruvian pepper tree, South American, Zone 9 (west)

Schinus terebinthifolius Raddi
  Brazilian pepper tree, Florida holly, South American, Zone 9 (east)

INDIGENOUS SPECIES

Rhus aromatica Aiton (syn. *R. trilobata* Nuttall)
fragrant sumac

Rhus copallina L.
dwarf or shining sumac

Rhus glabra L.
scarlet or smooth sumac

Cotinus obovatus Rafinesque
American chittamwood or smoketree

Rhus integrifolia (Nuttall) Brewer & Watson
lemonade sumac

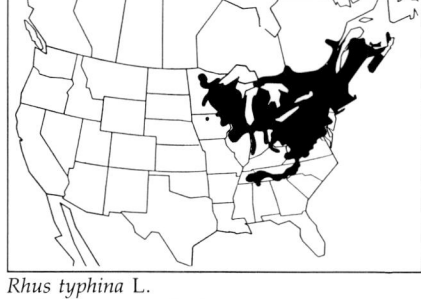

Rhus ovata Watson
sugar bush

Rhus typhina L.
staghorn or velvet sumac

Toxicodendron diversilobum (Torrey & Gray) Greene
western poison oak

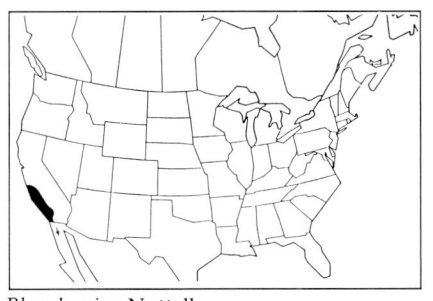

Rhus laurina Nuttall
laurel sumac

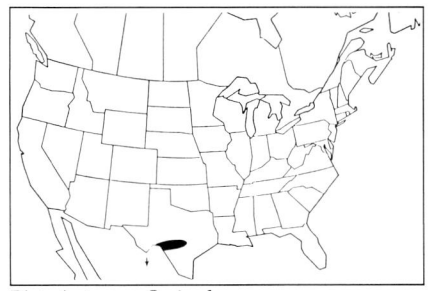

Pistacia texana Swingle
American or Mexican pistachio

Toxicodendron rydbergii (Rydberg) Greene
Rydberg's poison ivy

Toxicodendron radicans (L.) Kuntze
poison ivy

APPENDIX 1

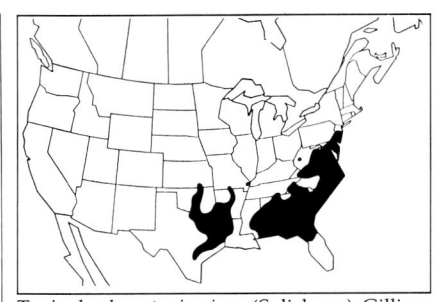
Toxicodendron toxicarium (Salisbury) Gillis
eastern poison oak

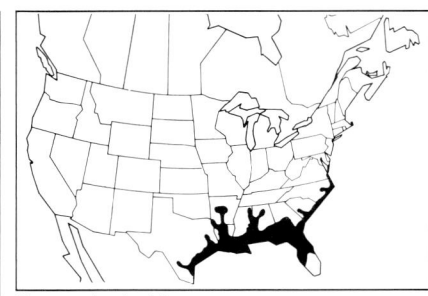
Ilex decidua Walter
possum haw

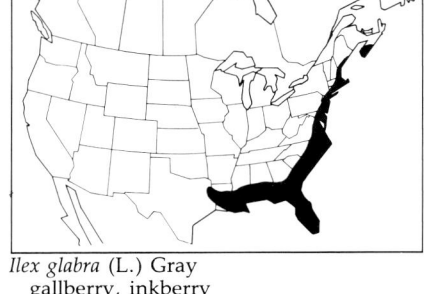
Ilex vomitoria Aiton
cassine, yaupon

AQUIFOLIACEAE

INTRODUCED SPECIES

Ilex aquifolium L.
English or European holly, European, Zone 7

Ilex cornuta Lindley & Paxton
Chinese or horned holly, Asian, Zone 7

Ilex crenata Thunberg
box-leaved or Japanese holly, Asian, Zone 6

INDIGENOUS SPECIES

Ilex ambigua (Michaux) Torrey (syn. *I. montana* Torrey & Gray)
large-leaved holly, mountain winterberry

Ilex cassine L.
cassine, dahoon, yaupon

Ilex coriacea (Pursh) Chapman
large or sweet gallberry

Ilex glabra (L.) Gray
gallberry, inkberry

Ilex laevigata (Pursh) Gray
smooth winterberry

Ilex opaca Aiton
American holly

Ilex verticillata (L.) Gray
common winterberry

ARALIACEAE

INTRODUCED SPECIES

Hedera helix L.
ivy, Eurasian, Zone 4

ARECACEAE

INTRODUCED SPECIES

Chamaedorea erumpens H.E. Moore
bamboo palm, Central American, Zone 10

Cocos nucifera L.
coconut palm, tropical Old World, Zone 10

Phoenix canariensis Chabaud
Canary Island date palm, Canary Is., Zone 9

Phoenix dactylifera L.
date palm, Afro-Asian, Zone 9

Phoenix reclinata Jacquin
Senegal date palm, African, Zone 9 (south)

Roystonea borinquena O.F. Cook
Puerto Rican royal palm, Zone 10

Roystonea regia (HBK) O.F. Cook
Cuban royal palm, tropical American, Zone 10

Trachycarpus fortunei (Hooker) H. Wendland
Chinese or windmill palm, Asian, Zone 8 (south)

Washingtonia robusta H. Wendland
thread palm, Mexican, Zone 9 (south)

DISTRIBUTION MAPS OF SPECIES

INDIGENOUS SPECIES

Acoelorrhaphe wrightii (Grisebach & H. Wendland) Becarri
Everglades palm

Pseudophoenix sargentii H. Wendland
Florida cherry palm

Rhapidophyllum hystrix (Pursh) H. Wendland & Drude
needle or porcupine palm, blue palmetto

Roystonea elata (Bartram) F. Harper
Florida royal palm

Sabal mexicana Martius
Texas palmetto

Sabal minor (Jacquin) Persoon
dwarf palmetto

Sabal palmetto (Walter) Schultes & Schultes f.
cabbage palmetto

Serenoa repens (Bartram) Small
saw palmetto

Thrinax morrisii H. Wendland
brittle thatch or Key palm

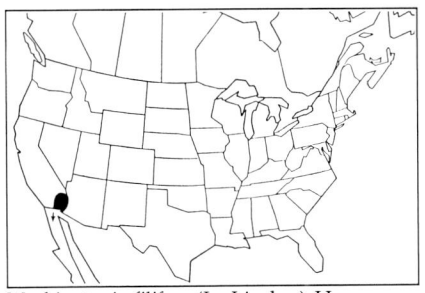

Washingtonia filifera (L. Linden) H. Wendland
desert fan palm

BERBERIDACEAE

INTRODUCED SPECIES

Berberis bealei Fortune (syn. *Mahonia bealei* (Fortune) Carrière)
  Asian, Zone 6
Berberis japonica Thunberg (syn. *Mahonia japonica* (Thunberg) DC.)
  Asian, Zone 6
Berberis thunbergii DC.
  Japanese barberry, Asian, Zone 6
Berberis vulgaris L.
  common barberry, European, Zone 4

INDIGENOUS SPECIES

Berberis aquifolium Pursh (syn. *Mahonia aquifolium* (Pursh) Nuttall)
Oregon grape

Berberis canadensis Miller
American or Alleghany barberry

BETULACEAE

INTRODUCED SPECIES

Alnus glutinosa (L.) Gaertner
  black alder, Eurasian, Zone 5

Betula pendula Roth (syn. *B. alba* L. in part)
  European white birch, Eurasian, Zone 2

Betula pubescens Ehrhart (syn. *B. alba* L. in part)
  white birch, Eurasian, Zone 2

Corylus avellana L.
  European hazelnut or filbert, European, Zone 4

Corylus maxima Miller
  giant hazelnut or filbert, Eurasian, Zone 4

Ostrya carpinifolia Scopoli
  European hop hornbeam, Mediterranean, Zone 5

APPENDIX 1
INDIGENOUS SPECIES

Alnus incana (L.) Moench) (syn. *A. rugosa* (DuRoi) Sprengel, *A. tenuifolia* Nuttall) speckled or white alder

Alnus serrulata (Aiton) Willdenow
hazel alder

Betula nigra L.
red or river birch

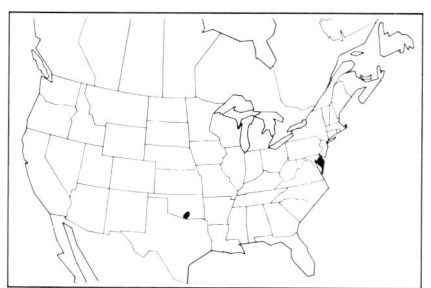

Alnus maritima Nuttall

Alnus viridis (Villars) Lamarck & DC. (syn. *A. cripsa* (Aiton) Pursh, *A. sinuata* (Regel) Rydberg) green, mountain or Sitka alder

Betula occidentalis Hooker
water birch

Alnus oblongifolia Torrey
Arizona alder

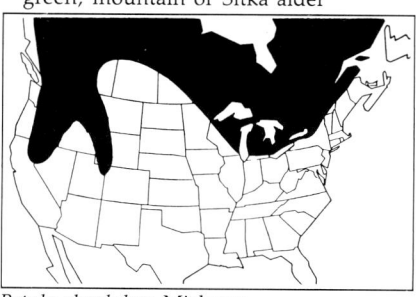

Betula glandulosa Michaux
shrub birch

Betula papyrifera Marshall
paper birch

Alnus rhombifolia Nuttall
white alder

Betula lenta L.
cherry or sweet birch

Betula populifolia Marshall
gray birch

Alnus rubra Bongard
red alder

Betula lutea Michaux f. (syn. *B. alleghaniensis* Britton)
gray or yellow birch

Betula pumila L.
low or swamp birch

DISTRIBUTION MAPS OF SPECIES

INDIGENOUS SPECIES

Carpinus caroliniana Walter
American hornbeam

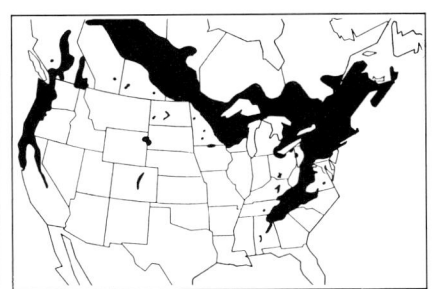
Corylus americana Walter
American hazelnut or filbert

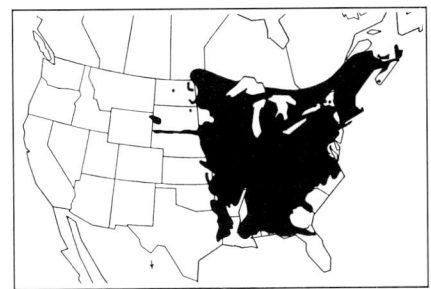
Corylus cornuta Marshall
beaked hazelnut or filbert

Catalpa bignonioides Walter
southern catalpa or cigar tree

Catalpa speciosa Warder
northern catalpa or cigar tree

CAPRIFOLIACEAE

INTRODUCED SPECIES

Lonicera caerulea L.
  Eurasian, Zone 3

Lonicera fragrantissima Lindley & Paxton
  Asian, Zone 6

Lonicera japonica Thunberg
  Japanese honeysuckle, Asian, Zone 5

Sambucus nigra L.
  European elder, Eurasian, Zone 6

Viburnum dilatatum Thunberg
  Asian, Zone 5

Viburnum lantana L.
  wayfaring tree, Eurasian, Zone 4

Weigela florida (Bunge) A. DC.
  Asian, Zone 5

INDIGENOUS SPECIES

Lonicera sempervirens L.
coral or trumpet honeysuckle

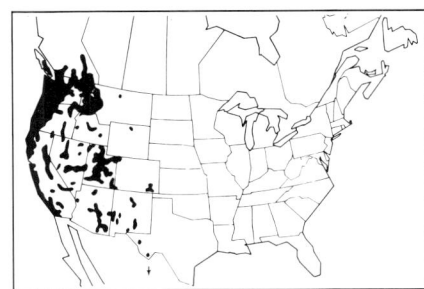
Sambucus caerulea Rafinesque (syn. *S. glauca* Nuttall)
blue elder or elderberry

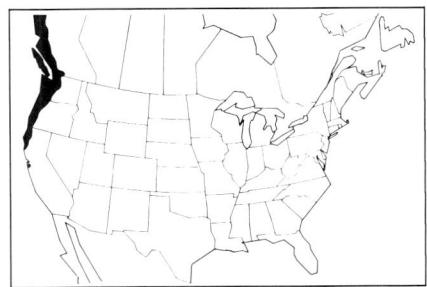
Sambucus callicarpa Greene
Pacific Coast red elder

Chilopsis linearis (Cavanilles) Sweet
desert-willow

BUXACEAE

INTRODUCED SPECIES

Buxus microphylla Siebold & Zuccarini
  Japanese boxwood, Asian, Zone 6

Buxus sempervirens L.
  common boxwood, Mediterranean, Zone 6

INDIGENOUS SPECIES

Ostrya virginiana (Miller) K. Koch
American hop hornbeam

Simmondsia chinensis (Link) C.K. Schneider
jojoba, goatnut

BIGNONIACEAE

APPENDIX 1

Sambucus canadensis L.
American elder or elderberry

Sambucus racemosa L.
red-berried or stinking elder

Viburnum lentago L.
nannyberry

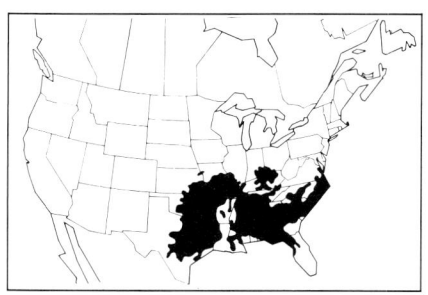

Viburnum rufidulum Rafinesque
rusty or southern blackhaw

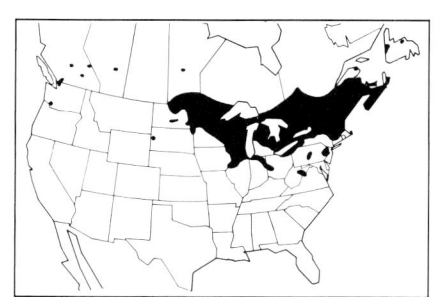

Viburnum trilobum Marshall
American cranberry bush

CASUARINACEAE

INTRODUCED SPECIES

Casuarina cunninghamiana Miquel
  Australian pine, horsetail tree, beefwood, Australian, Zone 10

Casuarina equisetifolia J.R. Forster & G. Forster
  Australian pine, Australian, Zone 10

CELASTRACEAE

INTRODUCED SPECIES

Celastrus orbiculatus Thunberg
  Oriental bittersweet, Asian, Zone 5

Euonymus alatus (Thunberg) Siebold
  winged euonymus or spindle tree, Asian, Zone 4

Euonymus fortunei (Turczaninow) Handel-Mazzetti
  Asian, Zone 6

Euonymus japonicus L.
  Japanese spindle tree, Asian, Zone 8

INDIGENOUS SPECIES

Celastrus scandens L.
American or false bittersweet

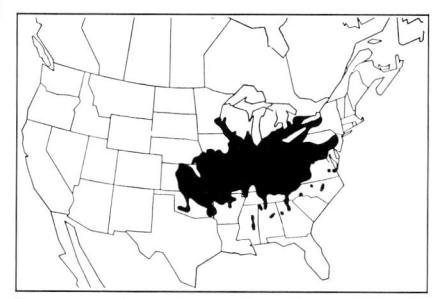

Euonymus atropurpureus Jacquin
eastern burning bush or wahoo

Euonymus occidentalis Torrey
western burning bush or wahoo

CISTACEAE

INDIGENOUS SPECIES

Hudsonia ericoides L.
golden-heather

CLUSIACEAE

INTRODUCED SPECIES

Hypericum calycinum L.
  St. John's wort, e. Mediterranean, Zone 5

Hypericum perforatum L.
  common St. John's wort, European, Zone 4

CORNACEAE

INTRODUCED SPECIES

Cornus alba L.
  Tartarian dogwood, Asian, Zone 3

Cornus mas L.
  Cornelian cherry, Eurasian, Zone 5

DISTRIBUTION MAPS OF SPECIES

INDIGENOUS SPECIES

Cornus alternifolia L. f.
alternate-leaved dogwood

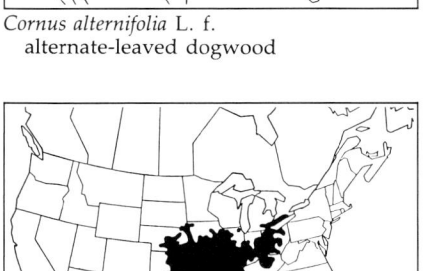

Cornus drummondii C.A. Meyer
rough-leaved dogwood

Cornus florida L.
flowering dogwood

Cornus glabra Bentham
brown dogwood

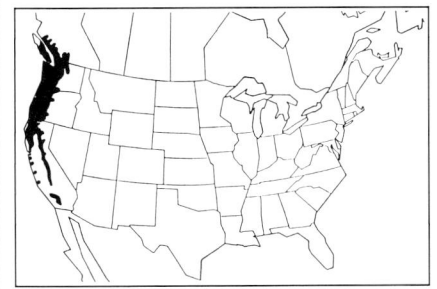

Cornus nuttallii Audubon
Pacific dogwood

Cornus occidentalis (Torrey & Gray) Coville
western dogwood

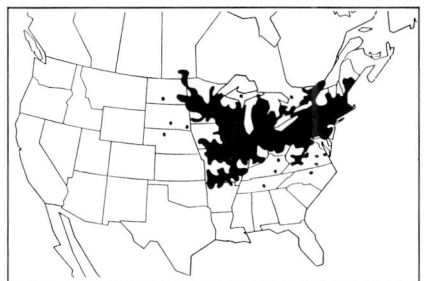

Cornus racemosa Lamarck
gray dogwood

Cornus rugosa Lamarck
round-leaved dogwood

Cornus stolonifera Michaux
red-osier dogwood

Cornus stricta Lamarck
swamp dogwood

ELAEAGNACEAE

INTRODUCED SPECIES

Elaeagnus angustifolia L.
  Russian or wild olive, oleaster, silverberry, Eurasian, Zone 3

Elaeagnus umbellata Thunberg
  Chinese oleaster, Asian, Zone 4

Hippophae rhamnoides L.
  sea buckthorn, Eurasian, Zone 4

INDIGENOUS SPECIES

Shepherdia argentea (Pursh) Nuttall
silver buffaloberry

Shepherdia canadensis (L.) Nuttall
buffalo or soapberry

EMPETRACEAE

INDIGENOUS SPECIES

Ceratiola ericoides Michaux
rosemary

APPENDIX 1

INDIGENOUS SPECIES

Corema conradii Torrey
poverty-grass

Arbutus menziesii Pursh
Pacific madrone

Rhododendron macrophyllum G. Don
West Coast rhododendron, California rosebay

Empetrum nigrum L.
black crowberry

Kalmia latifolia L.
mountain laurel

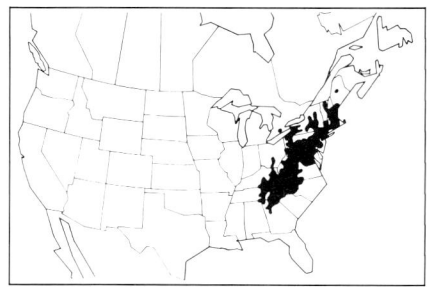

Rhododendron maximum L.
great laurel, rosebay rhododendron

ERICACEAE

INTRODUCED SPECIES

Arbutus × *andrachnoides* Link
hybrid (*A. andrachne* × *A. unedo*), Zone 8

Arbutus unedo L.
cane apples, strawberry tree, European, Zone 8

Bruckenthalia spiculifolia (Salisbury) Reinchenbach
spike heath, Mediterranean, Zone 6

Calluna vulgaris (L.) Hull
common heather, Eurasian, Zone 5

Erica carnea L.
spring heath, European, Zone 6

Erica cinerea L.
Scotch or twisted heath, European, Zone 6

Erica tetralix L.
bog or cross-leaved heath, European, Zone 4

Rhododendron mucronatum (Blume) G. Don
snow azalea, Asian, Zone 8

Rhododendron indicum (L.) Sweet
macranthum azalea, Asian, Zone 6

Rhododendron obtusum (Lindley) Planchet
Hirya azalea, Asian, Zone 7

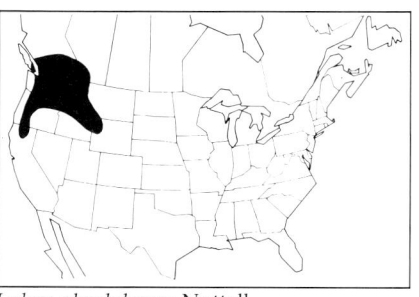

Ledum glandulosum Nuttall
trapper's tea

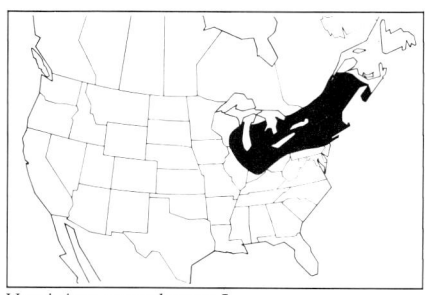

Vaccinium corymbosum L.
highbush blueberry or whortleberry

EUPHORBIACEAE
(see also Weeds and Herbs)

INTRODUCED SPECIES

Ricinus communis L.
castor bean, castor oil plant, tropics, Zone 9

Sapium sebiferum (L.) Roxburg
Chinese tallow tree, Asian, Zone 4

Sapium japonicum (Siebold & Zuccarini) Pax & Hoffman
Asian, Zone 8

INDIGENOUS SPECIES

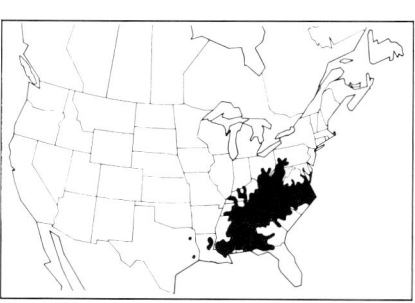

Oxydendrum arboreum (L.) DC.
sourwood

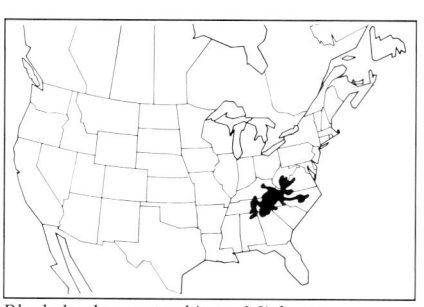

Rhododendron catawbiense Michaux
catawba rhododendron, mountain rosebay

Ateramnus lucidus (Sweet) Rothmann (syn. *Gymnanthes lucida* Sweet)
crabwood

FABACEAE
(see also Weeds and Herbs)

Subfamily *Mimosoideae*

INTRODUCED SPECIES

Acacia baileyana F.J. Mueller
  golden mimosa, Australian, Zone 10

Acacia decurrens Willdenow (syn. *A. dealbeta* Link)
  green, black or silver wattle, Australian, Zone 8

Acacia farnesiana (L.) Willdenow
  sweet acacia, tropical American, Zone 9

Acacia podalyriifolia A. Cunningham
  Queensland silver wattle, Australian, Zone 10

Albizia julibrissin Durazzini
  mimosa or silk tree, Afroasian, Zone 6

Calliandra haematocephala Hasskarl
  red powderpuff tree, South American, Zone 10(9)

Leucaena leucocephala (Lamarck) de Wit
  lead tree, tropical American, Zone 10

Mimosa pudica L.
  sensitive plant, tropical American, Zone 9

Pithecellobium dulce (Roxburgh) Bentham
  haumuchil, tropical American, Zone 10

INDIGENOUS SPECIES

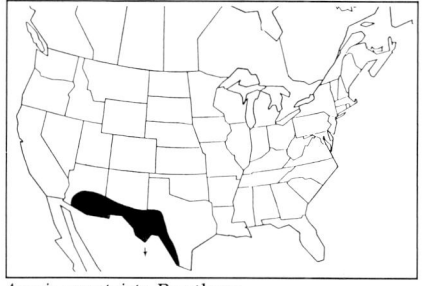
Acacia constricta Bentham
mescat acacia

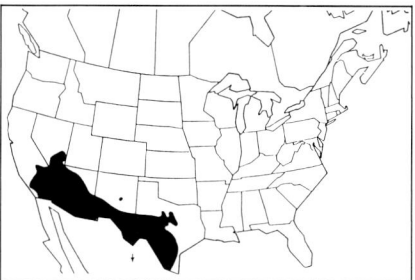
Acacia greggii Gray
cat's claw acacia

Acacia smallii Isely
Small's acacia

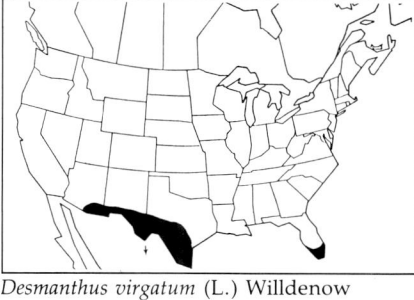
Desmanthus illinoensis (Michaux) MacMillan
prairie mimosa

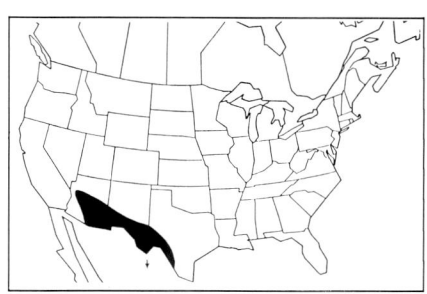
Desmanthus virgatum (L.) Willdenow

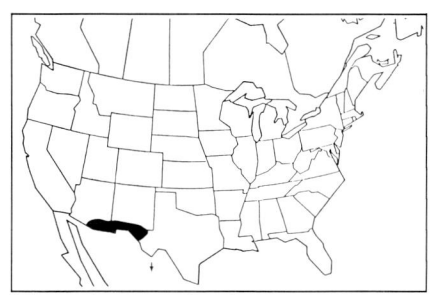
Mimosa biuncifera Bentham
cat's claw mimosa

Acacia angustissima (Miller) O. Kuntze
fern or prairie acacia

Acacia wrightii Bentham
Wright's acacia

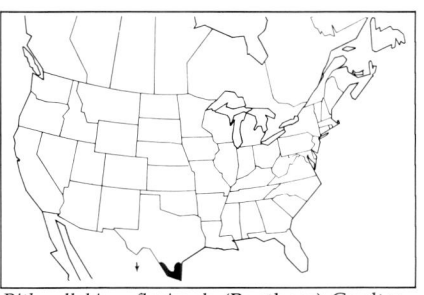
Mimosa dyscocarpa Bentham
gratiño

Acacia berlandieri Bentham
guajillo

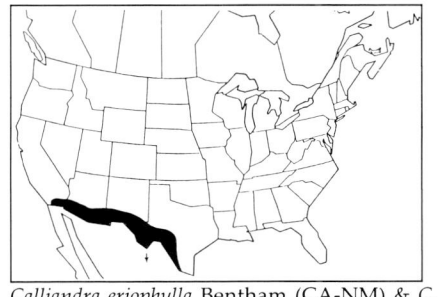
Calliandra eriophylla Bentham (CA-NM) & *C. conferta* Gray (TX)

Pithecellobium flexicaule (Bentham) Coulter
ebony blackbead

APPENDIX 1

Pithecellobium guadalupense (Persoon) Chapman
blackbead

Pithecellobium pallens (Bentham) Standley
huajillo

Pithecellobium ungus-cati (L.) Bentham
blackbead, cat's claw

Prosopis glandulosa Torrey
glandular mesquite

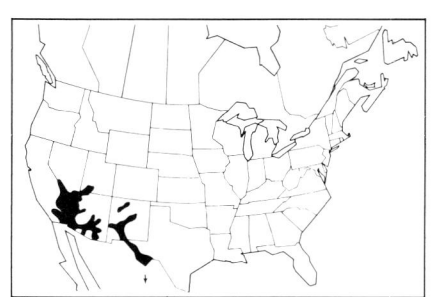

Prosopis pubescens Bentham
screwbean mesquite

Prosopis velutina Wooton
velvet mesquite

Subfamily *Caesalpinioideae*

INDIGENOUS SPECIES

Cassia fasciculata Michaux
partridge pea

Cercis canadensis L.
redbud

Gleditsia triancanthos L.
honey locust

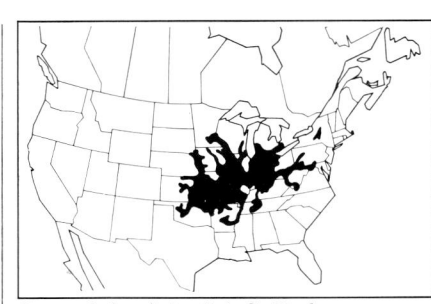

Gymnocladus dioica (L.) C. Koch
Kentucky coffee tree

Subfamily *Faboideae*

INDIGENOUS SPECIES

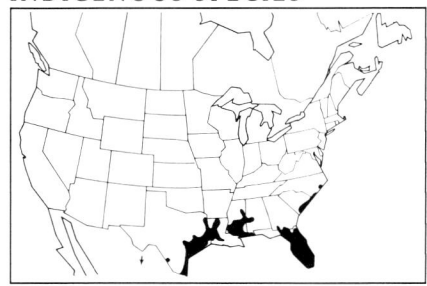

Erythrina herbacea L.
eastern coral bean

Robinia pseudoacacia L.
black locust

FAGACEAE

INTRODUCED SPECIES

Castanea mollissima Blume
Chinese chestnut, Asian, Zone 5

Castanea sativa Miller
European or Spanish chestnut, Mediterranean, Zone 6

Fagus sylvatica L.
European beech, European, Zone 5

INDIGENOUS SPECIES DISTRIBUTION MAPS OF SPECIES

Castanea alnifolia Nuttall
Florida chestnut (chinquapin)

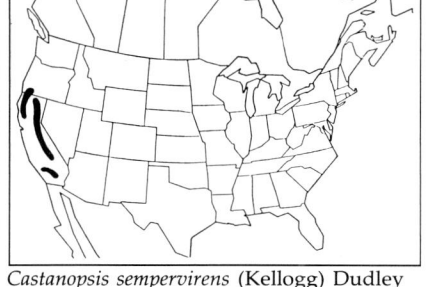
Castanopsis sempervirens (Kellogg) Dudley
bush or Sierra chinquapin

Quercus arizonica Sargent
Arizona white oak

Castanea dentata (Marshall) Borkhausen
American chestnut (uncommon in natural range)

Fagus grandifolia Ehrhart
American beech

Quercus arkansana Sargent
Arkansas oak

Castanea ozarkensis Ashe
Ozark chestnut (chinquapin)

Lithocarpus densiflorus (Hooker & Arnott) Rehder
tanbark oak

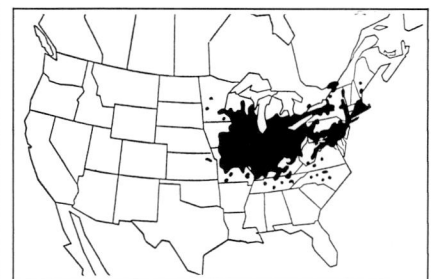
Quercus bicolor Willdenow
swamp white oak

Castanea pumila (L.) Miller
Allegheny chestnut (chinquapin)

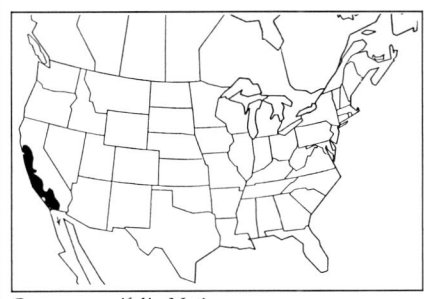
Quercus agrifolia Neé
California live oak

Quercus chapmanii Sargent
Chapman's oak

Castanopsis chrysophylla (Douglas) A. DC.
giant or golden chinquapin

Quercus alba L.
white oak

Quercus coccinea Muenchhausen
scarlet oak

APPENDIX 1

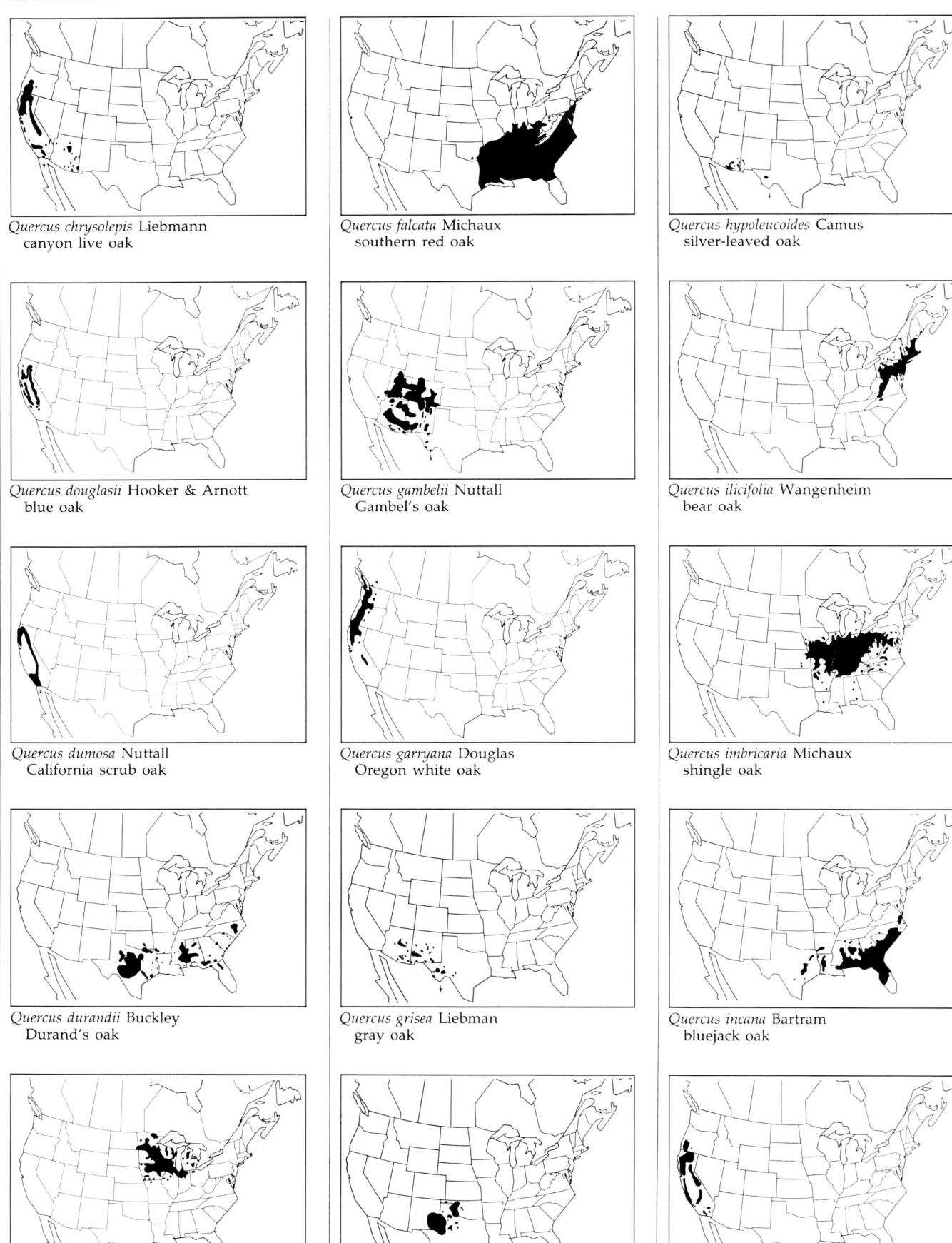

DISTRIBUTION MAPS OF SPECIES

Quercus laevis Walter (syn. *Q. catesbaei* Michaux)
turkey oak

Quercus marilandica Muenchhausen
blackjack oak

Quercus nuttallii Palmer
Nuttall's oak

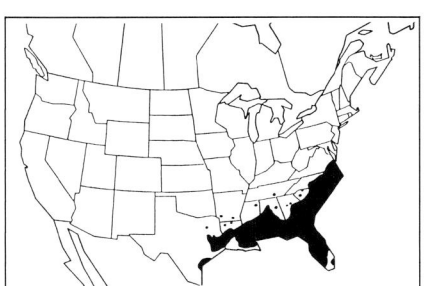
Quercus laurifolia Michaux
laurel oak

Quercus michauxii Nuttall
swamp chestnut oak

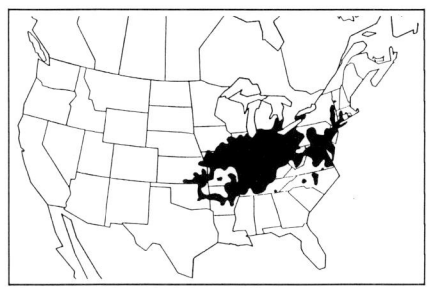
Quercus palustris Muenchhausen
pin oak

Quercus lobata Neé
California white oak

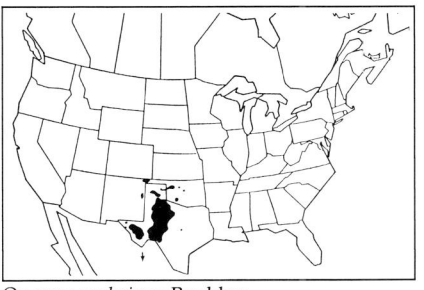
Quercus mohriana Buckley
Mohr's oak

Quercus phellos L.
willow oak

Quercus lyrata Walter
overcup oak

Quercus myrtifolia Willdenow
myrtle oak

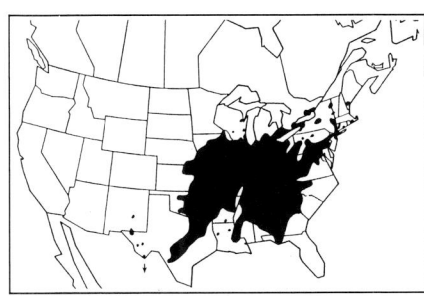
Quercus prinoides Willdenow (syn. *Q. muehlengergii* Engelmann)
chinquapin oak, dwarf chestnut oak

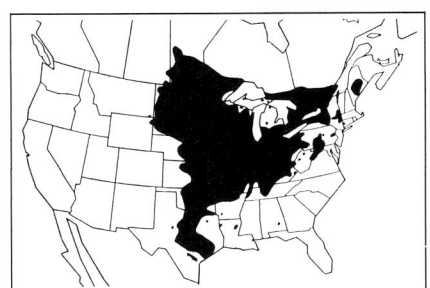
Quercus macrocarpa Michaux
bur oak

Quercus nigra L.
water oak

Quercus prinus L.
chestnut oak

APPENDIX 1

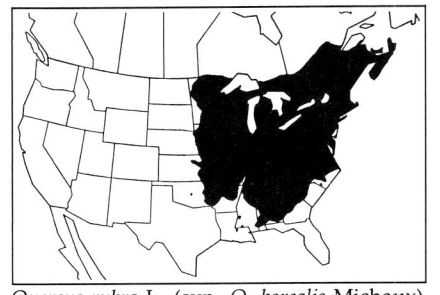

Quercus rubra L. (syn. *Q. borealis* Michaux)
 northern red oak

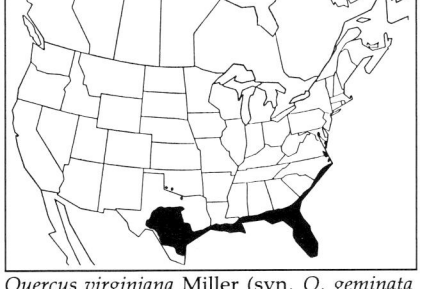

Quercus virginiana Miller (syn. *Q. geminata* Small)
 live oak

Garrya flavescens Watson

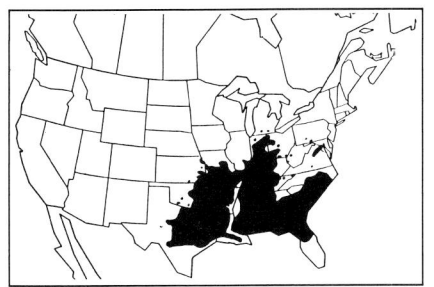

Quercus shumardii Buckley
 Shumard's oak

Quercus wislizenii A. DC.
 interior live oak

Garrya fremontii Torrey
 Fremont's silk-tassel

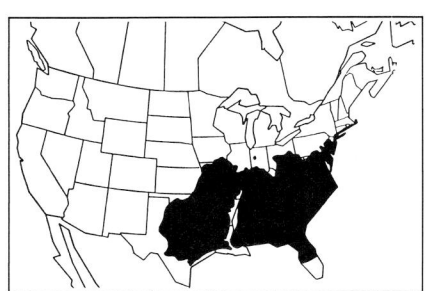

Quercus stellata Wangenheim
 post oak

GARRYACEAE

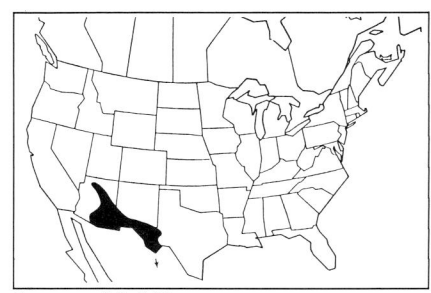

Garrya wrightii Torrey
 Wright's silk-tassel

INDIGENOUS SPECIES

HAMAMELIDACEAE

INTRODUCED SPECIES

Hamamelis japonica Siebold & Zuccarini
 Japanese witch-hazel, Asian, Zone 6

Hamamelis mollis D. Oliver
 Chinese witch-hazel, Asian, Zone 5

Garrya buxifolia Gray
 box-leaved silk-tassel

INDIGENOUS SPECIES

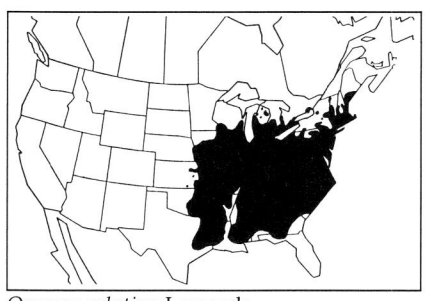

Quercus turbinella Greene
 shrub live oak

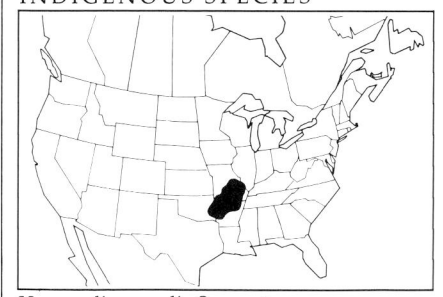

Hamamelis vernalis Sargent

Quercus velutina Lamarck
 black oak

Garrya elliptica Douglas
 waxy-leaved silk-tassel

DISTRIBUTION MAPS OF SPECIES
INDIGENOUS SPECIES

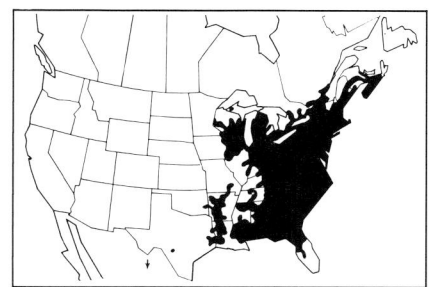

Hamamelis virginiana L.
witch-hazel

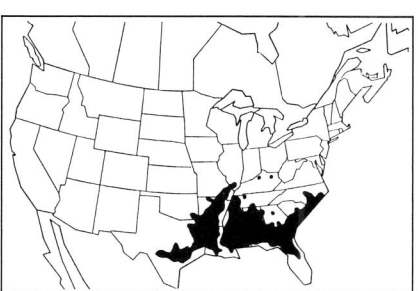

Aesculus octandra Marshall
sweet or yellow buckeye

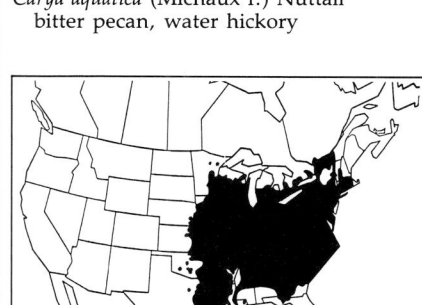

Carya aquatica (Michaux f.) Nuttall
bitter pecan, water hickory

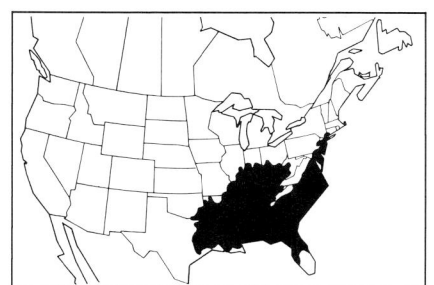

Liquidambar styraciflua L.
sweet gum

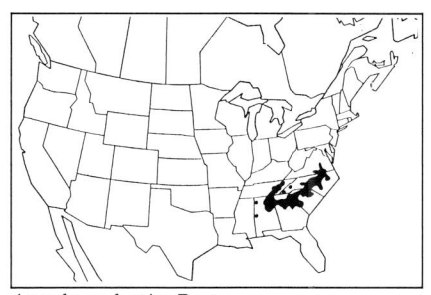

Aesculus pavia L.
red buckeye

Carya cordiformis (Wangenheim) K. Koch
bitternut hickory

HIPPOCASTANACEAE

INTRODUCED SPECIES

Aesculus hippocastanum L.
  horse chestnut, European, Zone 4

Aesculus sylvatica Bartram
dwarf or painted buckeye

Carya floridana Sargent
scrub hickory

INDIGENOUS SPECIES

Aesculus californica (Spach) Nuttall
California buckeye

JUGLANDACEAE

INTRODUCED SPECIES AND HYBRIDS

Carya × *laneyi* Sargent (*C. ovata* × *C. cordiformis*)

Juglans ailanthifolia Carrière
  Japanese walnut, Asian, Zone 5

Juglans cathayensis Dode
  Chinese walnut, Asian, Zone 6

Juglans regia L.
  English, Madeira or Persian walnut, Eurasian, Zone 7

Pterocarya × *rehderiana* C.K. Schneider (*P. fraxinifolia* × *P. stenoptera*)
  hybrid, Zone 5

Pterocarya stenoptera DC.
  Chinese wingnut, Asian, Zone 7

Carya glabra (Miller) Sweet
pignut hickory

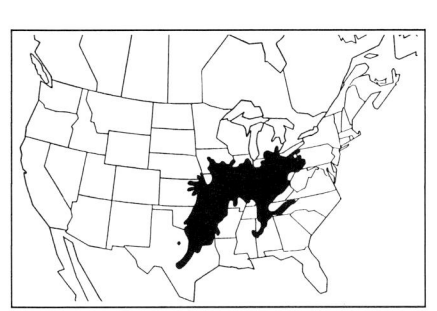

Aesculus glabra Willdenow
fetid or Ohio buckeye

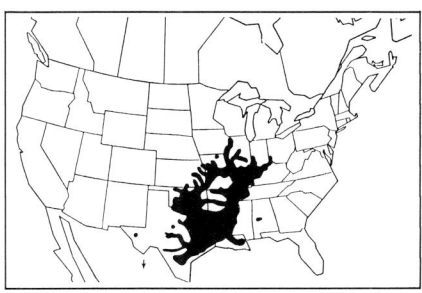

Carya illinoensis (Wangenheim) K. Koch
pecan

APPENDIX 1

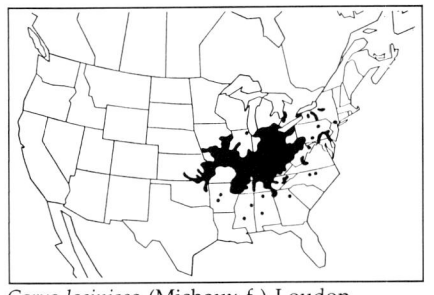

Carya laciniosa (Michaux f.) Loudon
shellback hickory

Carya tomentosa Nuttall
mockernut hickory

Juglans microcarpa Berlandier
little walnut

Carya myristicaeformis (Michaux f.) Nuttall
nutmeg hickory

Juglans californica Watson
California walnut

Juglans nigra L.
black walnut

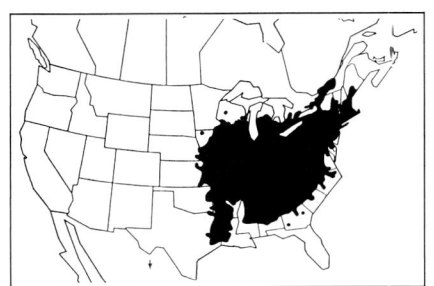

Carya ovata (Miller) K. Koch
shagbark hickory

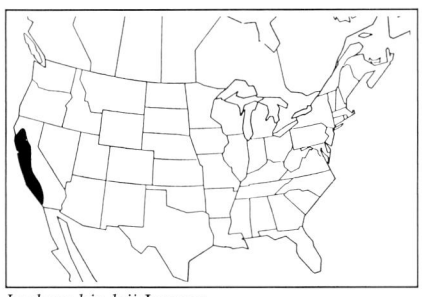

Juglans cinerea L.
butternut

LAURACEAE

INTRODUCED SPECIES

Cinnamomum camphora (L.) Presl
camphor tree, Asian, Zone 9

Cinnamomum zeylandicum Blume
cinnamon tree, Asian, Zone 10

Persea americana Miller
avocado, tropical American, Zone 10

INDIGENOUS SPECIES

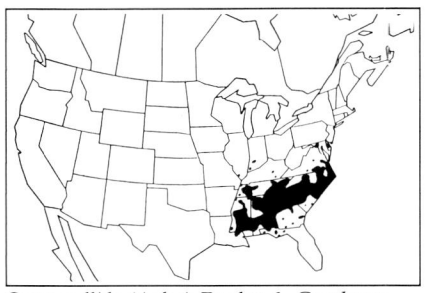

Carya pallida (Ashe) Engler & Graebner
sand hickory

Juglans hindsii Jepson
Hinds' walnut

Carya texana Buckley
black hickory

Juglans major (Torrey) Heller
Arizona walnut

Lindera benzoin (L.) Blume
spicebush

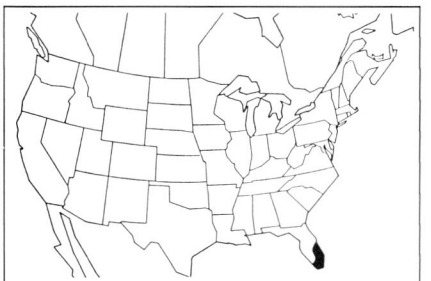

Nectandra coriacea (Sweet) Grisebach
Jamaica nectandro

DISTRIBUTION MAPS OF SPECIES

Persea barbonia (L.) Sprengel
red or swamp bay

Sassafras albidum (Nuttall) Nees
sassafras

Umbellularia californica (Hooker & Arnott) Nuttall
California bay or laurel, Pacific myrtle

LEITNERIACEAE

INDIGENOUS SPECIES

Leitneria floridana Chapman
Florida corkwood

MAGNOLIACEAE

INTRODUCED SPECIES AND HYBRID

Magnolia kobus DC. (syn. *M. stellata* Seibold & Zuccarini) Maximowicz
star magnolia, Asian, Zone 5 (south)

Magnolia × *soulangiana* Soulange-Bodin
Chinese or saucer magnolia

INDIGENOUS SPECIES

Liriodendron tulipifera L.
tulip tree, yellow poplar

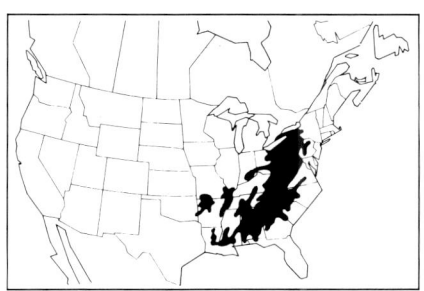

Magnolia acuminata L.
cucumber tree

Magnolia grandiflora L.
bull bay, southern magnolia

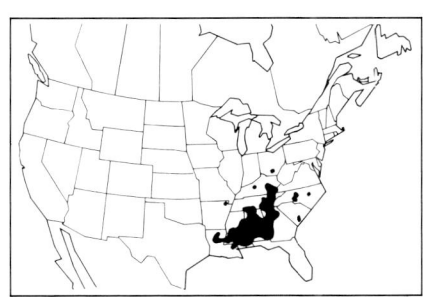

Magnolia macrophylla Michaux
large-leaved cucumber tree

Magnolia tripetala L.
umbrella magnolia

Magnolia virginiana L.
sweet bay

MORACEAE

INTRODUCED SPECIES

Broussonetia papyrifera (L.) Ventenat
paper mulberry, Asian, Zone 6

Cudrania tricuspidata (Carrière) Lavallée
Asian, Zone 7

Morus alba L.
white mulberry, Asian, Zone 5

Morus nigra L.
black mulberry, Asian, Zone 5

INDIGENOUS SPECIES

Maclura pomifera (Rafinesque) Schneider
Osage orange

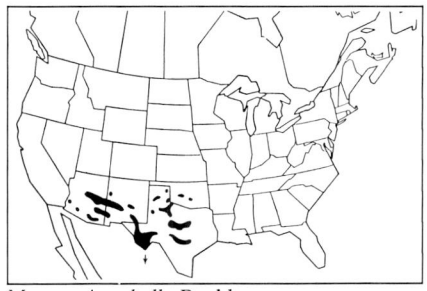

Morus microphylla Buckley
mountain or Texas mulberry

APPENDIX 1

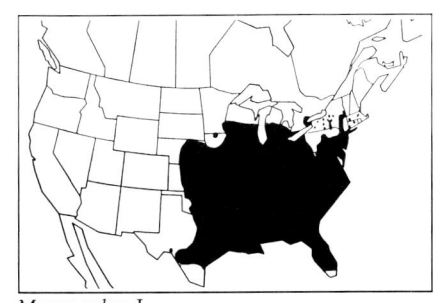

Morus rubra L.
red mulberry

MYRICACEAE

INTRODUCED SPECIES

Myrica faya Aiton
candleberry myrtle, Mediterranean and Canary Is., Zone 10

Myrica rubra Siebold & Zuccarini
Asian, Zone 8

INDIGENOUS SPECIES

Comptonia peregrina (L.) J. Coulter
sweet fern

Myrica californica Chamisso & Schlechtendal
California bayberry or wax-myrtle

Myrica cerifera L.
southern bayberry or wax-myrtle

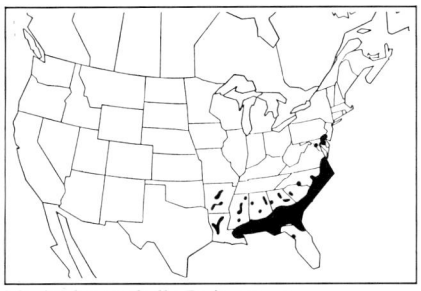

Myrica gale L.
sweet gale

Myrica heterophylla Rafinesque
evergreen bayberry

Myrica inodora Bartram
odorless bayberry

Myrica pensylvanica Loiseleur-Deslongschamps
northern bayberry

MYRTACEAE

INTRODUCED SPECIES

Callistemon citrinus (Curtis) Stapf
crimson bottlebrush, Australian, Zone 9

Callistemon viminalis (Gaertner) Cheel
weeping bottlebrush, Australian, Zone 9

Eucalyptus camaldulensis Dehnhardt (syn. *E. rostrata* Schlechtendal)
Murray red gum, Australian, Zone 9

Eucalyptus globulus Labillardière
blue gum, Australian, Zone 9 (many additional species cultivated Zones 9 and 10)

Eucalyptus rudis Endlicher
desert gum, Australian Zone 9

Eucalyptus torquata Luehmann
coral gum, Australian, Zone 9

Melaleuca decora (Salisbury) Britton (syn. *M. genistifolia* Smith)
Australian, Zone 9

Melaleuca quinquenervia (Cavanilles) S.T. Blake
cajeput, paperbark tree, Australian, Zone 9

NYSSACEAE

INDIGENOUS SPECIES

Nyssa sylvatica Marshall
black or sour gum, black tupelo

OLEACEAE

INTRODUCED SPECIES

Chionanthus retusus Lindley & Paxton
Chinese fringe tree, Asian, Zone 6

Forsythia × *intermedia* Zabel
forsythia, golden bells, Asian, Zone 5

Forsythia suspensa (Thunberg) Vahl
forsythia, golden bells, Asian, Zone 5

DISTRIBUTION MAPS OF SPECIES

Fraxinus angustifolia Vahl
 narrow-leaved ash, Mediterranean, Zone 7

Fraxinus excelsor L.
 European ash, Eurasian, Zone 4

Fraxinus ornus L.
 flowering or manna ash, Eurasian, Zone 6

Fraxinus uhdei (Wenzig) Lingelsheim
 evergreen ash, Mexican, Zone 8

Jasminum beesianum Forrest & Diels (and other species)
 jasmine, Asian, Zone 7

Ligustrum japonicum Thunberg
 Japanese or wax-leaved privet, Asian, Zone 7

Ligustrum lucidum Aiton
 Chinese or glossy privet, Asian, Zone 8

Ligustrum obtusifolium Siebold & Zuccarini
 Asian, Zone 4

Ligustrum ovalifolium Hasskarl
 California privet, Asian, Zone 6

Ligustrum sinensis Louriero
 Chinese privet, Asian, Zone 7

Ligustrum vulgare L.
 common privet, Mediterranean, Zone 5

Olea europaea L.
 common olive, Mediterranean, Zone 9

Osmanthus heterophyllus (G. Don) P.S. Green (syn. *O. ilicifolius* (Hasskarl) Carrière)
 Chinese or holly olive, Asian, Zone 7

Syringa reticulata (Blume) Hara
 Japanese tree lilac, Asian, Zone 4

Syringa vulgaris L.
 common lilac, Asian, Zone 4

INDIGENOUS SPECIES

Forestiera phillyreoides (Bentham) Torrey
 desert olive forestiera

Forestiera segregata (Jacquin) Krug & Urban
 Florida privet

Fraxinus americana L.
 white ash

Chionanthus virginicus L.
 fringe tree, old man's beard

Fraxinus anomala Torrey
 single-leaved ash

Forestiera acuminata (Michaux) Poiret
 swamp privet

Fraxinus berlandierana A. DC.
 Berlandier's ash

Fraxinus caroliniana Miller
 Carolina ash

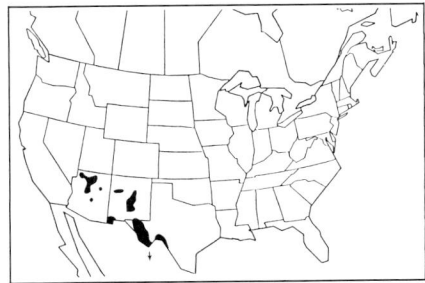

Fraxinus cuspidata Torrey
 fragrant ash

Fraxinus dipetala Hooker & Arnott
 two-petal ash

Fraxinus latifolia Bentham
 Oregon ash

Fraxinus nigra Marshall
 black ash

201

APPENDIX 1

Fraxinus pennsylvanica Marshall
green ash

Osmanthus americanus (L.) Gray
devilwood, American olive

PLATANACEAE

INTRODUCED SPECIES

Platanus × *acerifolia* (Aiton) Willdenow
London planetree, European (hybrid),
Zone 5

Platanus orientalis L.
Oriental planetree, Asia, Zone 6

RHAMNACEAE

INDIGENOUS SPECIES

Ceanothus americanus L.
New Jersey tea

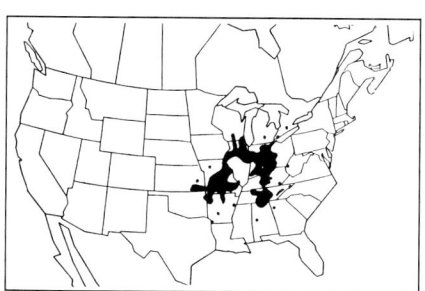
Fraxinus profunda (Bush) Bush
pumpkin ash

INDIGENOUS SPECIES

Platanus occidentalis L.
American sycamore

Fraxinus quadrangulata Michaux
blue ash

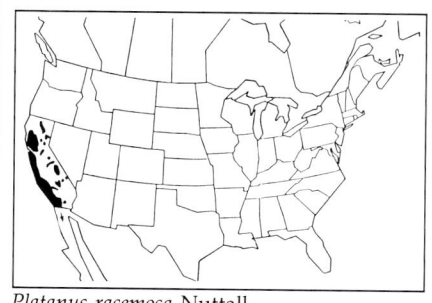
Platanus racemosa Nuttall
California sycamore

Ceanothus sanguineus Pursh
wild lilac, Oregon tea-tree

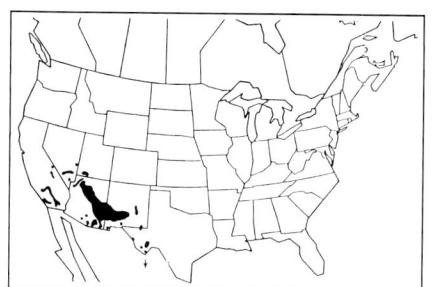
Fraxinus texensis Sargent
Texas ash

Platanus wrightii Watson
Arizona sycamore

Ceanothus thyrsiflorus Eschscholtz
blueblossom

Fraxinus velutina Torrey
velvet ash

Ceanothus velutinus Douglas
tobacco brush

202

ROSACEAE

INTRODUCED SPECIES AND HYBRIDS

Amelanchier ovalis Medicus
  European, Zone 5

Crataegus laevigata (Poiret) DC.
  English hawthorn, quick-set thorn, Zone 5

Crataegus monogyna Jacquin
  English hawthorn, Eurasian, Zone 5

Malus baccata (L.) Borkhausen
  Siberian crabapple, Asian, Zone 2

Malus floribunda Van Houtte
  showy crabapple, Asian, Zone 5

Malus pumila Miller
  common apple, Eurasian, Zone 4

Prunus armenica L.
  apricot, Asian, Zone 6

Prunus avium (L.) L.
  sweet cherry, Eurasian, Zone 4

Prunus domestica L.
  plum, Eurasian, Zone 5

Prunus dulcis (Miller) D.A. Webb
  almond (bitter and sweet), Asian, Zone 7

Pyrus calleryana Decnesne
  Callary pear, Asian, Zone 6

Pyrus communis L.
  common pear, Eurasian, Zone 5

Rosa alba L.
  white rose, Eurasian, Zone 5

Rosa canina L.
  brier or dog rose, European, Zone 4

Rosa eglanteria L.
  sweetbrier, Eurasian, Zone 6

Rosa rugosa Thunberg
  Japanese or Turkestan rose, Asian, Zone 2

Rosa, garden roses
  hybrid teas and floribundas, Eurasian, Zone 5

Sanguisorba minor Scopoli
  burnet, Eurasian, Zone 4

Sorbus aucuparia L.
  European mountain ash, rowan, European, Zone 2

Sorbus domestica L.
  service tree, European, Zone 6

Spiraea prunifolia Siebold & Zuccarini
  bridal wreath, Asian, Zone 5

Spiraea × vanhouttei (C. Briot) Zabel
  hybrid bridal wreath, Zone 5

INDIGENOUS SPECIES

DISTRIBUTION MAPS OF SPECIES

Acaena californica Bitter

Amelanchier sanguinea (Pursh) DC.
round-leaved serviceberry

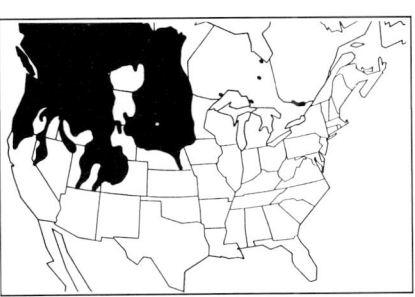
Amelanchier alnifolia Nuttall
Saskatoon serviceberry

Amelanchier utahensis Koehne
Utah serviceberry

Amelanchier arborea (Michaux f.) Fernald
downy serviceberry

Cercocarpus betuloides Torrey & Gray

Amelanchier canadensis (L.) Medicus
Canada serviceberry

Cercocarpus ledifolius Nuttall
curly-leaved mountain mahogany

Amelanchier interior Nielson
inland serviceberry

Cercocarpus montanus Rafinesque

203

APPENDIX 1

Cowania mexicana D. Don
cliffrose

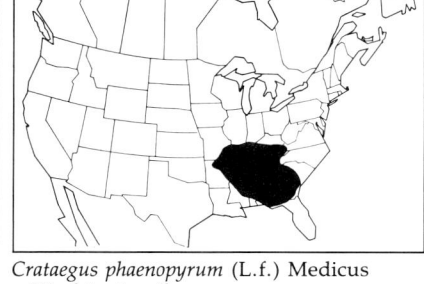
Crataegus phaenopyrum (L.f.) Medicus
Washington thorn

Holodiscus discolor (Pursh) Maximowicz
creambush, ocean spray

Crataegus calpodendron (J.E. Ehrhart) Medicus
blackthorn

Crataegus pruinosa (H.L. Wendland) C. Koch
frosted hawthorn

Malus angustifolia (Aiton) Michaux
southern crabapple

Crataegus crus-galli L.
cockspur thorn

Crataegus punctata Jacquin
dotted thorn

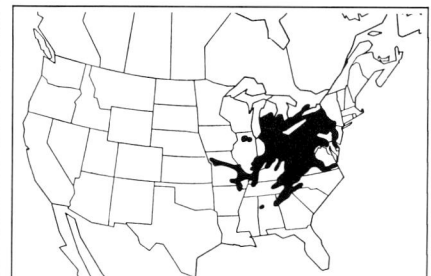
Malus coronaria (L.) Miller
sweet crabapple

Crataegus douglasii Lindley
black hawthorn

Crataegus succulenta Link
fleshy hawthorn

Malus fusca (Rafinesque) C.K. Schneider
Oregon crabapple

Crataegus mollis (Torrey & Gray) Scheele
dawny hawthorn

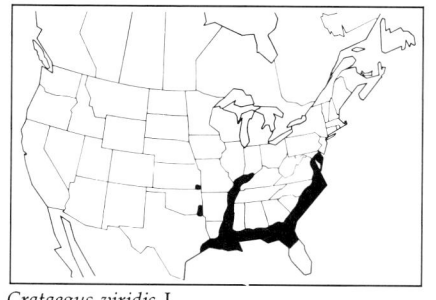
Crataegus viridis L.
green hawthorn

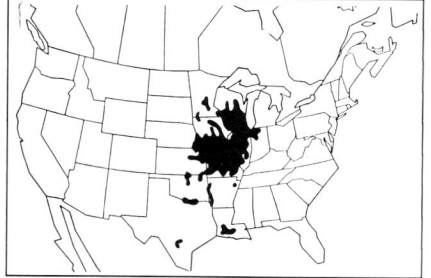
Malus ioensis (A. Wood) Britton
prairie crabapple, wild crab

DISTRIBUTION MAPS OF SPECIES

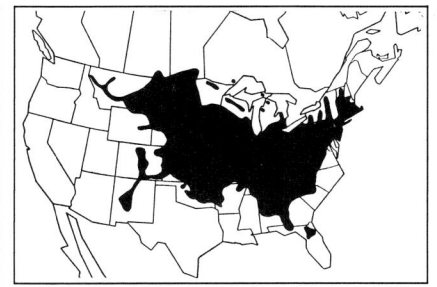

Prunus americana Marshall
American or wild plum

Prunus ilicifolia (Nuttall) Walpers
holly-leaved cherry

Prunus serotina Ehrhart
black cherry

Prunus angustifolia Marshall
chickasaw or sand plum

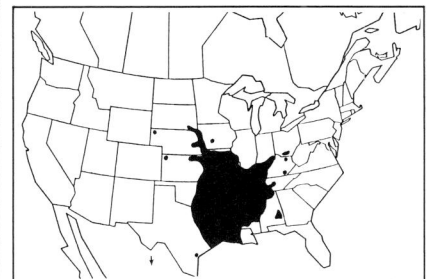

Prunus mexicana Watson
Mexican plum

Prunus subcordata Bentham
Pacific or Sierra plum

Prunus caroliniana (Miller) Aiton
Carolina laurelcherry

Prunus munsoniana F.W. Wight & Hedrick
wildgoose plum

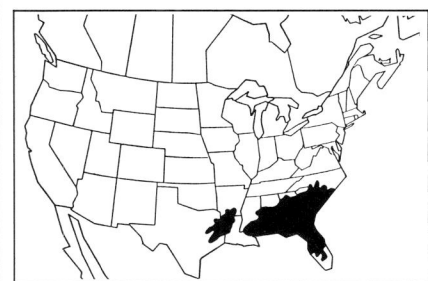

Prunus umbellata Elliott
flatwood plum

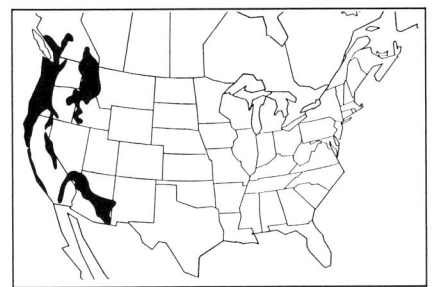

Prunus emarginata (Hooker) Walpers
bitter or Oregon cherry

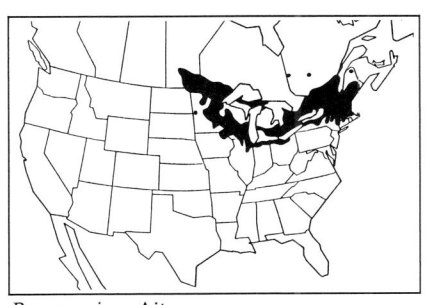

Prunus nigra Aiton
Canada plum

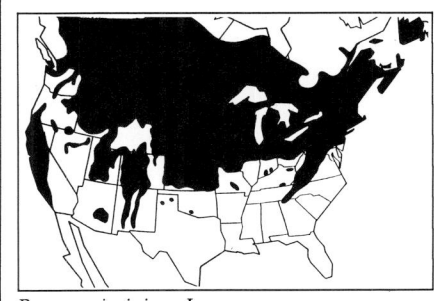

Prunus virginiana L.
chokecherry

Prunus hortulana Bailey
hortulan plum

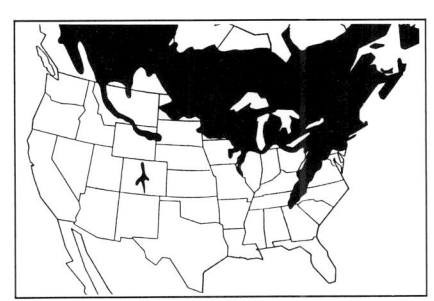

Prunus pensylvanica L.f.
pin or wild-red cherry

Rosa setigera Michaux
climbing or prairie rose

APPENDIX 1

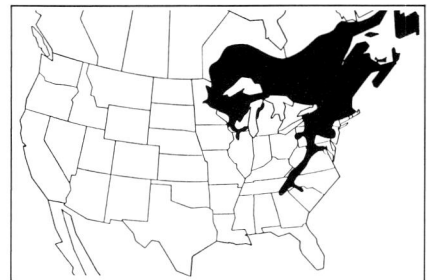

Sorbus americana Marshall
American mountain ash

Sorbus decora (Sargent) C.K. Schneider
showy mountain ash

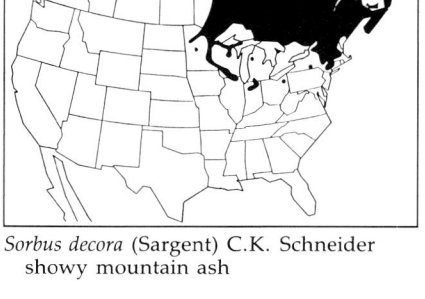

Sorbus sitchensis M.J. Roemer
Sitka mountain ash

Spiraea alba Du Roi
meadowsweet

Spiraea corymbosa Rafinesque

Spiraea densiflora Rydberg

Spiraea douglasii Hooker

Spiraea tomentosa L.
hardback, steeplebush

RUTACEAE

INTRODUCED SPECIES

Citrus aurantiifolia (Christmann) Swingel (lime), *C. aurantium* L. (sour, bitter or Seville orange), *C. limon* (L.) Burman f. (lemon), *C. medica* L. (citron), *C.* × *paradisi* Macfady (grapefruit), *C. sinensis* (L.) Osbeck (sweet orange) Asian, Zone 10 (9)

Phellodendron amurense Ruprecht
cork tree, Asian, Zone 4

Phellodendron chinensis Schneider
cork tree, Asian, Zone 4

INDIGENOUS SPECIES

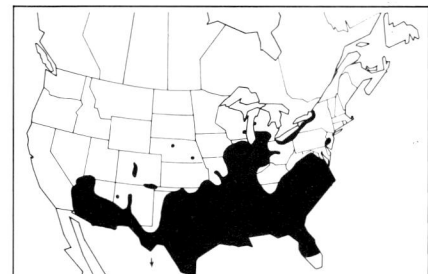

Ptelea trifoliata L.
stinking or water ash, hop tree

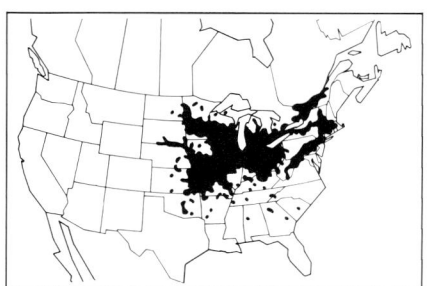

Zanthoxylum americanum Miller
prickly ash, toothache tree

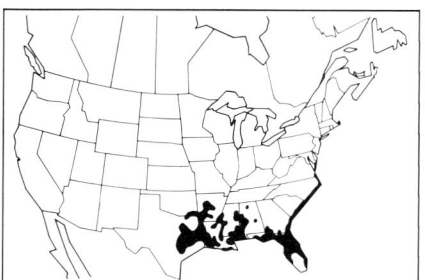

Zanthoxylum clava-herculis L.
Hercules'-club

Zanthoxylum fagara (L.) Sargent

Zanthoxylum hirsutum Buckley
Texas Hercules'-club

DISTRIBUTION MAPS OF SPECIES

Zanthoxylum pterota HBK

SALICACEAE

INTRODUCED SPECIES AND HYBRIDS

Populus × *acuminata* Rydberg (*P. angustifolia* × *P. deltoides*)
  hybrid, Zone 3

Populus alba L.
  silver or white poplar, abele, Eurasian, Zone 4

Populus × *canadensis* Moench (*P. deltoides* × *P. nigra*)
  hybrid, Zone 4

Populus nigra L.
  black and Lombardy poplar, Eurasian, Zone 3

Salix alba L.
  white willow, Mediterranean & Asian, Zone 2

Salix babylonica L.
  weeping willow, Asian, Zone 5

Salix fragilis L.
  brittle or crack willow, Eurasian, Zone 5

Salix purpurea L.
  purple osier, Mediterranean & Asian, Zone 5

INDIGENOUS SPECIES

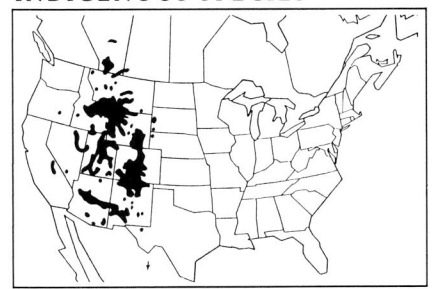

Populus angustifolia James
  narrow-leaved cottonwood

Populus arizonica Sargent
  Arizona cottonwood

Populus balsamifera L.
  balsam poplar

Populus deltoides Bartram (syn. *P. sargentii* Dode)
  eastern and plains cottonwood

Populus fremontii Watson
  Fremont's cottonwood

Populus grandidentata Michaux
  bigtooth aspen

Populus heterophylla L.
  swamp cottonwood

Populus tremuloides Michaux
  quaking aspen

Populus trichocarpa Torrey & Gray
  black cottonwood

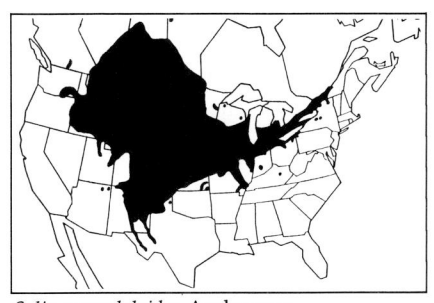

Salix amygdaloides Andersson
  peach-leaved willow

Salix arbusculoides Andersson
  littletree willow

APPENDIX 1

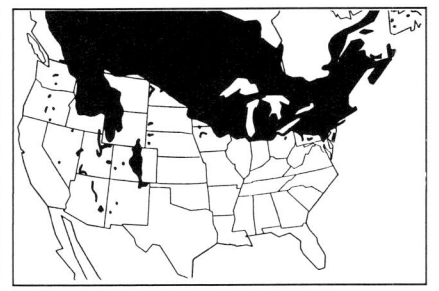
Salix bebbiana Sargent
long-beaked willow

Salix gooddingii Ball
Goodding's willow

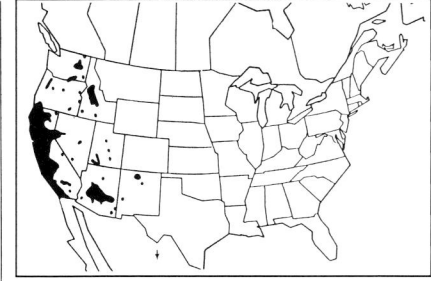
Salix lasiolepis Bentham
arroyo willow

Salix caroliniana Michaux
coastal plain willow

Salix hindsiana Bentham
Hinds' willow

Salix lucida Muhlenberg
shining willow

Salix discolor Muhlenberg
pussy willow

Salix hookerana Barratt
coastal or Hooker's willow

Salix mackenzieana (Hooker) Barratt
Mackenzie's willow

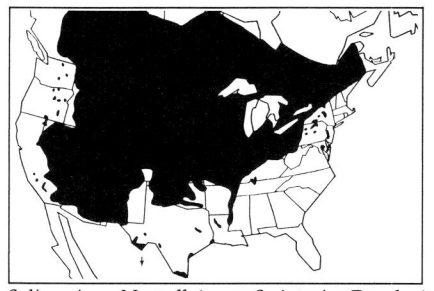
Salix exigua Nuttall (syn. *S. interior* Rowlee)
coyote willow

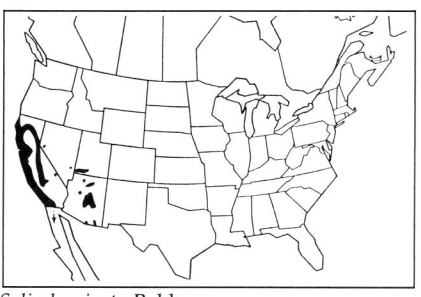
Salix laevigata Bebb
red willow

Salix monticola Bebb
serviceberry willow

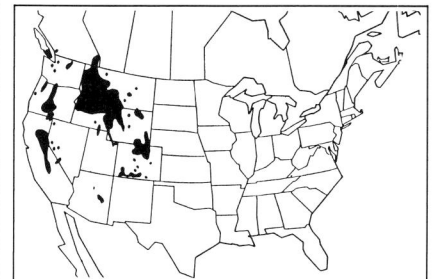
Salix geyeriana Andersson
Geyer's willow

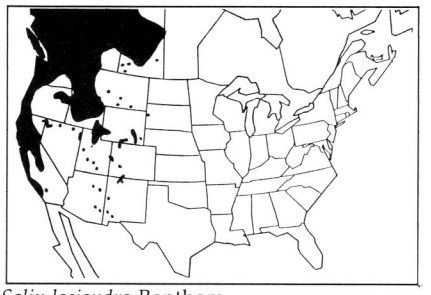
Salix lasiandra Bentham
western shining willow

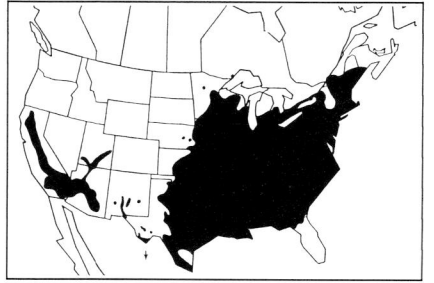
Salix nigra Marshall
black willow

DISTRIBUTION MAPS OF SPECIES

SAXIFRAGACEAE

INTRODUCED SPECIES

Hydrangea macrophylla (Thunberg) Seringe
French hydrangea, hortensia, Asian,
Zone 7

Hydrangea paniculata Siebold
peegee hydrangea, Asian, Zone 4

Philadelphus coronarius L.
mock-orange, Eurasian, Zone 5

Ribes nigrum L.
black currant, Eurasian, Zone 5

Ribes sativum (Reinchenback) Syme
common currant, European, Zone 5

Ribes uva-crispa L.
English gooseberry, European, Zone 5

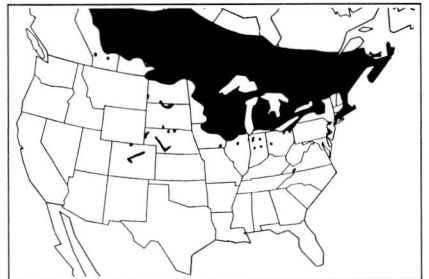

Salix pellita Andersson
satiny willow

Salix petiolaris Smith
meadow willow

Salix scouleriana Barratt
Scouler's willow

Salix sitchensis Sanson
Sitka willow

Salix taxifolia HBK
yew-leaved willow

Salix tracyi Ball
Tracy's willow

INDIGENOUS SPECIES

Hydrangea arborescens L. (syn. *H. cinerea*
Small, *H. radiata* Walter)
wild hydrangea

SAPINDACEAE

INTRODUCED SPECIES

Koelreuteria elegans (Seemann) A.C. Smith
flamegold tree, Pacific Is., Zone 9

Koelreuteria paniculata Laxmann
golden-rain or varnish tree, Asian,
Zone 5

INDIGENOUS SPECIES

Salix sericea Marshall
silky willow

Sapindus drummondii Hooker & Arnott
western soapberry

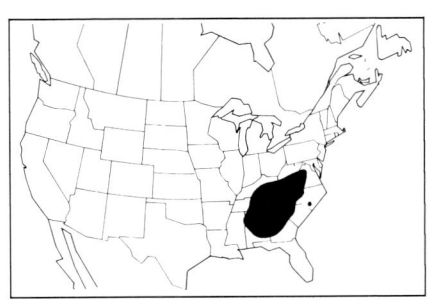

Hydrangea quercifolia Bartram
oak-leaved hydrangea

Philadelphus inodorus L. (syn. *P. grandiflorus*
Willdenow)

Salix sessilifolia Nuttall
northwest willow

APPENDIX 1

Philadelphus lewisii Pursh (syn. *P. californicus*
  Bentham)
  Lewis' mock-orange

Philadelphus pubescens Loiseleur

Ribes aureum Pursh
  golden or Missouri currant

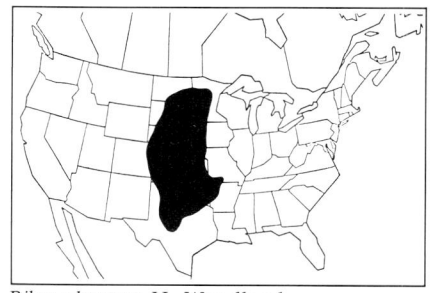

Ribes odoratum H. Wendland
  buffalo currant

Ribes viscossissimum Pursh
  sticky currant

SIMAROUBACEAE

INTRODUCED SPECIES

Ailanthus altissima (Miller) Swingle
  tree-of-heaven, Asian, Zone 5

INDIGENOUS SPECIES

Alvaradoa amorphoides Liebmann

TAMARICACEAE

INTRODUCED SPECIES

Tamarix aphylla (L.) Karsten
  Eurasian, Zone 7

Tamarix chinensis Louriero
  Chinese tamarisk or salt cedar, Asian,
  Zone 6

Tamarix gallica L.
  French tamarisk, manna, Mediterranean,
  Zone 6

Tamarix parviflora DC. (syn. *T. tetrandra*
  Pallas)
  European, Zone 7

Tamarix ramosissima Ledebour (syn. *T.
  pentandra* Pallas)
  Eurasian, Zone 6

TILIACEAE

INTRODUCED SPECIES

Tilia cordata Miller
  small-leaved European linden, European,
  Zone 4

Tilia platyphyllos Scopoli
  large-leaved linden, European, Zone 4

INDIGENOUS SPECIES

Tilia americana L.
  American basswood or linden

Tilia caroliniana Miller
  Carolina basswood or linden

Tilia floridana Small
  Florida basswood or linden

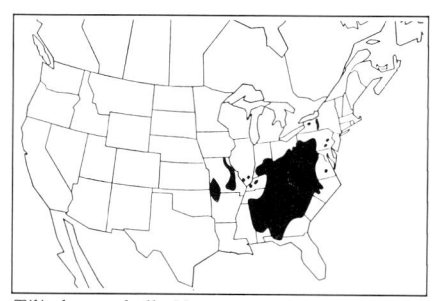

Tilia heterophylla Ventenat
  white basswood or linden

ULMACEAE

INTRODUCED SPECIES

Celtis australis L.
  European hackberry, Mediterranean,
  Zone 7

Ulmus glabra Hudson
  Scotch or Wych elm, Eurasian, Zone 5

DISTRIBUTION MAPS OF SPECIES

Ulmus parvifolia Jacquin
  Chinese or evergreen elm, Asian, Zone 6 (fall-flowering)
Ulmus procera Salisbury (syn. *U. campestris* of authors)
  English elm, European, Zone 6
Ulmus pumila L.
  Chinese dwarf or Siberian elm, Asian, Zone 3

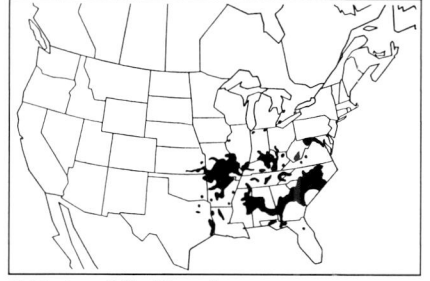
Celtis tenuifolia Nuttall
Georgia hackberry

Ulmus crassifolia Nuttall
cedar elm (fall-flowering)

INDIGENOUS SPECIES

Celtis laevigata Willdenow (syn. *C. mississippiensis* Bosc, *C. smallii* Beadle)
sugar or southern hackberry

Planera aquatica J.F. Gmelin
planer tree, water elm

Ulmus rubra Muhlenberg (syn. *U. fulva* Michaux)
red or slippery elm

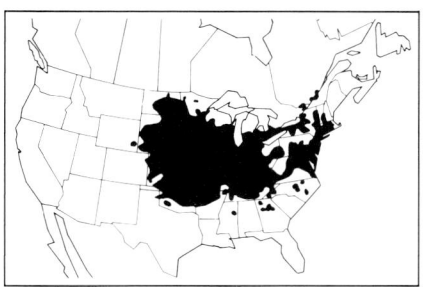
Celtis occidentalis L.
hackberry

Trema micrantha (L.) Blume (syn. *T. floridana* Britton)
Florida trema

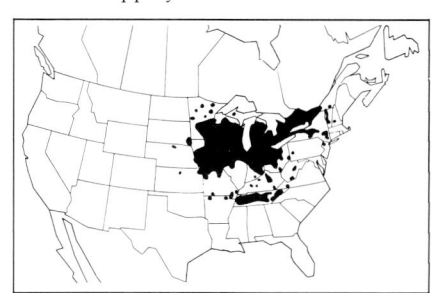
Ulmus thomasii Sargent (syn. *U. racemosa* Thomas)
cork or rock elm

Celtis pallida Torrey
desert hackberry

Ulmus alata Michaux
small-leaved or winged elm

Ulmus serotina Sargent
September elm (fall-flowering)

VISCACEAE

Celtis reticulata Torrey (syn. *C. douglasii* Planchon)
netleaf hackberry

Ulmus americana L. (syn. *U. floridana* Chapman)
American or white elm

APPENDIX 1
INDIGENOUS SPECIES

Arceuthobium douglasii Engelmann
  Douglas' dwarf mistletoe

Arceuthobium vaginatum (HBK) Eichler

VITACEAE

INTRODUCED SPECIES

Parthenocissus tricuspidata (Siebold & Zuccarini) Planchon
  Boston or Japanese ivy, Asian, Zone 5

Vitis vinifera L.
  European or wine grape, Eurasian, Zone 6 (south)

INDIGENOUS SPECIES

Parthenocissus quinquefolia (L.) Planchon
  Virginia creeper

Vitis aestivalis Michaux
  pigeon or summer grape

Vitis labrusca L.
  fox grape

Vitis riparia Michaux
  frost grape

Vitis rotundifolia Michaux
  muscadine grape

ZYGOPHYLLACEAE

INDIGENOUS SPECIES

Larrea divaricata Cavanilles
  creosote bush, greasewood

Grasses and Grasslike Plants

POACEAE

INTRODUCED SPECIES
■ (gray)

INDIGENOUS SPECIES
■ (black)

Agrostis elliottiana Schultes

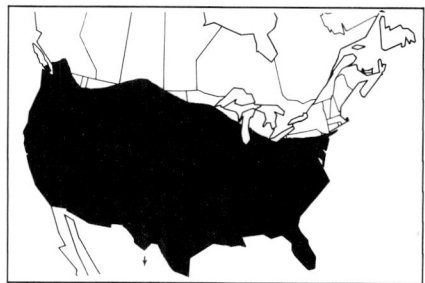

Alpopecurus carolinianus Walter
Carolina foxtail

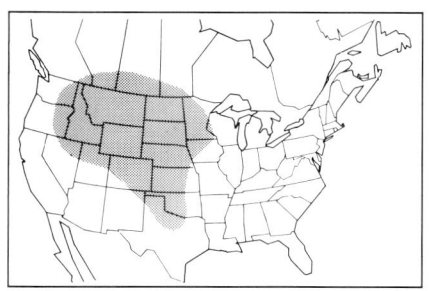

Agropyron cristatum (L.) Gaertner
crested wheatgrass, Eurasian

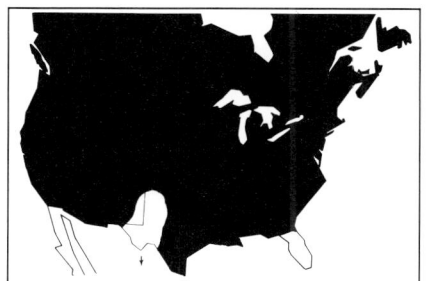

Agrostis hyemalis (Walter) Britton, Sterns & Poggenberg (syn. *A. scabra* Willdenow)
hairgrass, ticklegrass

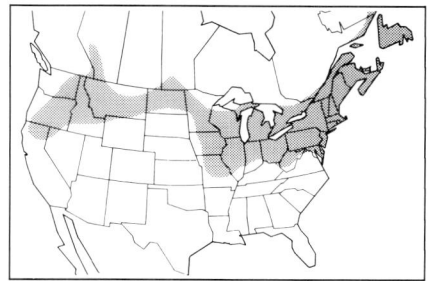

Alopecurus pratensis L.
meadow foxtail, Eurasian

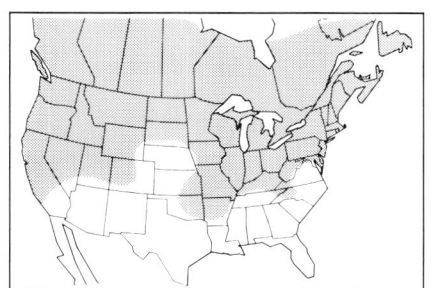

Agropyron repens (L.) Beauvois
couchgrass, quackgrass, Eurasian

Agrostis perennans (Walter) Tuckerman
autumn bent, upland bent

Ammophila arenaria (L.) Link
European beachgrass, European

Agropyron smithii Rydberg
western wheatgrass

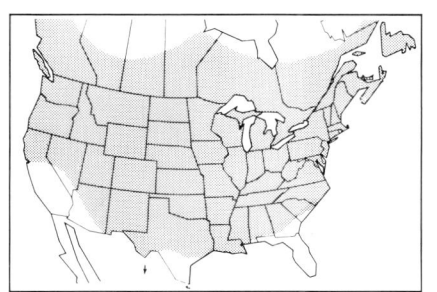

Agrostis stolonifera L. (syn. *A. alba* L. misapplied, *A. gigantea* Roth, *A. palustris* Hudson)
creeping bentgrass, redtop

Ammophila breviligulata Fernald
American beachgrass

APPENDIX 1

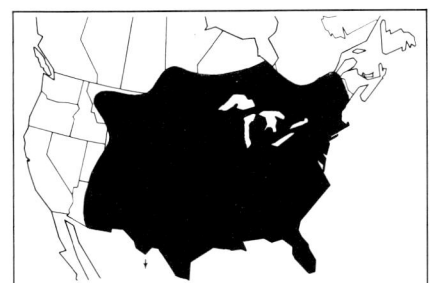
Andropogon gerardii Vitman
big bluestem

Avena barbata Brotero
slender oats, Old World

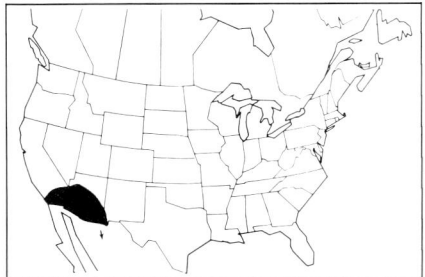
Bouteloua rothrockii Vasey
Rothrock grama

Andropogon virginicus L.
Virginia beardgrass

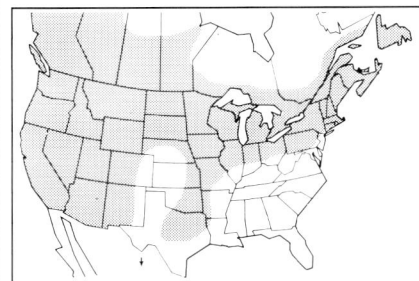
Avena fatua L.
wild oats, Eurasian

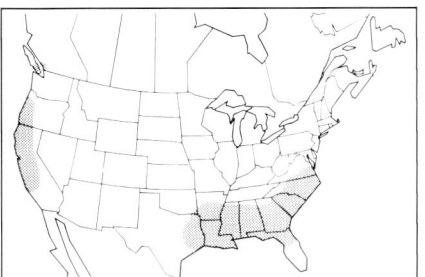
Briza minor L.
little quaking grass, European

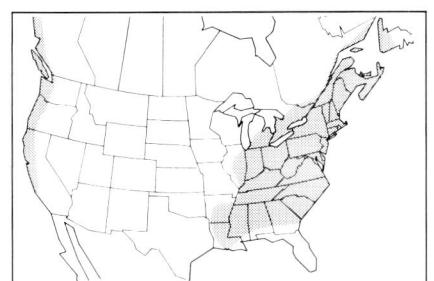
Anthoxanthum odoratum L.
sweet vernalgrass, Eurasian

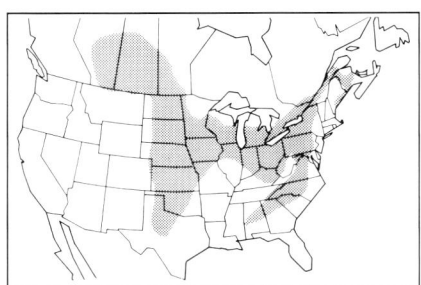
Avena sativa L.
oats (cultivated), Eurasian

Bromus carinatus Hooker & Arnott
California brome

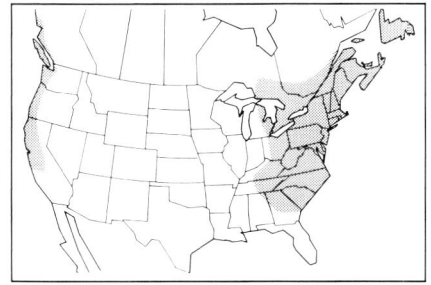
Arrhenatherum elatius (L.) Presl
tall oatgrass, Eurasian

Bouteloua curtipendula (Michaux) Torrey
side-oats or tall grama

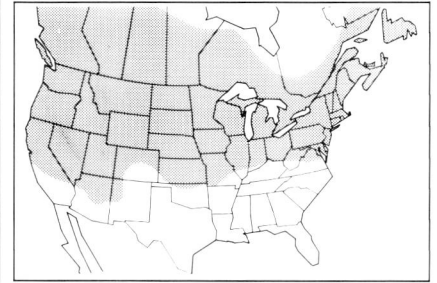
Bromus inermis Leysser
smooth or Hungarian brome, Eurasian

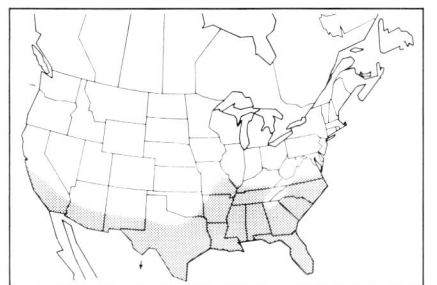
Arundo donax L.
giant reed, Old World

Bouteloua gracilis (HBK) Steudel
blue grama

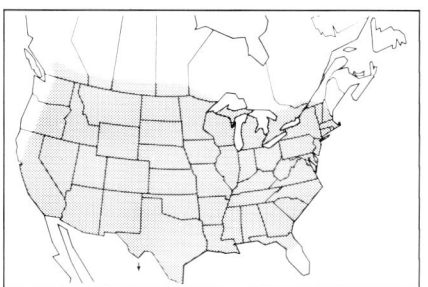
Bromus japonicus Thunberg
Japanese chess, Eurasian

DISTRIBUTION MAPS OF SPECIES

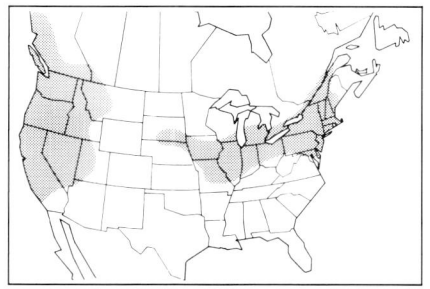

Bromus mollis L.
soft chess, Eurasian, northern North America

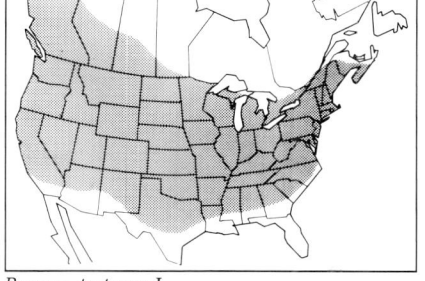

Bromus tectorus L.
downy chess, Eurasian

Chloris gayana Kunth
Rhodes grass, African

Bromus purgans L.
Canada brome

Buchloa dactyloides (Nuttall) Engelmann
buffalo grass

Cinna arundinacea L.
stout woodreed

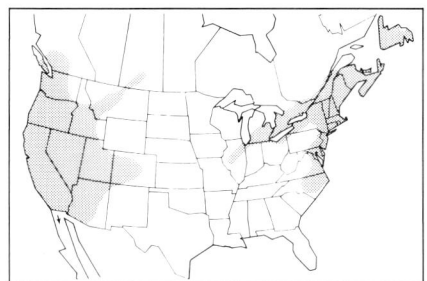

Bromus racemosus L.
Eurasian

Calamagrostis canadensis (Michaux) Nuttall
bluejoint

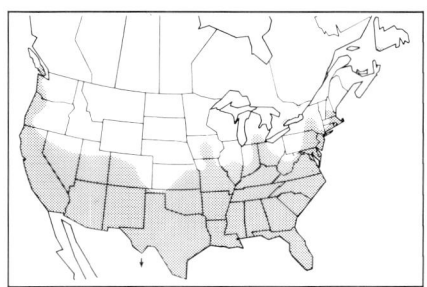

Cynodon dactylon (L.) Persoon
Bermuda grass, Eurasian

Bromus rigidus Roth
ripgut grass, European

Calamagrostis rubescens Buckley
pinegrass

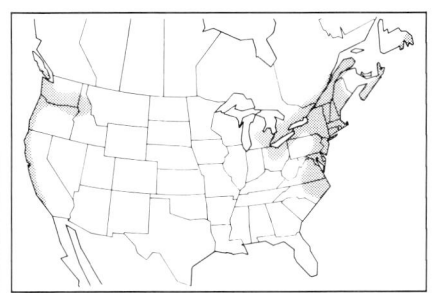

Cynosurus cristatus L.
crested dogtail, Eurasian

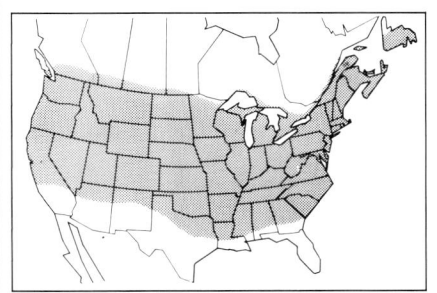

Bromus secalinus L.
chess, Eurasian

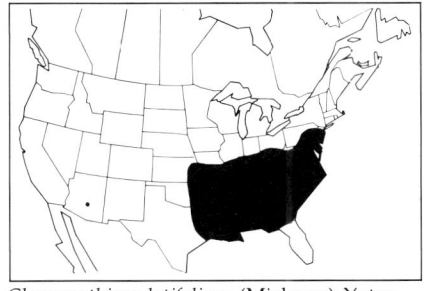

Chasmanthium latifolium (Michaux) Yates
broad-leaved sea oats

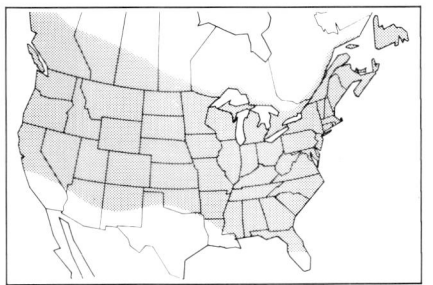

Dactylis glomerata L.
orchard grass, Eurasian

APPENDIX 1

Deschampsia caespitosa (L.) Beauvois
tufted hairgrass

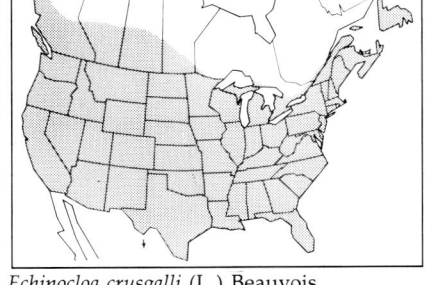
Echinocloa crusgalli (L.) Beauvois
barnyard grass, Eurasian

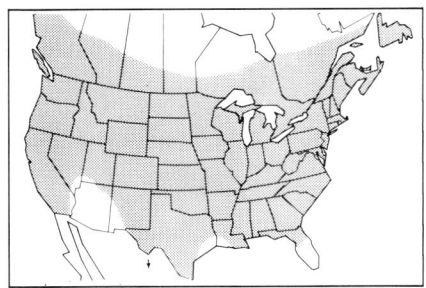
Festuca elatior L. (syn. *F. arundinacea* Schreber)
meadow fescue, Eurasian

Diarrhena americana Beauvois

Elymus canadensis L.
Canada wild-rye

Festuca obtusa Biehler
nodding fescue

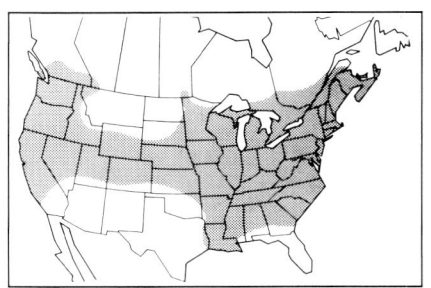
Digitaria ischaemum (Schreber) Muhlenberg
smooth crabgrass, Eurasian

Elymus villosus Muhlenberg
shaggy-haired wild-rye

Festuca octoflora Walter (syn. *Vulpia octoflora* (Walter) Rydberg)
six-weeks fescue

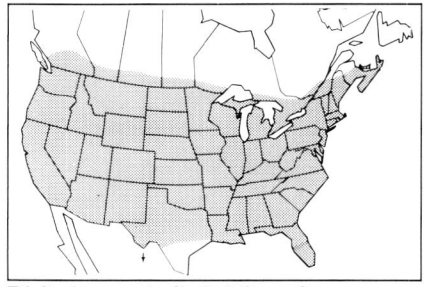
Digitaria sanguinalis (L.) Scopoli
crabgrass, Eurasian

Elymus virginicus L.
Virginia wild-rye

Festuca ovina L. (syn. *F. brachyphylla* Schultes)
sheep fescue

Distichlis spicata (L.) Greene
seashore saltgrass

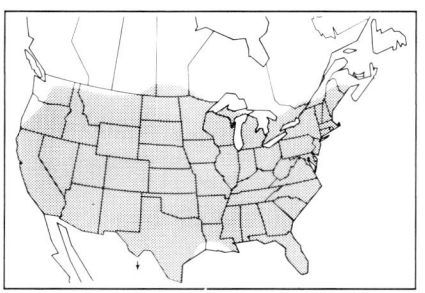
Eragrostis cilianensis (Allioni) Mosher (syn. *E. megastachya* (Koeler) Link)
stinkgrass, Eurasian

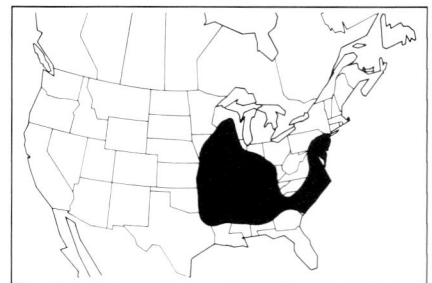
Festuca paradoxa Desvaux (syn. *F. shortii* Kunth)

DISTRIBUTION MAPS OF SPECIES

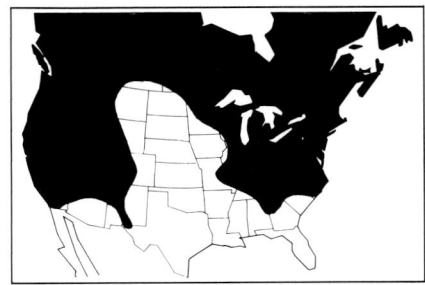
Festuca rubra L.
red fescue

Hordeum pusillum Nuttall
little barley

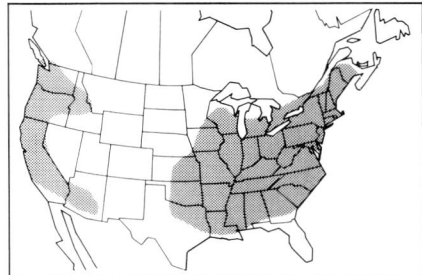
Lolium temulentum L.
bearded or poison darnel, Eurasian

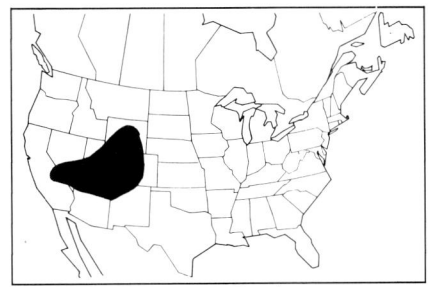
Hilaria jamesii (Torrey) Bentham
galleta

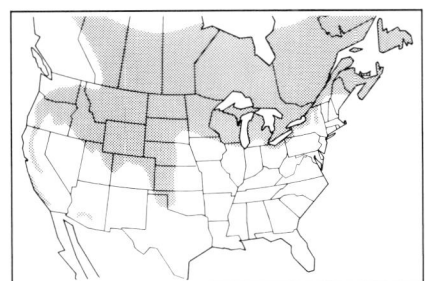
Hordeum vulgare L.
barley (cultivated), Eurasian

Melica nitens (Scribner) Nuttall
three-flowered melic

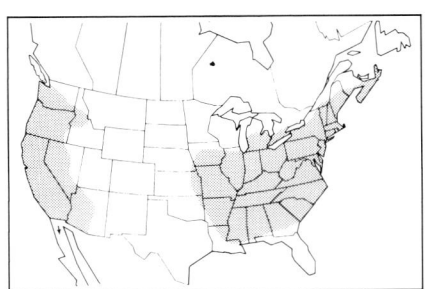
Holcus lanatus L. (syn. *Nothoholcus lanatus* (L.) Nash)
velvetgrass, Eurasian

Hystrix patula Moench
bottlebrush

Muhlenbergia sobolifera (Muhlenberg) Trinius
muhly

Hordeum jubatum L.
foxtail barley, squirrel-tail grass

Koeleria cristata (L.) Persoon (syn. *K. gracilis* Persoon)
crested hairgrass

Oryza sativa L.
rice (cultivated), Asian

Hordeum murinum L.
mouse barley, European

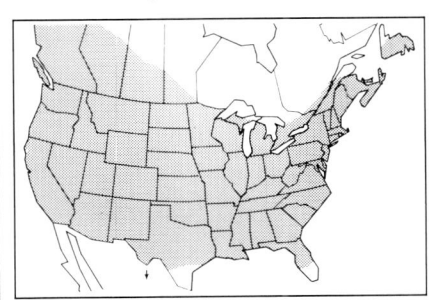
Lolium perenne L. (syn. *L. multiflorum* Lamarck)
common or English ryegrass, Eurasian

Oryzopsis hymenoides (Roemer & Schultes) Ricker
Indian ricegrass

217

APPENDIX 1

Panicum capillare L.
witchgrass

Panicum dichotomiflorum Michaux
fall panicum

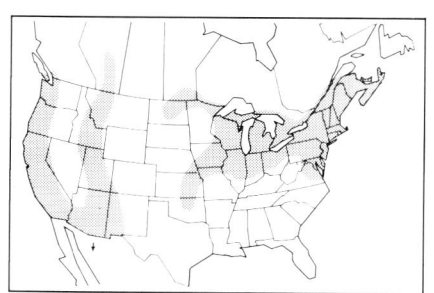

Panicum miliaceum L.
broomcorn or common millet, Eurasian

Panicum rigidulus Nees (syn. *P. agrostoides* Sprengel)

Panicum virgatum L.
switchgrass

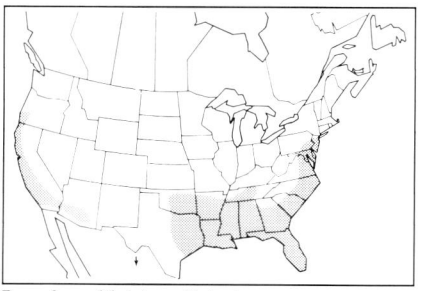

Paspalum dilatatum Poiret
Dallis grass, South American

Paspalum pubiflorum Fournier

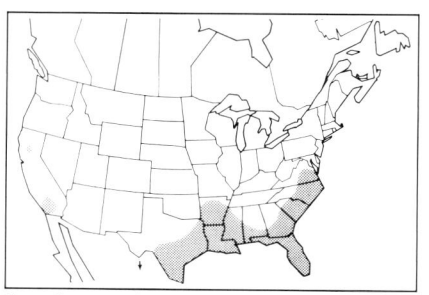

Paspalum urvillei Steudel (syn. *P. vaseyanum* Scribner)
Vasey grass, South American

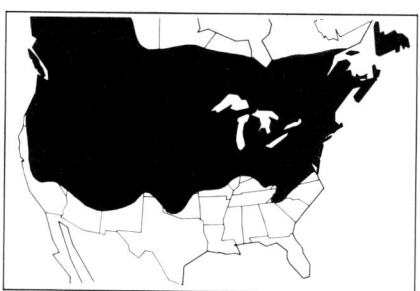

Phalaris arundinacea L.
reed canary grass

Phalaris minor Retzius
Mediterranean canary grass, Mediterranean

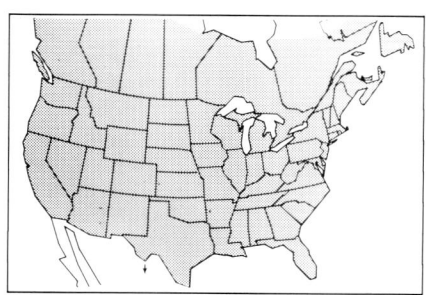

Phleum pratense L.
timothy, Eurasian

Phragmites communis Trinius
common reed

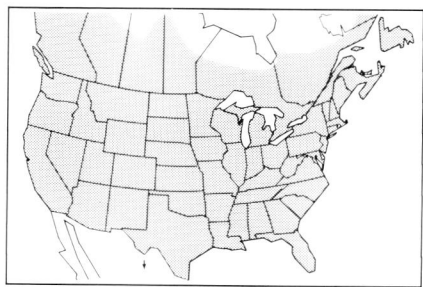

Poa annua L.
annual bluegrass, Eurasian

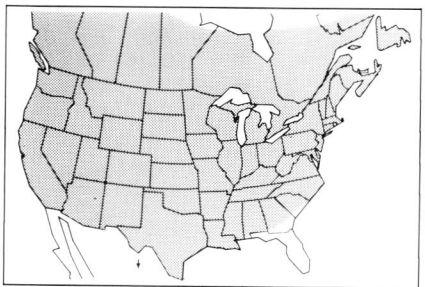

Poa compressa L.
Canada bluegrass, wiregrass, Eurasian

Poa pratensis L.
Kentucky bluegrass, June grass

DISTRIBUTION MAPS OF SPECIES

Poa sandbergii Vasey (syn. *P. secunda* of authors)
Sandberg bluegrass

Secale cereale L.
rye (cultivated), Eurasian

Sorghastrum nutans (L.) Nash
Indian grass

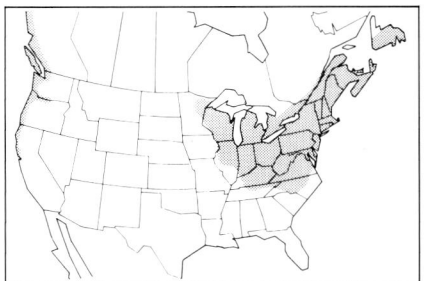

Poa trivialis L.
rough-stalked bluegrass, Eurasian

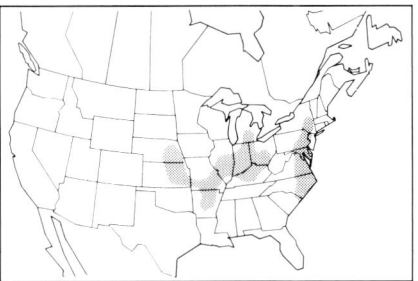

Setaria faberi Herrmann
Faber bristlegrass, Asian

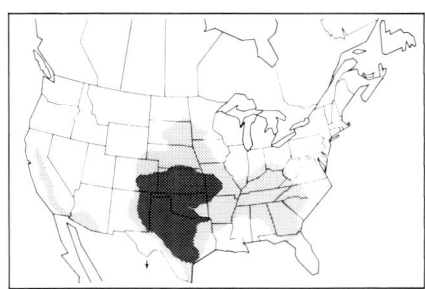

Sorghum bicolor (L.) Moench (syn. *Holcus sorghum* L., *S. vulgare* Persoon, *S. sudanense* (Piper) Stapf) (with cultivated belt)
sorghum, Sudan grass, African

Rhynchelytrum repens (Willdenow) Hubbard (syn. *R. roseum* (Nees) Stapf & Hubbard) Natal grass, South Africa

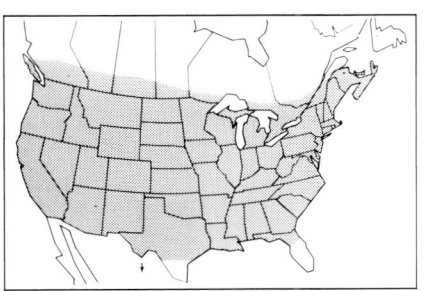

Setaria glauca (L.) Beauvois (syn. *S. lutescens* (Weigel) Hubbard)
yellow bristlegrass, Eurasian

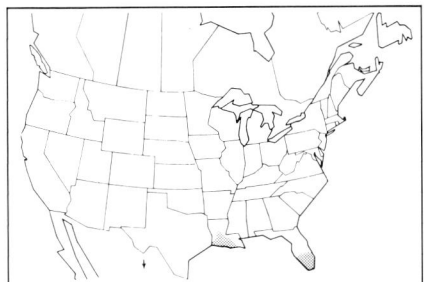

Saccharum officinarum L.
sugarcane, Asian

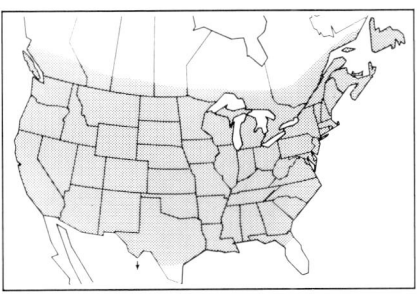

Setaria viridis (L.) Beauvois
green bristlegrass, Eurasian

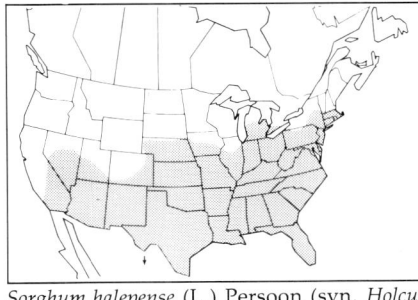

Sorghum halepense (L.) Persoon (syn. *Holcus halpensis* L.)
Johnson grass, European

Schizachyrium scoparium (Michaux) Nash (syn. *Andropogon scoparius* Michaux)
little bluestem, broom beardgrass

Sitanion jubatum J.G. Smith
big squirreltail

Spartina pectinata Link
prairie cordgrass

219

APPENDIX 1

Sporobolus cryptandrus (Torrey) Gray
sand dropseed

Tridens flavus (L.) Hitchock
purpletop

Zizania aquatica L. (syn. *Z. palustris* L.)
wildrice

Sporobolus heterolepis (Gray) Gray
prairie dropseed

Tripsacum dactyloides (L.) L.
eastern gamagrass

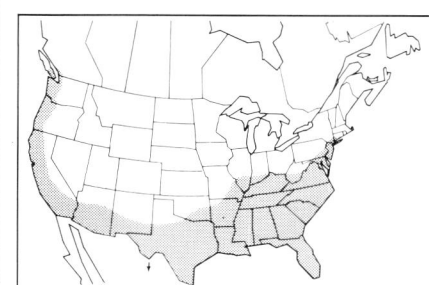
Zoysia japonica Steudel
Japanese lawngrass, zoysia, Asian

CYPERACEAE

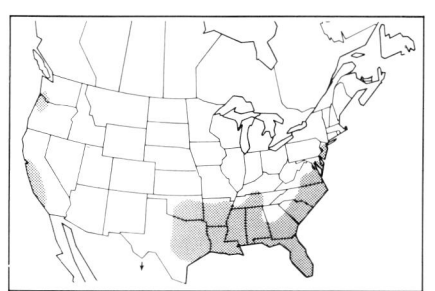
Sporobolus indicus (L.) R. Brown (syn. *S. poiretii* (Roemer & Schultes) Hitchcock)
smutgrass, Asian

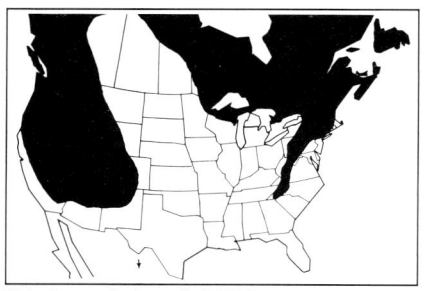
Trisetum spicatum (L.) Richter
spike trisetum

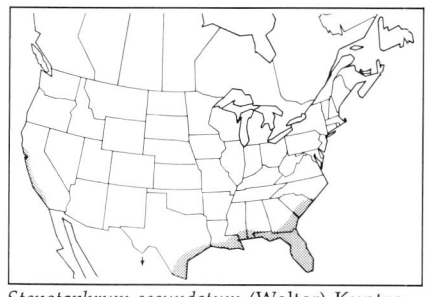
Stenotaphrum secundatum (Walter) Kuntze
St. Augustine grass, tropical American

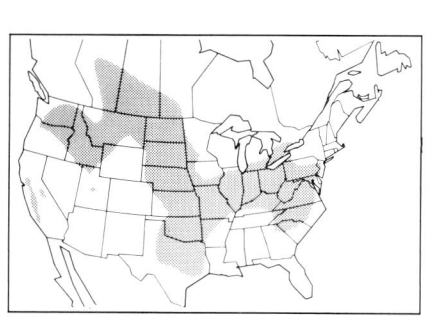
Triticum aestivus L.
wheat, Eurasian

Carex cephalophora Muhlenberg (syn. *C. leavenworthii* Dewey)

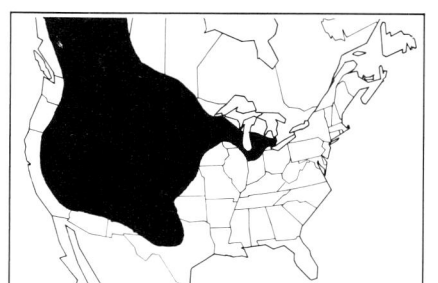
Stipa comata Trinius & Ruprecht
needle-and-thread

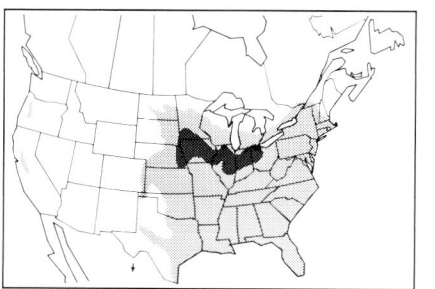
Zea mays L. (with cultivated belt)
corn, maize, tropical American

Carex crinita Lamarck
fringed or long-haired sedge

DISTRIBUTION MAPS OF SPECIES

Carex stricta Lamarck
erect sedge

Fimbristylis autumnalis (L.) Roemer & Schultes

Juncus bufonius L.
toad rush

Carex vesicaria L.
bladdery sedge

Scirpus maritimus L. (syn. *S. paludosus* Nelson, *S. robustus* Pursh)
saltmarsh bulrush

Juncus filiformis L.
thread-like rush

JUNCACEAE

Carex viridula Michaux
greenish sedge

Luzula acuminata Rafinesque
tapering woodrush

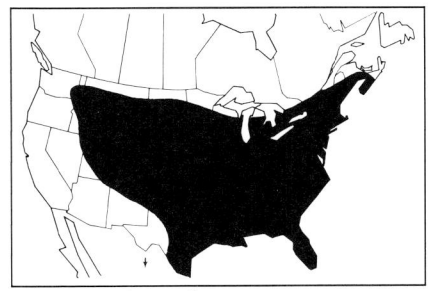

Cyperus esculentus L.
yellow nutgrass

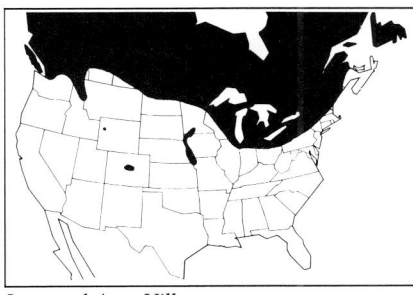

Juncus alpinus Villers
alpine rush

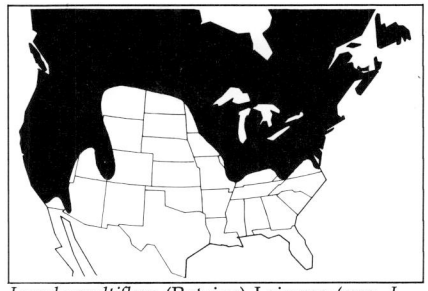

Luzula multiflora (Retzius) Lejeune (syn. *L. campestris* of authors, *Juncoides campestre* of authors)
many-flowered woodrush

Cyperus flavescens L.
yellowish umbrella-sedge

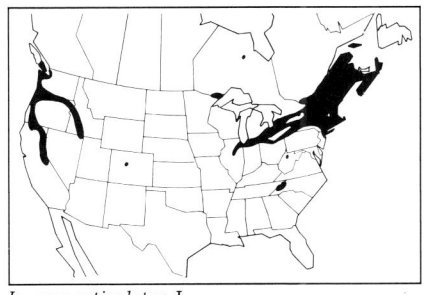

Juncus articulatus L.
jointed rush

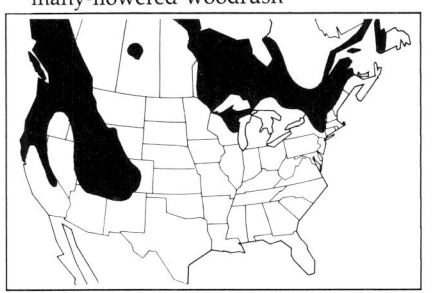

Luzula parviflora (Ehrhart) Desvaux
small-flowered woodrush

APPENDIX 1
JUNCAGINACEAE

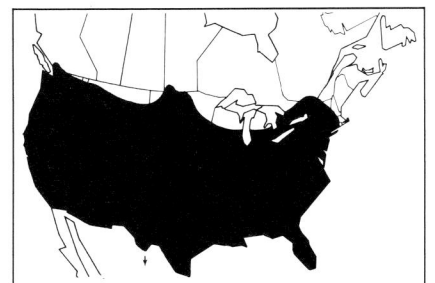
Potamogeton illinoensis Morong
Illinois pondweed

Typha angustifolia L.
narrow-leaved cattail

SPARGANIACEAE

Scheuchzeria palustris L. (syn. *S. americana* (Fernald) Jones)

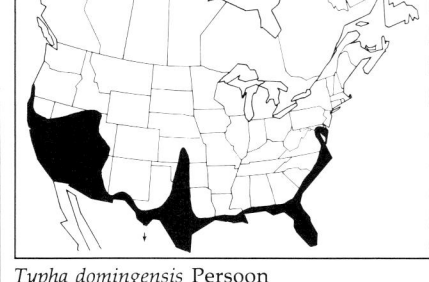
Typha domingensis Persoon
tule

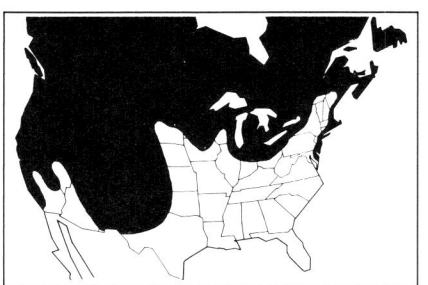
Triglochin maritimum L.
sea arrowgrass

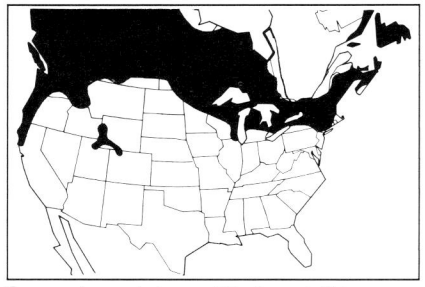
Sparganium minimum (Hartman) Fries
smallest bur-reed

Typha latifolia L.
common cattail

POTAMOGETONACEAE

Sparaganium multipedunculatum (Morong) Rydberg
many-peduncled bur-reed

TYPHACEAE

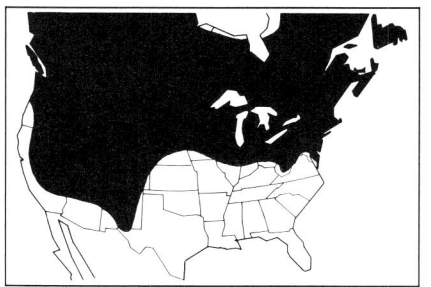
Potamogeton gramineus L.
grass-like pondweed

Weeds and Herbs

Many introduced annual species of weeds and herbs regardless of origin are able to reproduce freely throughout the continent to southern Canada or, if cultivated, are planted annually. Most major perennial herbs introduced from Eurasia are also found in the continent generally, while those from tropical areas survive only in southern latitudes (if mostly perennial, zone will be given). The most important introduced species are indicated (*).

Generalized distribution maps are given for the most common of the indigenous wind-pollinated genera. Typically insect-pollinated genera that may be implicated in pollinosis are represented by one or a few species only.

PTERIDOPHYTES

INDIGENOUS SPECIES

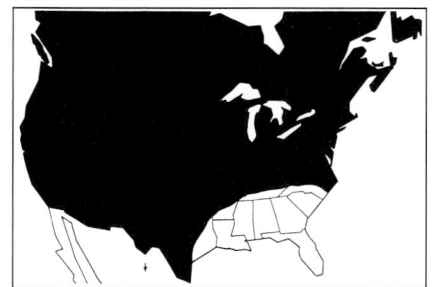

Equisetum arvense L.
  common horsetail (also Eurasian)

Equisetum hyemale L.
  winter scouring rush (also Eurasian)

Lycopodium clavatum L.
  running club-moss (also Eurasian)

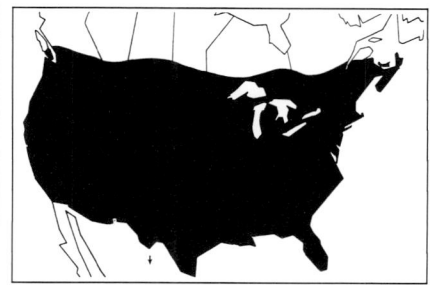

Pteridium aquilinum (L.) Kuhn
  bracken fern (also Eurasian)

AMARANTHACEAE

INTRODUCED SPECIES

Amaranthus caudatus L.
  purple amaranth, tropical American (cultivated annual)

*Amaranthus hybridus L.
  green or spleen amaranth, tropical American (annual)

*Amaranthus retroflexus L.
  redroot or rough pigweed, tropical American (annual)

*Amaranthus spinosus L.
  spiny amaranth, tropical American (annual)

Celosia argentea L.
  cockscomb, tropical American (annual, cultivated)

INDIGENOUS SPECIES

Alternanthera philoxeroides (Martius) Grisebach
  alligator weed

Amaranthus albus L.
  tumbleweed

Amaranthus arenicola I.M. Johnston
  sandhills amaranth

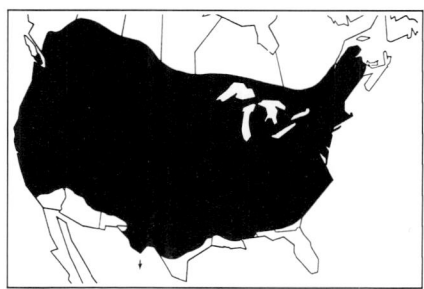

Amaranthus blitoides Watson (syn. *A. graecizans* of authors)
  prostrate pigweed, tumbleweed

223

APPENDIX 1

Amaranthus californicus (Moquin-Tandon) Watson

Amaranthus tuberculatus (Moquin-Tandon) Sauer

Iresine celosia L.
Juba's bush

Amaranthus cannabinus (L.) Sauer
waterhemp

Amaranthus wrightii Watson

Iresine rhizomatosa Standley

Amaranthus palmeri Watson
careless weed

Froelichia floridana (Nuttall) Moquin-Tandon
(syn. *F. campestris* Small)
Florida cottonweed

Tidestromia lanuginosa (Nuttall) Standley

Amaranthus tamariscinus Nuttall (syn. *Acnida tamariscina* (Nuttall) Wood)
western waterhemp

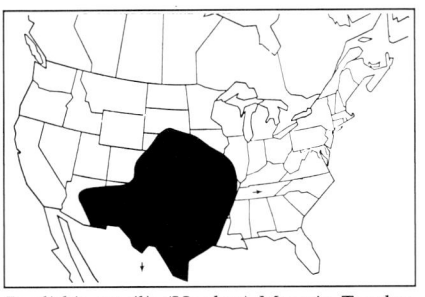

Froelichia gracilis (Hooker) Moquin-Tandon
slender cottonweed

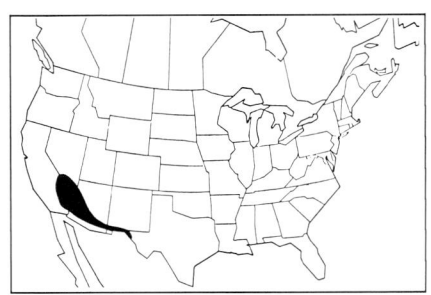

Tidestromia oblongifolia (Watson) Standley

Amaranthus torreyi (Gray) Bentham

Gomphrena caespitosa Torrey
globe amaranth

APIACEAE

INTRODUCED SPECIES

*Daucus carota L.
    wild carrot, Queen Anne's lace,
    Eurasian, Zone 4

Foeniculum vulgare Miller
    common fennel, Mediterranean region,
    Zone 5

ASTERACEAE

INTRODUCED SPECIES

**Artemisia absinthium* L.
  common wormwood, absinthium, European, Zone 4

Artemisia annua L.
  sweet wormwood, European, Zone 5

Artemisia stelleriana Besser
  beach wormwood, dusty miller, Asian, Zone 4

Artemisia vulgaris L.
  mugwort, Eurasian, Zone 4

Bellis perennis L.
  English daisy, Eurasian, Zone 4

Carduus nutans L.
  musk-thistle, Eurasian (biennial)

Cichorium intybus L.
  chicory, Eurasian, Zone 3

Tagetes patula L.
  French marigold, Mexican (annual)

Tanacetum vulgare L.
  tansy, Eurasian, Zone 2

Taraxacum officinale Wiggers
  dandelion, European, Zone 2

Xanthium spinosum L.
  spiny cocklebur, European (annual)

Xanthium strumarium L. (syn. *X. chinense* Miller, *X. orientale* L., *X. pensylvanicum* Wallroth, *X. speciosum* Kearney, and others)
  cocklebur, Eurasian (also native North American ?) (annual)

INDIGENOUS SPECIES

Ambrosia ambrosioides (Cavanilles) Payne
canyon ragweed, Sonora bur-sage

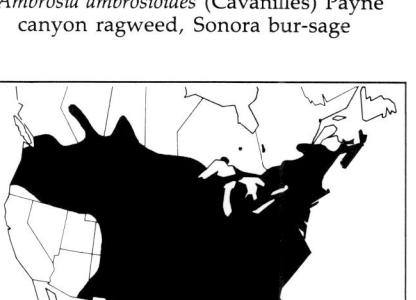

Ambrosia artemisiifolia L. (syn. *A. diversifolia* (Piper) Rydberg, *A. elatior* L., *A. glandulosa* Scheele, *A. longistylis* Nuttall, *A. media* Rydberg, *A. monophylla* (Walter) Rydberg)
short ragweed

Ambrosia chenopodiifolia (Bentham) Payne & *Ambrosia pumila* (Nuttall) Gray

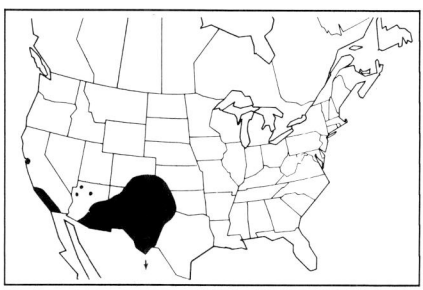

Ambrosia confertiflora DC. (syn. *F. tenuifolia* Harvey & Gray)
slender ragweed

Ambrosia cordifolia (Gray) Payne

Ambrosia (syn. *Acanthambrosia, Franseria, Gaertneria*)
  ragweed, false ragweed
Ambrosia acanthicarpa Hooker (syn. *F. montana* Nuttall)
  bur ragweed, sandbur

Ambrosia bidentata Michaux
southern ragweed

Ambrosia deltoidea (Torrey) Payne
Arizona bur-sage

Ambrosia chamissonis (Lessing) Greene (syn. *F. cuneifolia* Nuttall, *F. bipinnatifida* Nuttall, *F. villosa* Rydberg)

Ambrosia dumosa (Gray) Payne
burro-weed, white bur-sage

APPENDIX 1

Ambrosia eriocentra (Gray) Payne
woolly bur-sage

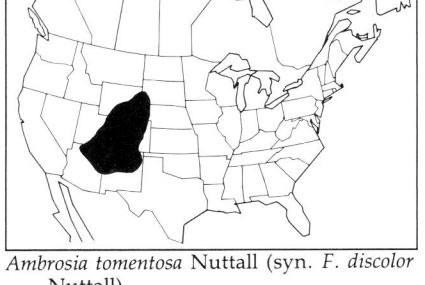
Ambrosia tomentosa Nuttall (syn. *F. discolor* Nuttall)

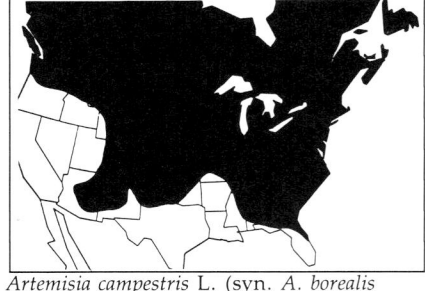
Artemisia campestris L. (syn. *A. borealis* Pallas, *A. canandensis* Michaux, *A. caudata* Michaux, *A. pacifica* Nuttall)
sagewort wormwood, western sagebrush

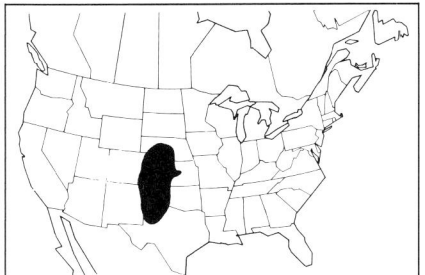
Ambrosia grayi (Nelson) Shinners

Ambrosia trifida L. (syn. *A. aptera* DC., *A. integrifolia* Willdenow, *A. variabilis* Rydberg)
giant ragweed

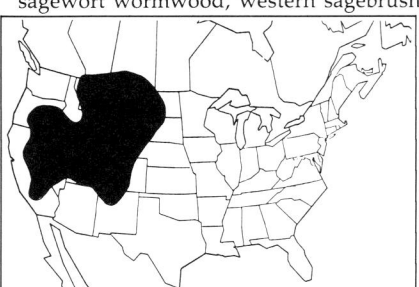
Artemisia cana Pursh
silvery sagebrush

Ambrosia hispida Pursh
coastal ragweed

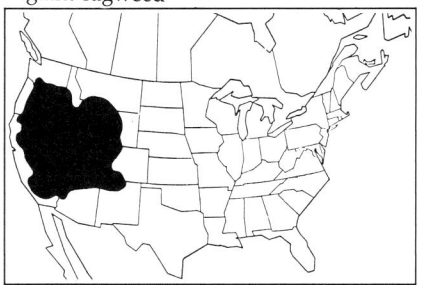
Artemisia arbuscula Nuttall (syn. *A. nova* Nelson)
black sage

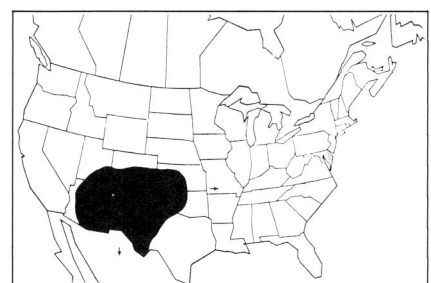
Artemisia carruthii Carruth

Ambrosia ilicifolia (Gray) Payne

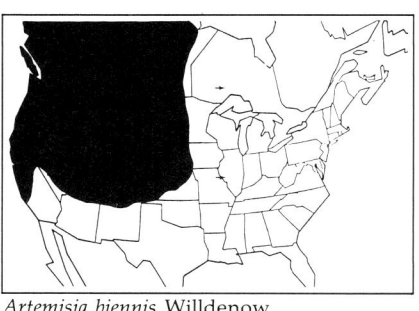
Artemisia biennis Willdenow
biennial wormwood

Artemisia douglasiana Besser

Ambrosia psilostachya DC. (syn. *A. californica* Rydberg, *A. coronopifolia* Torrey & Gray, *A. rugelii* Rydberg)
perennial or western ragweed

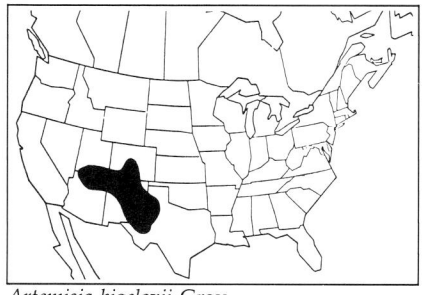
Artemisia bigelovii Gray
Bigelow sagebrush

Artemisia dracunculus L. (syn. *A. glauca* Pallas)
tarragon (also European)

DISTRIBUTION MAPS OF SPECIES

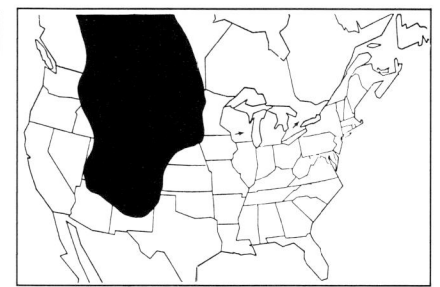
Artemisia frigida Willdenow
prairie sagewort

Artemisia tridentata Nuttall (syn. *A. bigelovii* Gray, *A. parishii* Gray)
common or basin sagebrush

Baccharis halimifolia L.
consumption weed, sea-myrtle

Artemisia ludoviciana Nuttall (syn. *A. albula* Wooton, *A. gnaphalodes* Nuttall, *A. incompta* Nuttall)
western mugwort, white sage

Baccharis angustifolia Michaux

Baccharis pilularis DC. (syn. *B. consanguinea* DC.)
chaparral broom, coyote bush

Artemisia michauxiana Beser

Baccharis brachyphylla Gray

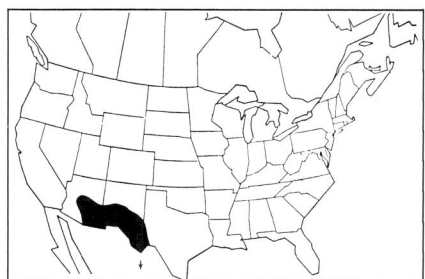
Baccharis pteronioides DC.

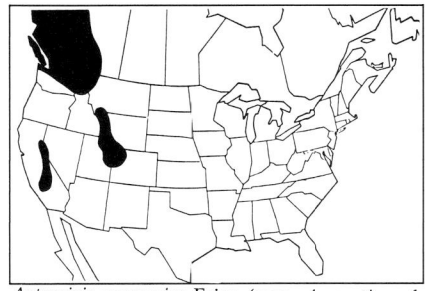
Artemisia norvegica Fries (syn. *A. arctica* of authors)

Baccharis glomeruliflora Persoon
groundsel tree

Baccharis salicina Torrey & Gray
willow baccharis

Artemisia spinescens D.C. Eaton

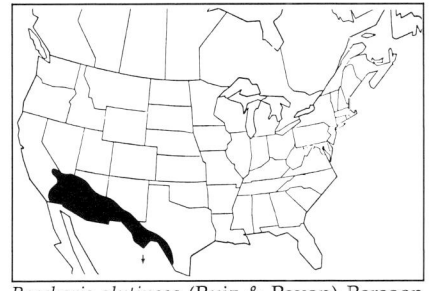
Baccharis glutinosa (Ruiz & Pavon) Persoon
seep-willow

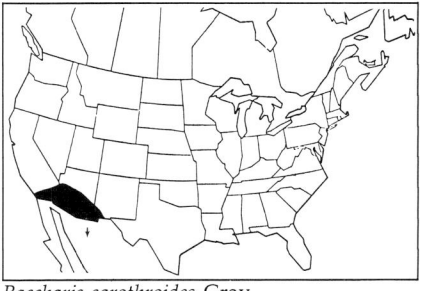
Baccharis sarothroides Gray
broom baccharis

227

APPENDIX 1

Baccharis viminea DC.
mule fat

Hymenoclea monogyra Torrey & Grey
burro-brush

Iva angustifolia DC. (syn. *I. texensis* Jackson)

Baccharis wrightii Gray

Hymenoclea salsola Torrey & Gray (syn. *H. pentalepis* Rydberg)

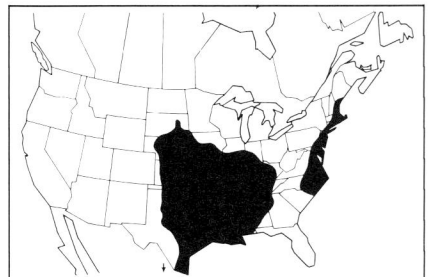
Iva annua L. (syn. *I. caudata* Small, *I. ciliata* Willdenow)
rough marsh-elder

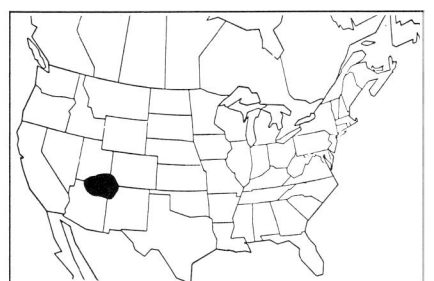
Dicoria brandegei Gray and *D. paniculata* Eastwood

Iva (syn. *Chorisiva, Cyclachaena, Leuciva, Oxytenia*)
marsh-elder, sump-weed
Iva acerosa (Nuttall) Jackson (syn. *Oxytenia acerosa* Nuttall)

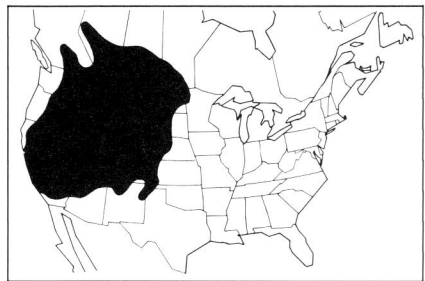
Iva axillaris Pursh
poverty weed

Dicoria canescens Torrey & Grey

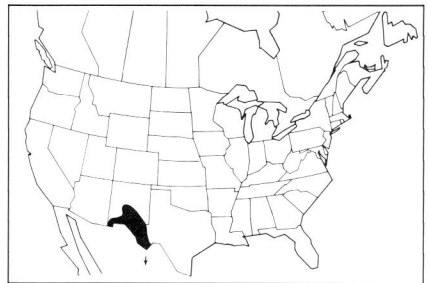
Iva dealbata Gray
woolly sump-weed

Eupatorium rugosum Houttuyn
white snakeroot

Iva ambrosiaefolia (Gray) Gray

Iva frutescens L. (syn. *I. oraria* Bartlett)

DISTRIBUTION MAPS OF SPECIES

Iva imbricata Walter

Iva microcephala Nuttall

Iva nevadensis Jones

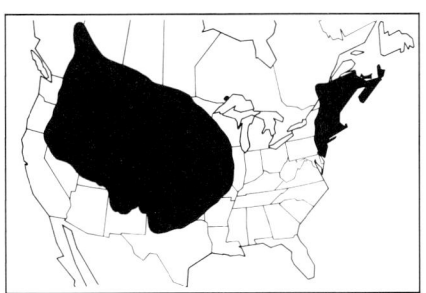
Iva xanthifolia Nuttall

Parthenice mollis Gray

Parthenium integrifolium L.
American feverfew

Solidago canadensis L.
Canada goldenrod

BRASSICACEAE (*Cruciferae*)

INTRODUCED SPECIES

Brassica geniculata (Desfontaines) J. Ball (syn. *Sinapis geniculata* Desfontaines)
European, Zone 7

Brassica hirta Moench (syn. *Sinapis alba* L.)
white mustard, Eurasian (annual)

Brassica juncea (L.) Cosson (syn. *Sinapis juncea* L.)
India mustard, Eurasian (annual)

*Brassica kaber (DC.) Wheeler (syn. *Sinapis kaber* DC.)
charlock, Mediterranean (annual)

*Brassica napus L.
rape, European (annual)

Brassica nigra (L.) Koch (syn. *Sinapis nigra* L.)
black mustard, Eurasian (annual)

Brassica oleracea L.
cabbage, European (annual to perennial)

*Brassica rapa L. (syn. *B. campestris* L.)
field mustard, European (annual or biennial)

Matthiola incana (L.) R. Brown
stock, s. European, Zone 8

CANNABACEAE

INTRODUCED SPECIES

Cannabis sativa L.
hemp, marijuana, Asian (annual)

Humulus lupulus L.
hops, Eurasian, Zone 4

Humulus scandens (Louriero) Merrill (syn. *H. japonicus* Siebold & Zuccarini)
Japanese hops, Asian, Zone 5

CHENOPODIACEAE

INTRODUCED SPECIES

Atriplex rosea L.
tumbling orach, European (annual)

Axyris amaranthoides L.
Russian pigweed, Asian (annual)

Bassia hyssopifolia (Pallas) Kuntze
smother weed, Eurasian (annual)

Beta vulgaris L.
sugar beet, Eurasian (annual and biennial)

*Chenopodium album L.
lamb's quarters, European (annual)

Chenopodium ambrosioides L.
Mexican tea, wormseed, tropical American, Zone 5 (annual or perennial)

Chenopodium botrys L.
Jerusalem-oak, Eurasian (annual)

Chenopodium glaucum L.
oak-leaved goosefoot, Eurasian (annual)

Chenopodium murale L.
nettle-leaved goosefoot, European (annual)

Chenopodium urbicum L.
European (annual)

Chenopodium valvaria L.
stinking goosefoot, European (annual)

Corispermum hyssopifolium L.
bugseed, Eurasian (annual)

Corispermum nitidum Schultes
Eurasian (annual)

Corispermum orientale Lamarck
Eurasian (annual)

Kochia scoparia (L.) Schrader
burning bush, summer-cypress, Eurasian (annual)

*Salsola kali L. (syn. *S. pestifer* Nelson)
Russian thistle, Eurasian (annual)

APPENDIX 1
INDIGENOUS SPECIES

Allenrolfea occidentalis (Watson) Kuntze
iodine bush, pickle weed

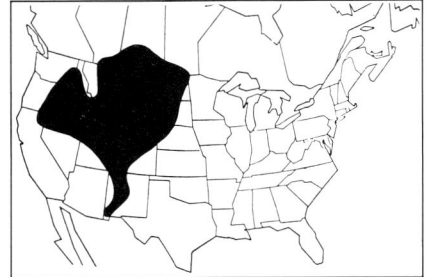

Atriplex nuttallii Watson
moundscale

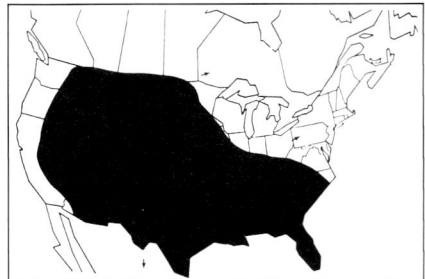

Chenopodium berlandieri Moquin-Tandon
pitseed goosefoot

Atriplex argentea Nuttall
saltbush, silverscale

Atriplex patula L.
orach, spearscale

Chenopodium capitatum (L.) Ascherson
strawberry-blite (also Eurasian)

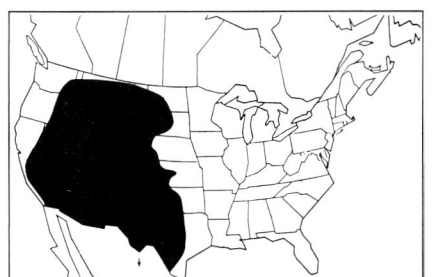

Atriplex canescens (Pursh) Nuttall
shadscale, wingscale

Atriplex phyllostegia (Torrey) Watson
arrowscale

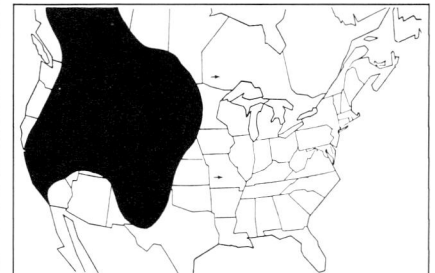

Chenopodium chenopodioides (L.) Aellen (syn.
C. *rubrum* of authors)
alkali- or coast-blite

Atriplex confertifolia (Torrey & Fremont)
Watson

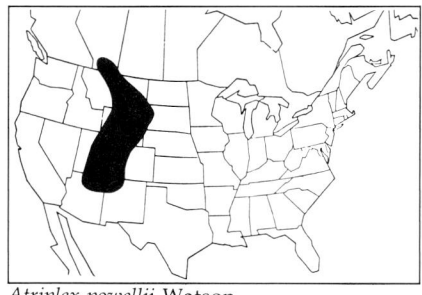

Atriplex powellii Watson

Chenopodium fremontii Watson
Fremont goosefoot

Atriplex dioica (Nuttall) MacBride
rillscale

Atriplex truncata (Torrey) Gray
wedgescale

Chenopodium hians Standley

DISTRIBUTION MAPS OF SPECIES

Chenopodium hybridum L. (syn. *C. gigantospermum* Aellen, *C. standleyanum* Aellen)
maple-leaved goosefoot (also Eurasian)

Chenopodium leptophyllum (Moquin-Tandon) Watson

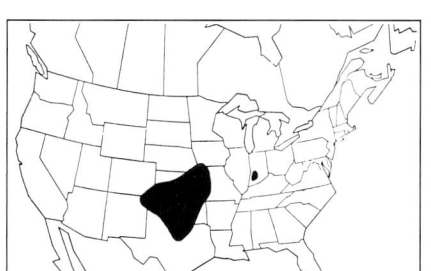

Chenopodium pallescens Standley

Cycloloma atriplicifolium (Sprengel) Coulter
winged pigweed

Eurotia lanata (Pursh) Moquin-Tandon
winter fat

Grayia spinosa (Hooker) Moquin-Tandon
hop-sage

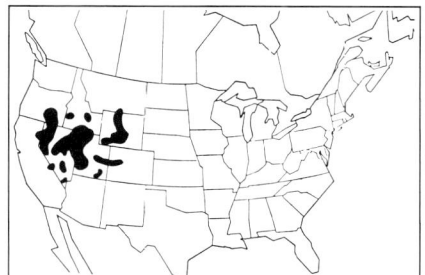

Halogeton glomeratus (Bieberstein) C.A. Meyer

Kochia americana Watson
red-sage, green-molly

Monolepis nuttalliana (Roemer & Schultes) Greene
poverty weed

Salicornia europaea L.
chicken claws (also Eurasian)

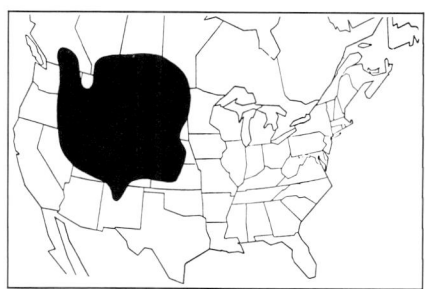

Salicornia rubra Nelson
red glasswort

Salicornia virginica L.
pickleweed, woody glasswort

Sarcobatus vermiculatus (Hooker) Torrey
greasewood

Suaeda americana (Persoon) Fernald
American seepweed, matted sea-blite

Suaeda calceoliformis (Hooker) Moquin-Tandon (syn. *S. depressa* (Pursh) Watson)
Pursh's sea-blite or seepweed

APPENDIX 1

Suaeda linearis (Elliott) Moquin-Tandon
southern sea-blite

Acalypha virginica L.

INDIGENOUS SPECIES

Myriophyllum exalbescens Fernald (syn. *M. spicatum* of authors)
water-milfoil

FABACEAE
(see also Trees and Shrubs)

INTRODUCED SPECIES

Medicago lupulina L.
black medick, Eurasian (annual)

Medicago sativa L.
alfalfa, lucerne, Asian, Zone 2

Melilotus alba Desrousseaux
white sweet clover, Eurasian (annual or biennial)

Melilotus officinalis (L.) Pallas
yellow sweet clover, Eurasian (annual or biennial)

Trifolium arvense L.
clover, European (annual)

Trifolium incarnatum L.
crimson clover, European (annual)

Trifolium pratense L.
red clover, European, Zone 2

Trifolium repens L.
white clover, European, Zone 2

Vicia angustifolia L.
common vetch, European (annual)

Vicia sativa L.
spring vetch, European (annual)

ONAGRACEAE

INDIGENOUS SPECIES

Epilobium angustifolium L.
fireweed

PAPAVERACEAE

INTRODUCED SPECIES

Bocconia frutescens L.
tropical American, Zone 9

Suaeda torreyana Watson (syn. *S. fruticosa* of authors, *S. intermedia* Watson)

EUPHORBIACEAE
(see also Trees and Shrubs)

INTRODUCED SPECIES

Mercurialis annua L.
mercury, European (annual)

INDIGENOUS SPECIES

Acalypha gracilens Gray
slender three-seeded mercury

Acalypha rhomboidea Rafinesque

HALORAGACEAE

INDIGENOUS SPECIES

Eschscholzia californica Chamisso
California poppy

PLANTAGINACEAE

INTRODUCED SPECIES

*Plantago lanceolata L.
    English plantain, European, Zone 2 or 3 (annual to perennial)

*Plantago major L.
    common plantain, European, Zone 2 (annual to perennial)

Plantago psyllium L.
    fleawort, Mediterranean (annual)

INDIGENOUS SPECIES

Littorella americana Fernald
American littorella

Plantago bigelovii Gray

Plantago cordata Lamarck
heart-leaved plantain

Plantago elongata Pursh

Plantago eriopoda Torrey
redwood plantain

Plantago heterophylla Nuttall

Plantago maritima L. (syn. *P. juncoides* Lamarck)
seaside plantain

Plantago patagonica Jacquin (syn. *P. aristata* Michaux, *P. purshii* Roemer & Schultes)
Patagonia Indian-wheat, bracted plantain, salt-and-pepper plant (also s. South America)

Plantago pusilla Nuttall

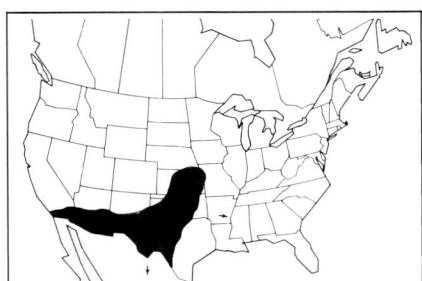

Plantago rhodosperma Decaisne
red-seeded plantain

Plantago rugelii Decaisne
Rugel's plantain

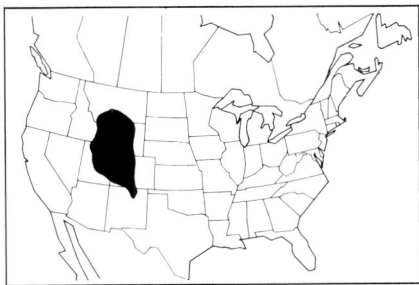

Plantago tweedyi Gray

Plantago virginica L.
hoary or pale-seeded plantain

APPENDIX 1

POLYGONACEAE

INTRODUCED SPECIES

Fagopyrum esculentum Moench
  buckwheat, Eurasian (annual)

Muehlenbeckia complexa (A. Cunningham) Meissner
  wire plant, New Zealand, Zone 9

Rheum rhabarbarum L. (syn. *R. rhaponticum* of authors)
  rhubarb, Asian, Zone 3

Rumex acetosa L.
  garden sorrel, sour dock, Eurasian, Zone 3

*Rumex acetosella L.
  common or sheep sorrel, Eurasian, Zone 2

*Rumex crispus L.
  curley or yellow dock, Eurasian, Zone 2 or 3

Rumex obtusifolius L.
  bitter dock, European, Zone 4

Rumex patientia L.
  patience dock, Eurasian, Zone 4

INDIGENOUS SPECIES

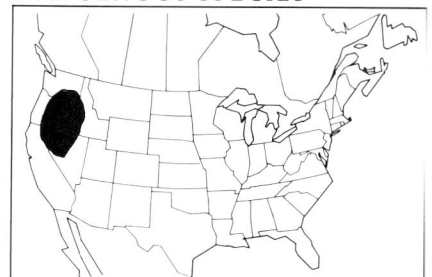

Eriogonum sphaerocephalum Bentham
wild buckwheat

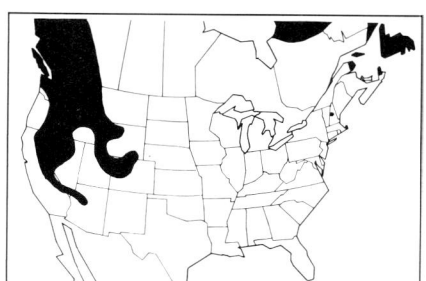

Oxyria digyna (L.) J. Hill
mountain sorrel

Polygonum coccineum Muhlenberg
knotweed, smartweed

Rumex altissimus Wood
pale dock

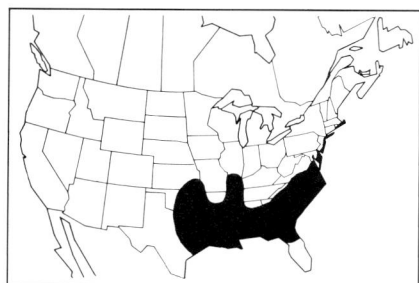

Rumex hastatulus Elliot
heart sorrel

Rumex hymenosepalus Torrey
canaigre, wild-rhubarb

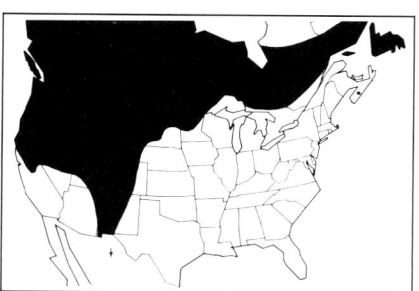

Rumex occidentalis Watson
western dock

Rumex paucifolius Nuttall
alpine sheep sorrel

Rumex salicifolius Weinmann (syn. *R. mexicanus* Meissner) (also European)

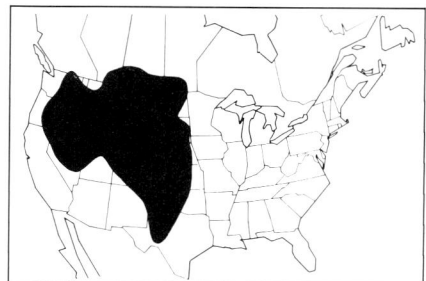

Rumex venosus Pursh
winged or veined dock

Rumex verticillatus L.
swamp dock

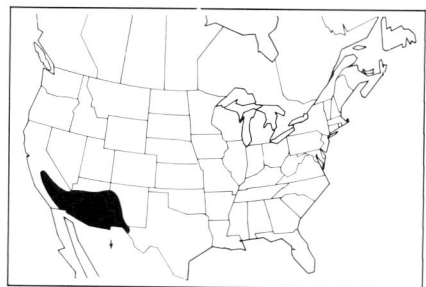

Rumex violascens Rechinger f.

RANUNCULACEAE

INTRODUCED SPECIES

Clematis dioscoreifolia Leveille & Vaniot
  Asian, Zones 4–5

Thalictrum aquilegifolium L.
  Eurasian, Zone 5

Thalictrum dipterocarpum Franchet
  Asian, Zone 5

Thalictrum minum L.
  Eurasian, Zone 5

INDIGENOUS SPECIES

Thalictrum alpinum L.
  alpine meadow-rue (also Eurasian)

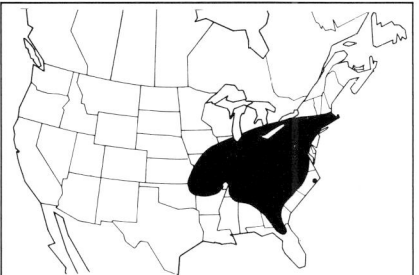

Thalictrum revolutum DC.
  wax-leaved meadow-rue

Laportea canadensis (L.) Weddell
  wood nettle

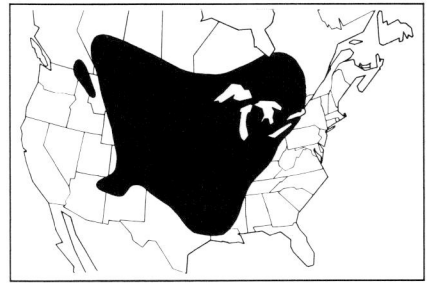

Thalictrum dasycarpum Fischer & Avé-Lallemant
  purple meadow-rue

Thalictrum venulosum Trelease

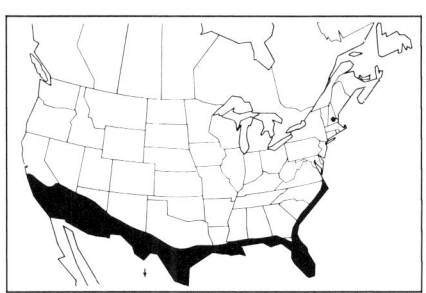

Parietaria floridana Nuttall (syn. *P. debilis* of authors)
  Florida pellitory

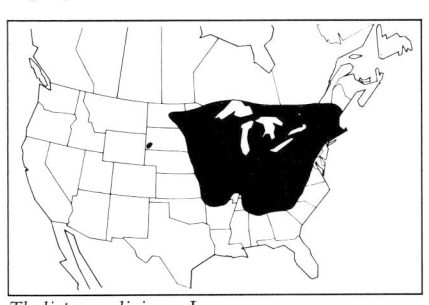

Thalictrum dioicum L.
  early meadow-rue

URTICACEAE

INTRODUCED SPECIES

Urtica pilulifera L.
  Roman nettle, s. European, Zone 7

Urtica urens L.
  burning, dog or dwarf nettle, Eurasian (annual)

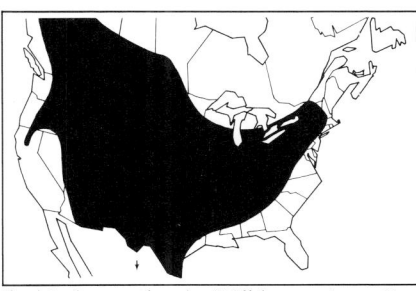

Parietaria pensylvanica Willdenow (syn. *P. obtusa* Rydberg)
  hammerwort, pellitory

INDIGENOUS SPECIES

Thalictrum fendleri Gray
  Fendler's meadow-rue

Boehmeria cylindrica (L.) Swartz
  bog-hemp

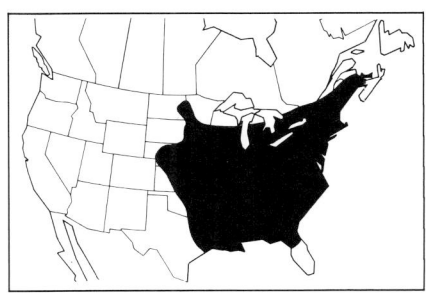

Pilea pumila (L.) Gray (syn. *P. fontana* (Lunell) Rydberg)
  clearweed, richweed

Thalictrum pubescens Pursh (syn. *T. polygamum* Muhlenberg)
  tall meadow-rue

Boehmeria nivea (L.) Gaudin

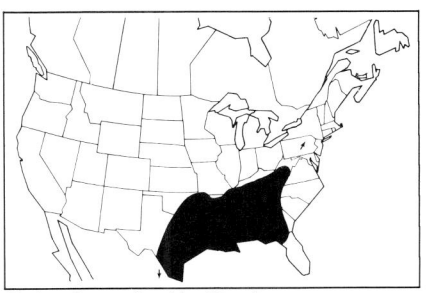

Urtica chamaedryoides Pursh

APPENDIX 1

Urtica dioica L. (syn. *U. gracilis* Aiton, *U. procera* Muhlenberg)
stinging or tall nettle (also Eurasian)

Urtica gracilenta Greene

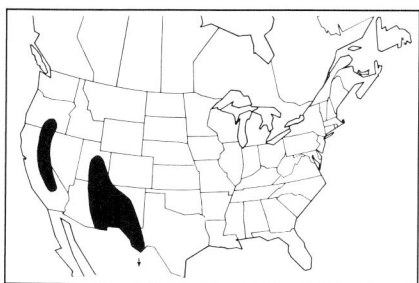

Urtica serra Blume

Appendix 2

Methodology

Pollen was obtained from either newly collected material or occasionally from herbarium specimens of the Missouri Botanical Garden. Determinations were made by the authors or by taxonomic specialists. Vouchers of new collections are deposited with the Missouri Botanical Garden herbarium (MO).

PREPARATION OF POLLEN GRAINS

Light Microscopy (LM)

Pollen of most collections was acetolyzed according to a modified procedure of Erdtman (1952, 1960) as follows. Fresh or dried anthers dissected from large flowers, or whole immature buds of small flowers, were placed in a centrifuge tube containing about a 2.5 ml mixture of 8 parts acetic anhydride and 1 part concentrated sulfuric acid. Dried material was pretreated overnight in Carnoy's solution (1 part glacial acetic acid and 3 parts absolute ethanol). The mixture was brought to 98–100° C in a water bath, and stirred gently for about 2 minutes, or less for thin-walled or large grains. The mixture was cooled for about 10 minutes, centrifuged up to 2 minutes at 1,000 rpm, and decanted. To the sediment was added about 2.5 ml of 95% ethanol and tap water to fill the centrifuge tube three-quarters. The sediment was rinsed through a bronze filter, 200 mesh, followed by centrifuging, decanting, and rinsing with tap water once or twice more. To the rinsed sediment was added about 10 drops of 50% glycerin for 15 minutes, after which the mixture was centrifuged, decanted, and the centrifuge tube inverted on filter paper overnight (or 2 hours at 60° C). Using a platinum needle, a 1 mm square of glycerin jelly* was gently rotated in the sediment and placed on a slide. The slide was placed on a warmer at 80° C; when the jelly melted a circular cover slip was placed over it. To make the slide permanent, melted paraffin was added to the edge of the cover slip until sealed. When cool, excess paraffin was removed with xylene.

Some pollen of most collections was not acetolyzed but was stained whole in basic fuchsin. Anthers from fresh or herbarium material were fixed in about 3 ml Carnoy's solution in centrifuge tube, gently crushed using a glass rod, and rinsed through a bronze filter, 200 mesh. The mixture was centrifuged for about 2 minutes, decanted, and rinsed repeatedly with tap water. About 2 ml dilute basic fuchsin* was added, thoroughly incorporated in the sediment, allowed to stand for 15 to 20 minutes, centrifuged, and decanted. To the stained sediment about 2 ml of 50% glycerin was then added for 15 minutes, centrifuged, decanted, and the centrifuge tube inverted on filter paper overnight. The material was mounted in glycerin jelly and the slide was made permanent as described above.

A Leitz Ortholux compound microscope with Plano 40× and 100× objectives was used to take the light micrographs using Kodak Technical Pan film. A Zeiss compound microscope with Plano 40× and 100× objectives was used for observation.

Scanning Electron Microscopy (SEM)

In preparing pollen for the scanning electron microscope, one of two techniques was followed, the choice depending on the strength of the pollen wall and its tendency to collapse during drying. Thick-walled pollen was air dried (ad) in a dessicator. Thin-walled pollen was acetolyzed (a), as outlined above, and dehydrated in an ethanol series of 50%, 75%, 95%, and 100%, and then critical-point dried.

Pollen was placed on double-stick Scotch tape which had been stuck on aluminum stubs. The material was then sputter-coated with a thin layer of gold and examined using a Hitachi S-450 at 15–25 Kv. Photomicrographs were taken on either Polaroid 55P/N or Kodak Ektapan film.

PHOTOGRAPHY OF PLANTS

Photographs of plants were taken with a Canon F-1 using a 100 mm macro-lens, a Calumet 4×5 with a Caltar 150 mm lens, or a Nikon FM with a 50 mm

---

* Glycerin jelly: 50 g gelatin, 175 ml distilled water, 150 ml glycerin, and 7 g phenol crystals heated until clear and stored at refrigerator temperatures.

* Dilute basic fuchsin: add 200 ml boiling distilled water to 1 g basic fuchsin, stir thoroughly, cool, and store at refrigerator temperatures. For staining pollen, dilute saturated aqueous solution $\frac{1}{10}$ (1 part stain and 10 parts distilled water).

Micro-Nikor lens. The film used was primarily Kodachrome 64 or, less frequently, Panatomic X or Ektapan.

PREPARATION OF POLLEN EXTRACTS

Although pollen extracts were not made for this project, a brief statement of their preparation is appropriate (for details, see Sheldon et al. 1967). Pollen is defatted with anhydrous ethyl ether. For small amounts of pollen, cover the pollen with ether, swirl for several minutes, let the mixture settle, decant, and repeat until the ether lacks color. Filter with a Whatman #4 in a funnel, evaporating the ether before beginning the extraction. Extract for 20 to 24 hours at room temperature. A usual extraction ratio is 1 g of pollen extracted in 10 ml of extracting fluid (Coca's solution),* 1:10 w/v. The liquid portion of the extract should be filtered (Whatman #4). The pH of the extract should be adjusted to the range 6.8 to 7.4 with concentrated HCl or NaOH, sterilized by filtering through a sterile 0.2 μm membrane filter using appropriate sterile procedures. To insure nontoxicity to humans animal tests should be completed.

For the prick, or scratch, test the concentrated extract is diluted usually to a ratio of 1:20 w/v in 50% glycerin in Coca's solution. For the intradermal tests the extract is commonly diluted to a final concentration of 1:1000 w/v in Coca's solution.

---

* Coca's solution: 5 g sodium chloride, 5.4 g sodium bicarbonate, 4.49 ml 89% solution liquified phenol, and distilled water for injection to fill 1 Liter.

Glossary

Acetolysis: a method of preparing pollen to illustrate the walls (sexine, nexine) resistant to hot acid treatment, the nonresistant contents of pollen being destroyed (intine, protoplasm, wall lipids, etc.). See Appendix 2 for methodology.
Achene: a simple, small, dry, 1-seeded indehiscent fruit.
Aeroallergen: an allergen distributed by the ambient air.
Aerobiology: the study of airborne cells or organisms, e.g., pollen, spores, bacteria.
Allergen: a substance capable of inducing allergy or specific hypersensitivity.
Amb: the outline of pollen viewed with one of the poles exactly uppermost.
Amphiphily: pollination by wind and vectors (usually insects).
Anemophily: pollination by wind.
Angulaperturate: describing the amb when the apertures are situated at its angles.
Angiosperm: a flowering plant whose seeds are borne within a mature ovary (fruit).
Annulus: a ring, referring to the thickened exine surrounding a pore; annulate (adj.).
Anthesis: the opening of a floral bud; generally, the time of flowering.
Antigen: a substance that provokes an immune response when introduced into the body.
Aperture: a weak, preformed opening of the exine that permits the exit of intra-exinous contents (e.g., pollen tube, intinous-held antigens).
Apocolpium (pl. **apocolpia**): the polar area.
Apomixis: in plants, reproduction from seeds in which there is no fusion of male and female gametes.
Arcus (pl. **arci**): a bandlike, thickened part of the sexine that extends from aperture to aperture.
Areolate: pertaining to small, usually circular, sexinous areas separated by grooves that form a negative reticulum (insulate).
Aspidate: referring to ± circular, protruding domes extending from around the apertures.
Atopic: allergic; the inherent potential to be allergic.
Baculum (pl. **bacula**): a rodlike or columnar structure, forming a zone of sexine between the outermost tectum or sexinous sculpturing and the inner nexine (Fig. 4).
Berry: a pulpy, indehiscent, few or many-seeded fruit, such as grapes, tomatoes.
Bilateral symmetry: symmetry having two vertical planes.
Bisexual: of an organism having both male and female reproductive structures and producing male and female gametes, the individual being either monoecious or hermaphroditic.
Bladder: see Saccus.
Blade: the broad, expanded part of the leaf; commonly referred to as the leaf.
Brevicolpate: having ± short colpi.
Brochus (pl. **brochi**): a part of the reticulum that includes a lumen and adjoining half of the muri.

Calyx: see Sepal.
Capitulum: a dense inflorescence consisting of an aggregation of usually sessile flowers.
Capsule: a dry, dehiscent fruit that develops from a compound ovary (of more than one carpel).
Carpel: a leaflike organ that encloses one or more ovules; a single pistil consists of one carpel, a compound pistil of two or more carpels.
Catkin: a spikelike inflorescence of unisexual flowers, typical of many woody, wind-pollinated plants.
Clavate: referring to club-shaped, sexinous processes that are higher than they are broad.
Colporate: having a compound aperture with a colpus (furrow) and an os (pore). **Colporoidate:** with a colpus and an indistinct os.
Colpus (pl. **colpi**): an elongated aperture, usually with acute ends; a germinal furrow (ruga); colpate (rugate) (adj.). **Colpoid:** colpuslike or having a faintly demarked colpus. See Sulcus.
Corolla: see Petal.
Corpus (pl. **corpi**): the body of the gymnospermous pollen excluding the air sacs.
Crassinexinous: with a nexine at least twice as thick as a sexine.
Crassitegillate: with a tegillum as thick as or thicker than the baculum zone (or that zone between the tegillum and the nexine).
Cymbiform: boat-shaped.
Cyme: a broad, ± flattened, determinate inflorescence, with central flowers opening first. **Cymule:** a small cyme, usually few-flowered.
Deciduous: referring to plant parts that separate at an abscission layer in response to environmental change, such as cold or drought.
Disc flower: a radially symmetrical, bisexual, tubular flower comprising the central part of the floral heads of many Asteraceae.
Distal face or **pole:** the part or point of the pollen surface that is directed outward when in tetrads.
Drupe: a fleshy, 1-seeded, indehiscent fruit with the seed enclosed by a hard wall; a stone fruit, such as peaches.
Duplibaculate: pertaining to muri supported by two rows of bacula.
Echinolophate: see Lophate.
Elater: a hygroscopic band attached to the spores of *Equisetum* (horsetail).
Elliptic: with the widest axis at midpoint of the structure and with margins symmetrically curved. **Ellipsoidal:** a solid having an elliptic shape.
Entomophily: pollination by insects.
Equatorial view: among radially symmetrical pollen, the view in which an aperture is directed toward the observer, thus allowing the measurement between the two poles of the polar axis (P).
Excrescence: a process (tuber, wart, etc.) on the sexine surface.

Exine: the outer, resistant pollen wall consisting of sexine and nexine (Fig. 4).

Facultative: taking place under certain conditions but not under others.

Flabellate: fanlike.

Floret: a small flower that makes up the floral heads or inflorescences of Asteraceae, or the spikes of grasses.

Flower: the reproductive structure of Angiosperms. A complete flower includes sepals, petals, stamens with anthers that at maturity release pollen, and pistils.

Foveolate: pertaining to the pitted (lumina) sculpturing of a reticulated sexine. **Scrobiculate:** when pits (lumina) are minute.

Fruit: a matured, ripened ovary containing the seeds.

Gametophyte: see Ovule; Pollen.

Gemma (pl. **gemmae**): a process on the surface of the sexine that is constricted basally and as wide as or wider than the height; gemmate (adj.).

Grain: see Pollen grain.

Granule: a small, ± rounded process on the surface of the sexine; granulate (adj.).

Gymnosperm: a cone-bearing vascular plant having exposed seeds.

Heterobrochate: referring to brochi that are distinctly different in size, giving an irregular reticulum.

Heteropolar: pollen having different distal and proximal poles, particularly with regard to apertures, shape, etc. Cf. Isopolar.

Hypersensitivity: an increased activity of the body to an antigen that incites an allergic reation; type I is antibody-mediated, type IV is cellular-mediated.

Hyposensitization: desensitization.

Inaperturate: having no distinct aperture.

Incrassate: thickened.

Inflorescence: a flower cluster.

Infructescence: a fruit cluster.

Insulate: see Areolate.

Intine: the innermost, nonresistant pollen wall that lies between the nexine and the protoplasm (Fig. 4).

Isopolar: pollen having no difference between proximal and distal poles.

Paraisopolar or **subisopolar:** pollen having a slight difference between proximal and distal poles. Cf. Heteropolar.

Lacuna (pl. **lacunae**): a large pit or depression in the sexine of lophate or reticulate pollen.

Lalongate: referring to an os (pore) that is transversely elongated.

Leaf: a flattened outgrowth of a stem consisting of blade, petiole, and sometimes stipule, usually green and photosynthetic.

Ligule: a small appendage at the base of grass leaves.

Ligule flower: a small, bisexual flower having a bilaterally symmetrical, ribbonlike corolla with five terminal teeth, as in floral heads of Asteraceae, tribe Lactuceae (chicory).

Lira (pl. **lirae**): the narrow ridge between the striae of striated pollen.

Lolongate: referring to an os (pore) that is longitudinally elongated.

Lophate: referring to ridged sculpturing, connected or free, with the ridges usually higher than the muri. **Echinolophate:** having spines on the ridge crests.

LO-pattern: a pattern that at high focus of the microscope appears as "bright islands" separated by "dark channels" and on lower focus the reverse, "dark islands" separated by "bright channels." **OL-pattern:** reverse of LO-pattern.

Lumen (pl. **lumina**): a cavity between the ridges (muri) of a reticulum. **Luminoid:** a cavity that is indistinct.

Micropore: a minute perforation or hole in the tectum (tegillum), often irregular in size and spacing. See Punctum.

Microtuberculate: see Tuberculate.

Microverrucate: see Verrucate.

Monad: a single pollen grain.

Monoecious: producing male and female gametes in one individual, but in plants having the stamens and pistils in separate flowers.

Murus (pl. **muri**): a ridge separating the cavities (lumina) of a reticulum.

Nexine: the inner, nonsculptured, resistant wall of the exine lying between the sexine and the intine (Fig. 4).

Nut: a dry, indehiscent, hard, 1-seeded fruit often produced from a compound pistil.

Oblate: distinctly flattened, used to describe the shape of radially symmetrical pollen where the ratio between polar axis and equatorial diameter is 0.5–0.75 (4:8–6:8). **Suboblate:** less distinctly flattened, the ratio being 0.75–0.8.

Oblate-spheroidal: somewhat flattened to nearly spherical, used to describe the shape of radially symmetrical pollen where the ratio between polar axis and equatorial diameter is 0.88–0.99 (7:8–7.9:8) (subspheroidal).

Oblong: with widest axis at midpoint of the structure and with margins essentially parallel. Cf. Elliptic.

Oncus (pl. **onci**): a thickened portion of the intine below the apertures.

Operculum (pl. **opercula**): a lid or covering of exine over the apertures.

Os (pl. **ora**): the inner, porelike portion of a compound aperture.

Ovary: an enlarged, ovule-bearing, basal portion of the pistil; the ovary becomes the fruit.

Ovule: a structure of the ovary that contains the egg (within the female gametophyte); the ovule becomes the seed.

Pantocolpate: with colpi ± uniformly distributed over the pollen surface (syn. pantorugate, polyrugate).

Pantoporate: with pores ± uniformly distributed over the pollen surface (syn. pantoforate, polyforate).

Papillate: with nipplelike processes well-spaced over the surface of the sexine.

Paraisopolar: see Isopolar.

Parasyncolpate: see Syncolpate.

Peltate: shield-shaped.

Perianth: the sepals and petals considered together.

Perine: the outermost, extra-exinous wall layer in some spores of ferns (syn. perispore).

Peroblate: very flattened, used to describe the shape of radially symmetrical pollen where the ratio between polar axis and equatorial diameter is <0.5 (<4:8).

Perprolate: very elongated, used to describe the shape of radially symmetrical pollen where the ratio between polar axis and equatorial diameter is >2 (>8:4).

Pertectum: see Tectum.

Petal: a floral part that is usually conspicuously colored. **Corolla:** the petals collectively.

Petiole: the leaf stalk.

Phylogeny: the evolutionary history of an organism.

Pilate: referring to small, club-shaped processes on the surface of the sexine; pilum (n.). **Retipilate:** when the processes of the reticulum are free and not united into ridges.

Pistil: the central structure of the flower consisting of ovary, style, and stigma; a pistil consists of one or several fused carpels. **Pistillate:** referring to a flower with a pistil (one or more carpels), but no functional stamens.

Polar axis: see Equatorial view.

Polar view: among radially symmetrical pollen, the view in which a pole is directed toward the observer, thus allowing the measurement of the equatorial diameter (E).

Pole: among polar pollen, there are two poles, one (proximal) directed toward the center of the tetrad, the other (distal) facing in the opposite direction.

Pollen or **pollen grain:** the microgametophyte (male gametophyte) of Gymnosperms and Angiosperms that possesses two sperm at maturity; the haploid microspore.

Pollination: the transfer of pollen from where it was formed, usually in the anther, to a receptive surface, usually the stigma.

Pollinosis: an allergic reaction, usually to airborne pollen, resulting in type I, antibody-mediated hypersensitivity; hay fever due to pollen.

Polyad: an aggregation of more than four pollen.

Polyploid: having more than two complete sets of chromosomes, such as tetraploid with four sets ($4\times$).

Pore: a circular or ovoid germinal aperture (foramen); porate (adj.). **Poroid:** porelike. See Os; Ulcus.

Pororate: referring to a compound aperture with an outer and inner pore.

Prolate: distinctly elongated, used to describe the shape of radially symmetrical pollen where the ratio between polar axis and equatorial diameter is 2.0–1.33 (8:6–8:7). **Subprolate:** less distinctly elongated, the ratio being 1.33–1.14.

Prolate-spheroidal: somewhat elongated to nearly spherical, used to describe the shape of radially symmetrical pollen where the ratio between polar axis and equatorial diameter is 1.14–1.01 (8:7–8:7.9) (subspheroidal).

Proximal face or **pole:** the part or point of the pollen surface that is directed inward when in tetrad.

Punctum (pl. **puncti**): a minute perforation in the surface of the sexine, similar to micropore (punctitegillate); punctate (adj.).

Pyriform: pear-shaped.

Radial symmetry: symmetry with more than two planes.

Ray flower: a small, flattened, pistillate or sterile flower with the ribbonlike corolla having two or three teeth, as the margins of Asteraceae inflorescences.

Reticulum (reticulate): netlike sculpturing on the surface of the sexine, formed by connected ridges (muri) between cavities (lumina); reticulate (adj.).

Retipilate: see Pilate.

Rugulate: referring to the corrugated, ribbed, or wrinkled surface of the sexine.

Saccus (pl. **sacci**): a winglike air sac, e.g., formed when the sexine is loosened from the nexine as in some Gymnosperms.

Samara: a simple, dry, indehiscent fruit with winglike outgrowths.

Scabrate: having minute pits or projections not exceeding 1 µm, the sexine appearing flecked.

Scrobiculate: see Foveolate.

Seed: a structure formed following the maturation of the ovule after fertilization.

Sepal: the outermost floral structure, usually green and enclosing other parts in the floral bud. **Calyx:** the sepals collectively.

Sexine: the outer, sculptured, resistant wall of the exine (Fig. 4).

Simplibaculate: having muri supported by one row of bacula.

Spathe: a bract or leaf surrounding a flower cluster.

Spheroidal: spherical, used to describe the shape of pollen where polar axis and equatorial diameter are about equal. See Oblate-spheroidal, Prolate-spheroidal.

Spike: an inflorescence with a long main axis and sessile flowers.

Spikelet: a small group of grass flowers.

Spine or **spina** (pl. **spinae**): a long, pointed process exceeding 3 µm in length; spinose (adj.) **Spinule** or **spinula** (pl. **spinulae**): a long, pointed process not exceeding 3 µm in length; spinulose (adj.).

Spiraperturate: pertaining to one or several spiral apertures.

Sporosis: an allergic reaction to spores resulting in type I, antibody-mediated hypersensitivity.

Stamen: the structure of the flower that produces pollen, usually consisting of anther and filament. **Staminate:** referring to a flower with stamens, but no functional pistil.

Stellate: star-shaped.

Stephanaperturate: having more than three apertures in the equatorial area, i.e., the colpi are stephanocolpate and the pores are stephanoporate.

Stigma: the uppermost part of the pistil serving as a receptive surface for pollen on which they germinate.

Stipule: a leafy appendage at the base of some leaves.

Striate: having long, parallel grooves and ridges.

Striato-reticulate: having a sculpturing pattern intermediate between striate and reticulate.

Striato-rugulate: having a sculpturing pattern intermediate between striate and rugulate.

Style: the slender, often elongated part of a pistil (between the stigma and ovary) through which the pollen tube grows.

Subisopolar: see Isopolar.

Suboblate: see Oblate.

Subprolate: see Prolate.

Sulcus: a colpus of the distal face, such as the germinal furrow of many monocotyledons (palms) and primitive dicotyledons (*Magnolia*); sulcate (adj.).

Synclinorate: having transversely elongated ora forming a connected equatorial band.

Syncolpate: with colpi connecting at the poles. **Parasyncolpate:** with colpi almost connecting at the poles.

Taxon (pl. **taxa**): a general term used for a taxonomic group of any rank.

Tectum (pl. **tecta**): a ± homogenous, continuous zone of the sexine external to the bacula, the outermost part of the sexine; used interchangeably with Tegillum. **Pertectum:** a very complete tectum (Fig. 4).

Tegillum: see Tectum.

Tetrad: a union of four pollen grains or spores. **Rhomboidal tetrad:** with all grains or spores of a tetrad in one plane forming a quadrangle. **Tetrahedral tetrad:** with one grain or spore of a tetrad lying centrally over the other three.

Trilete: with a three-slit aperture (fissure) at the proximal face of a spore.

Truncate: ending abruptly as if cut across.

Tuberculate: having rounded, short, warty processes densely placed on the surface of the sexine; tubercle (n.). **Microtuberculate:** with very small processes.

Ulcus: an irregulate, often indistinct, porelike aperture; ulcerate (adj.). **Ulceroid:** a vaguely defined ulcus.

Umbel: an indeterminate, flattened inflorescence whose floral stalks arise from a common point. **Umbellate:** pertaining to umbels.

Unisexual: of an organism having separate sexes and producing only one kind of gamete, either male or female, the individual being dioecious.

Vermiculate: having grooves that resemble worm tracts.

Verrucate: having warty processes on the surface of the sexine, the basal diameter being greater than the height. **Microverrucate:** with very small, warty processes.

Warty: see Verrucate.

Zoophily: pollination by animals. Cf. Entomophily.

Literature Cited

Adams, D.E., W.E. Perkins, and J.R. Estes. 1981. Pollination systems in *Paspalum dilatatum* Poir. (Poaceae): an example of insect pollination in a temperate grass. Amer. J. Bot. 68: 389–94.

Adams, R.J. and J.K. Morton. 1972–79. An atlas of pollen of the trees and shrubs of eastern Canada and the adjacent United States. Parts I–IV. Univ. of Waterloo Biol. Series 8–11.

Airy Shaw, H.K. 1973. *Willis' A Dictionary of the Flowering Plants and Ferns*, 8th ed. Cambridge: Cambridge Univ. Press, 1245 p.

Alemany-Vall, R. 1955. Casos de polinosis en 1954–1955. Med. Clin. 26: 100–108.

Allessio, M.L. and J.R. Rowley. 1966. Atmospheric pollen in Amherst, Massachusetts. Bot. Gaz. 127: 35–40.

Adolphson, C., L. Goodfriend, and G.J. Gleich. 1978. Reactivity of ragweed allergens with IgE antibodies. J. Allergy Clin. Immunol. 62: 197–210.

Anderson, G.J. 1976. The pollination biology of *Tilia*. Amer. J. Bot. 63: 1203–12.

Andersen, S.T. and F. Bertelsen. 1972. Scanning electron microscope studies of pollen of cereals and other grasses. Grana 12: 79–86.

Ballestero, L.H. and V. Monticelli. 1944. El tratamiento de la polinosis mediante antigenos genéricos y antigenos especificos con especial referencia al genero *Atriplex*. Semina Méd. 1: 715–18.

Balyeat, R.M. and H.J. Rinkel. 1931. Distribution and importance of the paper mulberry (*Papyrus papyrifera* Kuntze) as a cause of hay fever and asthma in the United States. J. Allergy 3: 7–18, 97–99.

Balyeat, R.M. and T.R. Stemen. 1927. Distribution and importance of the acnidas as a cause of hay fever and asthma in the United States. Amer. J. Med. Sci. 174: 639–47.

Barkley, T.M. (ed.). 1977. *Atlas of the Flora of the Great Plains*. Ames: Iowa State University Press, 600 p.

Barrett, C.E. 1934. Studies in hay fever. J. Allergy 5: 406–26.

Bassett, I.J. and C.W. Crompton. 1967. Air-borne pollen surveys in British Columbia. Canad. J. Pl. Sci. 47: 251–61.

Bassett, I.J., C.W. Crompton, and J.A. Parmelee. 1978. *An Atlas of Airborne Pollen Grains and Common Fungus Spores in Canada*. Monograph 18. Ottawa: Canada Department of Agriculture, 321 p.

Bawa, K.S. and J.E. Crisp. 1980. Wind-pollination in the understory of a rain forest in Costa Rica. J. Ecology 68: 871–76.

Belin, L. 1972a. Studies of birch pollen antigens with special reference to the allergenic principle. Grana 12: 65–78.

Belin, L. 1972b. Immunological analyses of birch pollen antigens, with special reference to the allergenic components. Inter. Arch. Allergy Appl. Immunol. 42: 300–322.

Bernton, H.S. 1925. Plantain hay-fever and asthma. JAMA 84: 944–46.

Bernton, H.S. 1928. Hay fever and asthma caused by the pollen of the paper mulberry (*Papyrus papyrifera* Kuntze). J. Lab. Clin. Med. 13: 829–36.

Bieberdorf, F. and B. Swinny. 1952. Mesquite and related plants in allergy. Ann. Allergy 10: 720–24.

Black, J.H. 1929. Cedar hay fever. J. Allergy 1: 71–73, 87, 89.

Blackwell, W.H. and M.J. Powell. 1982. A preliminary note on pollination in the Chenopodiaceae. Ann. Missouri Bot. Gard. 68: 524–26.

Blumstein, G.I. 1943. Sensitivity to *Ailanthus* pollen. J. Allergy 14: 329–34.

Blumstein, G.I. and L. Tuft. 1937. Plantain hay fever: its incidence and importance. JAMA 108: 1500–1502.

Blue, J.A. 1955a. Oklahoma, pp. 218–23. In *Regional Allergy of the United States, Canada, Mexico and Cuba*. M. Samter and O.C. Durham (eds.). Springfield, Ill.: C.C. Thomas.

Blue, J.A. 1955b. Summer blooming lamb's-quarters: a factor in inhalant allergy. Ann. Allergy 13: 304–6.

Bookman, R. 1978. Allergenicity of lawn grasses (letter). Ann. Allergy 40: 364.

Bruggen, T. van. 1976. *The Vascular Plants of South Dakota*. Ames: Iowa State University Press, 538 p.

Carron, R.F. and H.E. Malvarez. 1941. Polinosis por tala. Prensa Med. Argent. 28: 1114–16.

Chamberlain, C.T. 1927. Observation of the treatment of hayfever in the Pacific Northwest. Ann. Otol. Rhinol. Laryngol. 36: 1083–92.

Charpin, J. and H. Charpin. 1980. *Parietaria officinalis* L. allergy. J. Allergy Clin. Immunol. 65: 80–81.

Chen, S.-U. and T.-C. Huang. 1980. Aeropalynological study of Taipei basin, Taiwan. Grana 19: 147–55.

Ciampolini, F. and M. Cresti. 1981. *Atlante dei Principali Pollini Allergenici presenti in Italia*. Siena, Italy: Università da Siena.

Colldahl, H. 1954. Rape pollen allergy. Acta Allergol. 7: 367–69.

Collins, B.S. and J.A. Quinn. 1982. Displacement of *Andropogon scoparius* on the New Jersey Piedmont by the successional shrub *Myrica pensylvanica*. Amer. J. Bot. 69: 680–89.

Correll, D.S. and M.C. Johnston. 1970. *Manual of the Vascular Plants of Texas*. Renner: Texas Research Foundation. 1881 p.

Cronquist, A. 1955. Phylogeny and taxonomy of the Compositae. Amer. Midl. Nat. 53: 478–511.

Cronquist, A. 1977. The Compositae revisited. Brittonia 29: 137–53.

Dalen, G. van and R. Voorhorst. 1981. Allergen community in pollen from certain tree species. Ann. Allergy 46: 276–78.

Derbes, V.J. 1941. The linden (*Tilia*) as a factor in season pollinosis. J. Allergy 12: 502–6.

LITERATURE CITED

Dumm, J.F. and O. Zarate. 1944. La *Artemisia verlotorum* Lamott como in un factor de polinosis en la ciudad de La Plata. Rev. Asoc. Med. Argentina 58: 590–93.

Durham, O.C. 1943. *Kochia scoparia* as a factor in inhalant allergy. J. Allergy 14: 160–70.

Edmonds, R.L. (ed.). 1979. *Aerobiology: The Ecological Systems Approach*. Stroudsburg, Penn.: Dowden, Hutchinson & Ross, 386 p.

Elias, T.S. 1980. *The Complete Trees of North America*. New York: Van Nostrand Reinhold, 948 p.

Ellis, R.V. and C.O. Rosendahl. 1933. A survey of the causes of hay fever in the state of Minnesota. Minn. Med. 16: 379–89.

Erdtman, G. 1952. *Pollen Morphology and Plant Taxonomy: Angiosperms*. Stockholm: Almqvist & Wiksell, 553 p.

Erdtman, G. 1960. The acetolysis method, a revised description. Svensk. Bot. Tidskr. 54: 561–64.

Erdtman, G. 1969. *Handbook of Palynology*. New York: Hafner, 486 p.

Fein, B.T. and P.B. Kamin. 1962. A 10 year survey of the hay fever plants and important atmospheric allergens in the San Antonio, Texas, metropolitan area. J. Allergy 33: 141–52.

Fernald, M.L. 1950. *Gray's Manual of Botany*, 8th ed. New York: American Book, 1632 p.

Fly, L.B. 1952. A preliminary pollen analysis of the Miami, Florida, area. J. Allergy 23: 48–57.

Freeman, J. 1955. Mass experiments on pollen during 1949–1954. Acta Allergol. 8: 156–57.

French, S.W. 1930. Pollens which cause hay fever in south central Texas. J. Allergy 1: 286–91.

Furlow, J.J. 1979. The systematics of the American species of *Alnus* (Betulaceae). Rhodora 81: 1–121, 151–248.

Giscafre, L. and A.E. Ragonese. 1946. Polinosis occasionade por el Ligustro (*Ligustrum licidum* Ait.) en la Ciudad le Sante Fe. Darwiniana (Buenos Aires) 7: 198–207.

Gutmann, M.J. 1950. Hay fever in Palestine, second report. Ann. Allergy 8: 345–49, 381.

Hall, H.M. 1922. Hay fever plants of California. Public Health Rep. 37: 803–22.

Heijer, A. and K. Göransson. 1965. The significance of testing and hyposensitization with several grass pollens for hay fever. Acta Allergol. 23: 146–60.

Heinberg, C.J. 1930. Spiny amaranth. J. Florida Med. Assoc. 17: 221.

Heslop-Harrison, J. 1979. Aspects of the structure, cytochemistry and germination of the pollen of rye (*Secale cereale* L.). Ann. Bot. 44 (suppl. 1): 1–47, 18 pl.

Hitchcock, A.S. 1971. *Manual of the Grasses of the United States*, 2d ed. (revised by A. Chase). New York: Dover. 1051 p.

Hitchcock, C.L., A. Cronquist, M. Ownbey, and J.W. Thompson. 1955–69. *Vascular Plants of the Pacific Northwest*, vols. 1–5. Seattle: University of Washington Press.

Homan, R.B. 1963. Lamb's quarters pollen. Ann. Allergy 21: 647–49.

Hortus Third. 1976. New York: Macmillan. 1290 p.

Hyde, H.A. 1950. Studies in atmospheric pollen. IV. Pollen deposition in Great Britain, 1943. New Phytol. 49: 398–420.

Hyde, H.A. 1959. Volumetric counts of pollen grains at Cardiff, 1954–57. J. Allergy 30: 219–34.

Hyde, H.A. 1972. Atmospheric pollen and spores in relation to allergy. I. Clin. Allergy 2: 153–79.

Hyde, H.A. and D. A. Williams. 1945. Pollen of lime (*Tilia* spp.). Nature 155: 457.

Hyde, H.A. and K.F. Adams. 1958. *An Atlas of Airborne Pollen*. London: Macmillan. 111 p.

Jackson, R.C. 1960. A revision of the genus *Iva* L. Univ. Kansas Sci. Bull. 41: 793–876.

Jimenez-Diaz, C. 1932. *El Asma y Otras Enfermedades Alergicas*. Madrid: Editorial España. 999 p.

Kahn, I.S. and E.M. Grothaus. 1931. Hay fever and asthma due to red cedar (*Juniperus virginiana*) and to mountain cedar (*Juniperus sabinoides*). S. Med. J. 24: 729–30.

Kahn, I.S. and E.M. Grothaus. 1936. *Parthenium hysterophorus*: antigenic properties, respiratory and cutaneous. Texas St. J. Med. 32: 284–88.

Käpylä, M. 1981. Diurnal variation of non-arboreal pollen in the air in Finland. Grana 20: 55–59.

Kearney, T.H. and R.H. Peebles. 1951 (1960). *Arizona Flora (and Supplement)*. Berkeley and Los Angeles: University of California Press. 1085 p.

Kessler, A. 1958. Sensitivity to olive pollen (*Olea europaea*) as the cause of allergic diseases. Dapim Refuiim 17: 3.

King, T.P. 1979. Immunochemical properties of some atopic allergens. J. Allergy Clin. Immunol. 64: 159–63.

Kircher, Jr., T.E. 1955. New Mexico, pp. 333–39. In *Regional Allergy of the United States, Canada, Mexico and Cuba*. M. Samter and O.C. Durham (eds.). Springfield, Ill.: C.C. Thomas.

Knuth, P. 1906–9. *Handbook of Flowering Pollination*, vol. 1 (1906), vol. 2 (1908), vol. 3 (1909). Translated by J.R.A. Davis. Oxford: Clarendon Press.

Köhler, E. and E. Lange. 1979. A contribution to distinguishing cereal from wild grass pollen grains by LM and SEM. Grana 18: 113–40.

Kupias, R., A. Koivikko, and Y. Mäkinen. 1981. Liberation of *Taraxacum* and *Leucanthemum* pollen in the air through mechanical agitation. Grana 20: 199–203.

Lambright, G.L. and R.P. Albaugh. 1934. Perennial hay fever from *Lycopodium*. J. Allergy 5: 590–91.

Lamson, R.W. 1931. Evidence of biologic relationships among species of Chenopodiales. Proc. Soc. Exper. Biol. Med. 28: 502–4.

Lamson, R.W. and A. Watry. 1933. The importance of the Chenopodiaceae in pollinosis: with special reference to Winslow and Holbrook, Arizona. J. Allergy 4: 255–81.

Langley, W.D. 1937. Sensitivity to *Piqueria* pollen with case report. J. Allergy 9: 60–61.

Leiferman, K.M., G.J. Gleich, and R.T. Jones. 1976. Cross-reactivity of IgE antibodies with pollen allergens. II. Analysis of various species of ragweed and other fall weed pollens. J. Allergy Clin. Immunol. 58: 140–48.

Leuschner, R.M. and G. Boehm. 1981. Pollen and inorganic particles in the air of climatically very different places in Switzerland. Grana 20: 161–67.

Levetin, E. and P. Buck. 1980. Hay fever plants in Oklahoma. Ann. Allergy 45: 26–32.

Lewis, W.H. and M.P.F. Elvin-Lewis. 1977. *Medical Botany: Plants Affecting Man's Health*. New York: Wiley-Interscience. 515 p.

Lewis, W.H. and W.E. Imber. 1975a. Allergy epidemiology in the St. Louis, Missouri, area. II. Grasses. Ann. Allergy 35: 42–50.

Lewis, W.H. and W.E. Imber. 1975b. Allergy epidemiology in the St. Louis, Missouri, area. III. Trees. Ann. Allergy 35: 113–19.

Lewis, W.H. and W.E. Imber. 1975c. Allergy epidemiology in the St. Louis, Missouri, area. IV. Weeds. Ann. Allergy 35: 180–87.

Lewis, W.H. and P. Vinay. 1979. North American pollinosis due to insect pollinated plants. Ann. Allergy 42: 309–18.

Lewis, W.H. and P. Vinay. 1980. The unique role of pollen in relation to allergy, pp. 101–14. In *Allergy: Immunology and Medical Treatment*, F. Johnston and J.T. Spencer, Jr. (eds.). Miami: Symposia Specialists.

Lichtenstein, L.M., D.G. Marsh, and D.H. Campbell. 1969. In vitro studies of rye grass pollen antigens. J. Allergy 44: 307–14.

Licitis, R. 1953. Air-borne pollen and spores sampled at five New Zealand stations, 1951–52. New Zealand J. Sci. Tech., section B, 34: 289–316.

Lieux, M.H. 1980–82. An atlas of pollen of trees, shrubs, and woody vines of Louisiana and other southeastern states. Parts 1–3. Pollen et Spores 22: 17–57, 191–43; 24: 21–64.

Little, Jr., E.L. 1971–78. *Atlas of United States Trees*, vols. 1–5. Washington, D.C.: U.S.D.A. Forest Service.

Lindenbaum, S.E. 1966a. Case report: pollinosis due to *Ricinus communis* or castor bean plant. Ann. Allergy 24: 23–25.

Lindenbaum, S.E. 1966b. Pollinosis due to Amaranthaceae-Chenopodiaceae in Israel. Israel J. Med. Sci. 2: 546–48.

Lockey, R.F., J.J. Stablein, and L.R.F. Binford. 1981. *Melaleuca* tree and respiratory disease, pp. 101–15. In *Proceedings of Melaleuca Symposium*, R.K. Geiger (ed.). Tallahassee, Fla.: Division of Forestry, Department of Agriculture and Consumer Services.

Long, R.W. and O. Lakela. 1976. *A Flora of Tropical Florida*, rev. ed. Miami: Banyan Books. 962 p.

Luippold, E.J. 1974. 25,000,000 annual visitors to Florida—vacation or ordeal. Ann. Allergy 33: 206–8.

Mabry, T.J. 1970. Infraspecific variation of sesquiterpene lactones in *Ambrosia* (Compositae): application of evolutionary problems at the populational level, Chapter 13, pp. 269–300. In *Phytochemical Phylogeny*, J.B. Harborne (ed.). New York: Academic Press.

MacQuiddy, E.L. 1955. Northern prairies and plains (Iowa, Nebraska, and the Dakotas), pp. 183–195. In *Regional Allergy of the United States, Canada, Mexico & Cuba*, M. Samter and O.C. Durham (eds.) Springfield, Ill.: C.C. Thomas.

Maloney, E.S. and M.H. Brodkey. 1940. Hemp pollen sensitivity in Omaha. Nebraska Med. J. 25: 190–91.

Malley, A. et al. 1975. The chemical characterization of a major allergen of timothy pollen. Immunochem. 12: 551–54.

Marchand, A.M. 1948. Hay fever plants of Puerto Rico. Bol. Assoc. Med. Puerto Rico 40: 20–23.

Markgraf, V. 1980. Pollen dispersal in a mountain area. Grana 19: 127–46.

Marsh, D.G., Z.H. Haddad, and D.H. Campbell. 1970. A new method of determining the distribution of allergenic fractions in biological materials: its application to grass pollen extracts. J. Allergy 46: 107–21.

Martin, P.S. and C.M. Drew. 1969–70. Scanning electron photomicrographs of southwestern pollen grains. J. Arizona Acad. Sci. 5: 147–76; 6: 140–61.

Matsumura, T. et al. 1969. Rice pollen asthma. J. Asthma Res. 7: 7–16.

Menon, M.P.S., A.K. Das, and A.B. Singh. 1977. Dual asthmatic responses to *Prosopis juliflora*. Ann. Allergy 39: 351–54.

Morton, J.F. 1978. Poisonous plants of Florida. J. Florida Med. Assoc. 65: 162–70.

Munz, P.A. 1959. *A California Flora and Supplement* (1968). Berkeley and Los Angeles: University of California Press. 1681 p, 224 p.

Native Trees of Canada, 4th ed. 1949. Bulletin 61. Ottawa: Canada Department of Mines & Resources. 293 p.

Newmark, F.M. and I.H. Itkin. 1967. Asthma due to pine pollen. Ann. Allergy 25: 251–52.

Nichol, E.S. and O.C. Durham. 1931. A pollen survey of Miami, Florida. S. Med. J. 24: 947–48.

Nilsson, S. and S. Persson. 1981. Tree pollen spectra in the Stockholm region (Sweden), 1973–1980. Grana 20: 179–82.

Nilsson, S., J. Praglowski, and L. Nilsson. 1977. *Atlas of Airborne Pollen Grains and Spores in Northern Europe*. Stockholm: Almqvist & Wiksell. 160 p.

Novey, H.S., M. Roth, and I.D. Wells. 1977. Mesquite pollen—an aeroallergen in asthma and allergic rhinitis. J. Allergy Clin. Immunol. 59: 359–63.

Nowicke, J.W. 1975. Pollen morphology in the order Centrospermae. Grana 15: 51–77.

Ogden, E.C. et al. 1974. *Manual for Sampling Airborne Pollen*. New York: Hafner Press. 182 p.

Ordman, D. 1945. Cypress pollinosis in South Africa. S. Afr. Med. J. 19: 142–46.

Ordman, D. 1958. Lucerne as a cause of respiratory allergy in South Africa. S. Afr. Med. J. 32: 1121–22.

Ordman, D. 1959. The *Prosopis* tree as a cause of seasonal hay fever and asthma in Southwest Africa and South Africa. S. Afr. Med. J. 33: 12–14.

Ordman, D. 1963. Seasonal respiratory allergy and the associated pollens in South Africa. S. Afr. Med. J. 37: 321–25.

O'Rourke, M.K. 1980. Pollen dispersal and its relationship to respiratory illness, pp. 81–87. In *Proc. First Inter. Confer. on Aerobiology*, Munich. Berlin: E. Schmidt.

Page, J.S. 1978. A scanning electron microscope survey of grass pollen. Kew Bull. 32: 313–19.

Panzani, R. 1956. L'asthme pollinique à la pariétaire de France. Presse Med. 64: 908–10.

Payne, W.W. 1964. A re-evaluation of the genus *Ambrosia* (Compositae). J. Arnold Arb. 45: 401–30, 8 pl.

Perlman, F. 1955. The Pacific Northwest—west of the Cascade Mountain range, pp. 296–306. In *Regional Allergy of the United States, Canada, Mexico and Cuba*. M. Samter and O.C. Durham (eds.). Springfield, Ill.: C.C. Thomas.

Phillips, E.W. 1932. Pollen incidence of central Arizona. J. Allergy 3: 489–94.

Phillips, E.W. 1939. Time required for the production of hay fever by a newly encountered pollen, sugar beet. J. Allergy 11: 28–31.

Phillips, E.W. 1955. Arizona, pp. 340–45. In *Regional Allergy of the United States, Canada, Mexico and Cuba*. M. Samter and O.C. Durham (eds.). Springfield, Ill.: C.C. Thomas.

LITERATURE CITED

Piness, G. 1955. Southern California, pp. 356–67. In *Regional Allergy of the United States, Canada, Mexico and Cuba*. M. Samter and O.C. Durham (eds.). Springfield, Ill.: C.C. Thomas.

Player, G. 1979. Pollination and wind dispersal of pollen in *Arceuthobium*. Ecol. Monogr. 49: 73–87.

Prince, H.E. and G.H. Meyer, 1977. Hay fever from southern wax-myrtle (*Myrica cerifera*): a case report. Ann. Allergy 38: 252–54.

Prince, H.E. and P.G. Secrest, Jr. 1939. Immunologic relationship of giant, western, common ragweed and marsh elder (*Iva ciliata*). J. Allergy 10: 537–50.

Pujevic, S. 1959. Les manifestations allergiques dues à la sensibilisation par pollen du peuplier. Acta Allergol. 14: 180–84.

Purseglove, J.W. 1972. *Tropical Crops: Monocotyledons*. London: Longman. 607 p.

Quintero, J.M. 1955. Cuba, pp. 135–42. In *Regional Allergy of the United States, Canada, Mexico and Cuba*. M. Samter and O.C. Durham (eds.). Springfield, Ill.: C.C. Thomas.

Radford, A.E., H.E. Ahles, and C.R. Bell. 1968. *Manual of the Vascular Flora of the Carolinas*. Chapel Hill: University of North Carolina Press. 1183 p.

Randolph, H. and M. McNeil. 1944. Pollen studies of the Phoenix area. J. Allergy 15: 125–37.

Raynor, G.S., E.C. Ogdon, and J.V. Hayes. 1972. Dispersion and deposition of timothy pollen from experimental sources. Agric. Meteorol. 9: 347–66.

Rehder, A. 1947. *Manual of Cultivated Trees and Shrubs Hardy in North America*, 2d ed. New York: Macmillan. 995 p.

Rogers, H.L. 1944. Sensitivity to minor pollens. Ann. Allergy 2: 125–28.

Roth, A. and J. Shia. 1966. Allergy in Hawaii. Ann. Allergy 24: 73–78.

Rowley, J.R. 1960. The exine structure of "cereal" and "wild" type grass pollen. Grana Palynol. 2: 9–15.

Rowley, J.R. 1964. Formation of the pore in pollen of *Poa annua*, pp. 59–69. In *Pollen Physiology and Fertilization*, H.F. Linskens (ed.). Amsterdam: North-Holland.

Rowe, A.H. 1927. A study of the atmospheric pollen and botanic flora of the east shore of San Francisco bay. J. Lab. Clin. Med. 13: 416–39.

Rowe, A.H. 1939. Pine pollen allergy. J. Allergy 10: 377–78.

Rowe, A.H. 1955. Northern California, pp. 346–55. In *Regional Allergy of the United States, Canada, Mexico and Cuba*. M. Samter and O.C. Durham (eds.). Springfield, Ill.: C.C. Thomas.

Salazar Mallen, M. 1955. Mexico, pp. 368–379. In *Regional Allergy of the United States, Canada, Mexico and Cuba*. M. Samter and O.C. Durham (eds.). Springfield, Ill.: C.C. Thomas.

Salén, E.G. 1951. *Lycopodium* allergy. Acta Allergol. 4: 308–19.

Scheppegrell, W. 1917. Hay fever: its cause and prevention in the Rocky Mountain and Pacific states. U.S. Public Health Rep. 412: 1135–52.

Scheppegrell, W. 1922. *Hayfever and Asthma*. New York and Philadelphia: Lea & Febiger. 274 p.

Scoggan, H.J. 1978-79. *The Flora of Canada*, vols. 1–4. Ottawa: National Museums of Canada.

Sellers, E.D. 1929. Mesquite tree pollen as a cause of hayfever. Texas St. J. Med. 25: 297–99.

Sellers, E.D. 1935. Pollinosis in the southwest. S. Med. J. 28: 710-14.

Sellers, E.D. and W.B. Adamson. 1932. A study of the apparent atopic similarity of certain Chenopodiales pollens. J. Allergy 3: 161–71.

Seymour, F.C. 1969. *The Flora of New England*. Rutland, Vt.: C.E. Tuttle. 596 p.

Serafini, U. 1957. Studies on hay fever with special regard to pollinosis due to *Parietaria officinalis*. Acta Allergol. 11: 3–27.

Sheldon, J.M., R.G. Lovell, and K.P. Mathews. 1967. *A Manual of Clinical Allergy*, 2d ed. Philadelphia: Saunders. 550 p.

Silva, Q.G.P. da. 1960. The incidence of *Olea* pollen in Portugal in five consecutive years. Acta Allergol. 15: 107–12.

Simpson, B.B. (ed.). 1977. *Mesquite: Its Biology in Two Desert Scrub Ecosystems*. Stroudsburg, Penn.: Dowden, Hutchinson & Ross.

Stingh, K. and D.N. Shivpuri. 1971. Studies in yet unknown allergenic pollens of Delhi state metropolitan (clinical aspects). Indian J. Med. Res. 59: 1397–1410.

Skvarla, J. and D.A. Larson. 1965. An electron microscopic study of pollen morphology in the Compositae with special reference to the Ambrosiinae. Grana Palynolog. 6: 210–69.

Small, W.S. 1952. Increasing castor bean allergy in southern California due to fertilizer. J. Allergy 23: 406–15.

Small, W.S. and G.M. Small. 1946. Botanical survey of southern California. Ann. Allergy 4: 352–71.

Smart, I.J. and R.B. Knox. 1979. Aerobiology of grass pollen in the city atmosphere of Melbourne: quantitative analysis of seasonal and diurnal changes. Austral. J. Bot. 27: 317–31.

Solomon, W.R. 1969. An appraisal of *Rumex* pollen as an aeroallergen. J. Allergy 44: 25–36.

Solomon, W.R. and O.C. Durham. 1967. Aeroallergens II. Pollens and the plants that produce them, pp. 340–397. In *A Manual of Clinical Allergy*, J.M. Sheldon, R.G. Lovell, and K.P. Mathews (eds.). Philadelphia: Saunders.

Stanley, R.G. and H.F. Linskens. 1974. *Pollen: Biology, Biochemistry, Management*. Berlin: Springer. 307 p.

Statistical Report of the Pollen and Mold Committee of the American Academy of Allergy. 1974–78. Columbus, O.: Ross Laboratories.

Steyermark, J.A. 1963. *Flora of Missouri*. Ames: Iowa State University Press. 1725 p.

Stone, D.E. and C.R. Broome. 1975. World Pollen and Spore Flora 4: Juglandaceae. Stockholm: Almqvist & Wiksell. 35 p.

Strausbaugh, P.D. and E.L. Core. 1978. *Flora of West Virginia*, 2d ed. Gransville, West Va.: Seneca Books. 1079 p.

Stuessy, T.F. 1973. A systematic review of the subtribe Melampodiinae (Compositae, Heliantheae). Contr. Gray Herb. (Harvard Univ.) 203: 65–80.

Swineford, Jr., C. 1940. *Catalpa* as a cause of hay fever. J. Allergy 11: 398–401.

Targow, A.M. 1971. The mulberry tree: a neglected factor in respiratory allergy in southern California. Ann. Allergy 29: 318–22.

Tas, J. 1956. *Ailanthus glandulosa* pollen as a cause of hay fever. Ann. Allergy 14: 47–49.

Tas, J. 1965. Hay fever due to the pollen of *Cupressus sempervirens*. Acta Allergol. 20: 405–7.

Thiberge, N. and G.H. Hauser. 1931. The value of atmospheric pollen plates in hayfever and asthma. S. Med. J. 24: 1049–53.

Tinsley, H.M. and R.T. Smith. 1974. Surface pollen studies across a woodland / heath transition and their application to the interpretation of pollen diagrams. New Phytol. 73: 547–65.

Tocker, A.M. 1956. The possible role of *Iva ciliata* in hay fever therapy. S. Med. J. 49: 445–52.

Tomlinson, P.B. 1980. *The Biology of Trees Native to Tropical Florida*. Allston, Mass.: Harvard University Printing Office. 480 p.

Unger, D.L. 1977. Allergy practice in the desert. Ann. Allergy 39: 300–301.

Ursing, B. 1968. Sugar beet pollen allergy as an occupational disease. Acta Allergol. 23: 396–99.

Vaughan, W.T. 1939. *Practice of Allergy*. St. Louis: Mosby. 1082 p.

Vaughan, W.T. and J.H. Black. 1948. *Practice of Allergy*, 2d ed. St. Louis: Mosby. 1132 p.

Vaughan, W.T. and R.W. Crockett. 1932. An assay of goldenrod as a cause of hay fever. Ann. Int. Med. 6: 789–94.

Vaughan, W.T., W.R. Graham, and R.W. Crockett. 1933. Hayfever pollen prevalence in Virginia; review of a six year's survey. Va. Med. Monthly 60: 158–62.

Wagatsuma, Y. and T. Shida. 1969. Ezo-Yomogi (*Artemisia montana*) pollinosis in Sapporo, Hokkaido, Japan. Allergy 18: 980–90.

Wagatsuma, Y. et al. 1972. Japanese white birch pollinosis in Sapporo, Japan. Allergy 21: 710–17.

Walkington, D.L. 1960. A survey of the hayfever plants and important atmospheric allergens in Phoenix, Arizona, metropolitan area. J. Allergy 31: 25–41.

Walton, C.H.A. and M. Dudley. 1947. A geographical study of hay fever plants in Manitoba. Canad. Med. Assoc. J. 56: 142–48.

Watson, L. and R.B. Knox. 1976. Pollen wall antigens and allergens: taxonomically-ordered variation among grasses. Ann. Bot. 40: 399–408.

Weber, R.W. 1981. Cross-reactivity among pollens. Ann. Allergy 46: 208–14.

Wodehouse, R.P. 1935. *Pollen Grains*. New York: McGraw-Hill. 574 p.

Wodehouse, R.P. 1957. Antigenic analysis by gel diffusion. III. Pollens of the Amaranth-Chenopod group. Ann. Allergy 15: 527–36.

Wodehouse, R.P. 1971. *Hayfever Plants*, 2d ed. New York: Hafner. 280 p.

Wolf, A.F. 1948. A fall-pollinating red berry juniper. Ann. Allergy 6: 431–34, 441.

Wright, G.L.T. and H.T. Clifford. 1965. The relationship of immediate grass pollen skin reactions to taxonomic groups in grasses. Med. J. Austral. 2: 74–75.

Ybert, J.-P. 1980. Le contenu pollinique de l'atmosphère en Côte d'Ivoire et au Tchad. Grana 19: 31–46.

Yoo, T.-J., E. Spitz, and J.L. McGerity. 1974. Allergy to Cupressaceae pollen. J. Allergy Clin. Immunol. 53: 71–72 (abstract).

Zivitz, N. 1942. Allergy to Australian pine. J. Allergy 13: 314–16.

Index

For page references of common (vernacular) names of plants, see their scientific names. Generic names only are in italics, including those considered synonyms (syn.). Names of higher taxa, such as tribe (-eae), subfamily (-oideae), and family (-aceae), are indexed with synonyms where relevant, but without italics. Species names are not indexed but may be found under appropriate genera either in the text or in greater detail in Appendix 1 (pp. 177–236). Colored plates of pollen, organized as an identification key, and of selected trees, grasses, and weeds found in the front of the volume are indexed according to plate number (I–XVI). Pages for plant distribution maps or occurrence in hardiness zones from Appendix 1 are followed by an "m."

Abelia, 31
Abies, 9–10, 179m
Acacia, XIII, 45, 48–49, 191m
Acaena, 81, 83, 203m
Acalypha, X, XVI, 158–60, 232m
Acanthambrosia (syn.). See *Ambrosia*
Acer, I, XV, 12–16, 18, 182m
Aceraceae, 12–16, 18
Acetolysis, 5, 237
Achillea, 139, 143
Achyranthes, 129
Acnida (syn.). See *Amaranthus*
Acoelorraphe, 22, 185m
Aconitum, 171
Aerobiology. See Pollen
Aesculus, 58, 62, 197m
Ageratum, 139, 143
Agropyron, 106–11, 113–14, 213m
Agrostideae, 106
Agrostis, 106–10, 113–15, 213m
Ailanthus, XVI, 87, 89, 93–94, 210m
Albizia, 45, 48–49, 191m
Alder. See *Alnus*
Alfalfa. See *Medicago*
Allenrolfea, 152, 155, 230m
Allergenicity, 5–6
Almond. See *Prunus*
Alnus, I, XIV, 23–28, 30, 185–86m
Alopecurus, 106–10, 113–16, 213m
Alternanthera, 129, 133, 223m
Alvoradoa, 89, 210m
Amaranth. See *Amaranthus*
Amaranth, globe. See *Gomphrena*
Amaranth family. See Amaranthaceae
Amaranthaceae, 129, 132–35, 155

Amaranthus, XIV, 129, 132–35, 155, 223–24m
Ambrosia, IX, XVI, 3, 110, 137–44, 146, 149–50, 158, 225–26m
Ambrosiinae, 139–40
Amelanchier, 82–83, 203m
American Academy of Allergy, Pollen and Mold Committee of the, 1, 4
Ammophila, 106–8, 110, 113–14, 116, 213m
Anacardiaceae, 15, 17, 19–20
Anacardium, 15, 183m
Andropogon, 106–11, 113–14, 116, 214m
Andropogoneae, 106
Angiosperms, 12–103, 105–27, 129, 132–76
Angiosperms, phylogenetic tree of, 1–2
Antennaria, 142
Anthemideae, 139–40, 149
Anthemis, 139, 143
Anthoxanthum, 106–8, 110, 112, 114, 116, 214m
Antigen E (AgE), 149–50
Antigen K (AgK), 149–50
Antigen Ra3 (AgRa3), antigen Ra4 (AgRa4), antigen Ra5 (AgRa5), 149–50
Aperture, 5
Apiaceae, 135–36
Apple. See *Malus*
Apricot. See *Prunus*
Aquifoliaceae, 17, 21
Araliaceae, 19, 21–22
Arborvitae. See *Thuja*
Arbutus, 40, 42–44, 190m
Arceuthobium, 98, 101, 212m
Arctium, XVI, 142
Arctotideae, 139–40
Arecaceae, 21–24
Arrhenatherum, 106–10, 114, 116, 214m
Arrowgrass. See *Triglochin*
Arrowgrass family. See Juncaginaceae
Artemisia, X, XV, 138–43, 146, 149, 225–27m
Artemisiastrum (syn.). See *Artemisia*
Arundineae, 106
Arundinoid subfamily. See Arundinoideae
Arundinoideae, 106
Arundo, 106–11, 113–14, 116, 214m
Ash. See *Fraxinus*
Aspen. See *Populus*
Aster, 139, 142–43
Aster tribe or family. See Astereae; Asteraceae
Asteraceae, 3, 135–50; tribes, 139–40

Astereae, 140, 149
Ateramnus, 45, 190m
Atriplex, XIV, 152, 154–56, 158, 229–30m
Australian pine. See *Casuarina*
Australian pine family. See Casuarinaceae
Avena, XIII, 105–11, 113–14, 116, 214m
Aveneae, 106
Avocado. See *Persea*
Axyris, 152, 155–56, 229m
Azalea. See *Rhododendron*

Baccharis, XVI, 137, 139–40, 142–43, 147, 149, 227–28m
Bahia grass. See *Paspalum*
Bald cypress. See *Taxodium*
Bald cypress family. See Taxodiaceae
Bamboo palm. See *Chamaedorea*
Bambusoid subfamily. See Bambusoideae
Bambusoideae, 106
Barberry. See *Berberis*
Barberry family. See Berberidaceae
Barley. See *Hordeum*
Barnyard grass. See *Echinochloa*
Basic fuchsin preparation, 237
Bassia, XIV, 152, 154–56, 158, 229m
Basswood. See *Tilia*
Bay, California. See *Umbellularia*
Bay, red. See *Persea*
Bay, swamp. See *Persea*
Beachgrass. See *Ammophila*
Beach-heath. See *Hudsonia*
Beardgrass. See *Andropogon*
Beech. See *Fagus*
Beech, southern. See *Nothofagus*
Beech family. See Fagaceae
Beefwood. See *Casuarina*
Bellis, 138–39, 225m
Bentgrass. See *Agrostis*
Berberidaceae, 23, 25
Berberis, XV, 23, 25, 185m
Bermuda grass. See *Cynodon*
Beta, XIV, 152, 155–56, 158, 229m
Betula, I, XIV, 23–28, 30, 68, 185–86m
Betulaceae, 23–30, 68, 159
Betuloideae, 25
Bignonia family. See Bignoniaceae
Bignoniaceae, 30–31
Birch. See *Betula*
Birch family. See Betulaceae
Bittersweet. See *Celastrus*
Black gum. See *Nyssa*
Blackbeard. See *Pithecellobium*
Bloodleaf. See *Iresine*
Blue palmetto. See *Rhapidophyllum*
Blueberry. See *Vaccinium*
Bluegrass. See *Poa*

249

INDEX

Bluestem. See *Andropogon*
Bocconia, 162, 165, 232m
Boehmeria, 173–75, 235m
Bog-hemp. See *Boehmeria*
Bogrush. See *Juncus*
Boneset. See *Eupatorium*
Boneset tribe. See Eupatorieae
Boston ivy. See *Parthenocissus*
Bottlebrush. See *Callistemon*; *Hystrix*
Bougueria, 164
Bouteloua, 105–10, 114, 166, 214m
Box-elder. See *Acer*
Boxwood. See *Buxus*
Boxwood family. See Buxaceae
Bracken fern. See *Pteridium*
Brassica, 150–51, 229m
Brassicaceae, 150–51
Bridal wreath. See *Spiraea*
Bristlegrass. See *Setaria*
Brittle bush. See *Encelia*
Briza, 106, 108–10, 113–14, 116, 214m
Broccoli. See *Brassica*
Bromeae, 106
Bromegrass. See *Bromus*
Bromus, VII, 106–10, 113–14, 116, 214–15m
Broussonetia, 64, 66, 69–70, 199m
Bruckenthalia, 42, 190m
Buchloe, 105–10, 114, 116, 215m
Buckeye. See *Aesculus*
Buckthorn family. See Rhamnaceae
Buckwheat. See *Fagopyrum*
Buckwheat, wild. See *Eriogonum*
Buckwheat family. See Polygonaceae
Buffalo grass. See *Buchloe*
Buffaloberry. See *Shepherdia*
Bulrush. See *Scirpus*
Burnet. See *Sanguisorba*
Burning bush. See *Euonymus*; *Kochia*
Bur-reed. See *Sparganium*
Bur-reed family. See Sparganiaceae
Burro-brush. See *Hymenoclea*
Bur-sage. See *Ambrosia*
Buttercup. See *Ranunculus*
Buttercup family. See Ranunculaceae
Buxaceae, 30–32
Buxus, 30–32, 187m

Cabbage. See *Brassica*
Caesalpinioideae, 51
Calamagrostis, 106–10, 114, 116, 215m
Calamondin. See X *Citrofortunella*
Calenduleae, 140
Calliandra, XIII, 45, 49, 191m
Callistemon, XVI, 69, 71, 73–74, 200m
Calluna, XIII, 40, 42–44, 190m
Camphor. See *Cinnamomum*
Campsis, 30
Canaigre. See *Rumex*
Canary grass. See *Phalaris*
Cannabaceae, 150–53
Cannabis, 150–53, 229m
Caprifoliaceae, 31, 33–34
Carduus, 139, 142–43, 147, 225m
Carex, VIII, XIII,, 121–22, 220–21m
Carob. See *Ceratonia*
Carpinus, XIV, 23, 25, 27, 29–30, 187m

Carrot. See *Daucus*
Carrot family. See Apiaceae
Carya, III, XIV, 58, 60, 63–65, 197–98m
Cashew. See *Anacardium*
Cashew family. See Anacardiaceae
Cassia, 48, 51, 192m
Castanea, XVI, 51–54, 56, 192–93m
Castanopsis, XVI, 51, 53–54, 56, 193m
Castor bean. See *Ricinus*
Casuarina, XIV, 34–36, 188m
Casuarinaceae, 34–36, 68
Catalpa, 30–31, 187m
Cat's claw. See *Mimosa*
Cattail. See *Typha*
Cattail family. See Typhaceae
Ceanothus, 80–81, 202m
Cedar. See *Cedrus*; *Juniperus*
Cedar, incense. See *Libocedrus*
Cedar, red. See *Juniperus*
Cedrus, 179m
Celastraceae, 36–37
Celastrus, 36, 188m
Celosia, 129, 133, 135
Celtideae, 96
Celtis, VI, XIV, 92–99, 210–11m
Centaurea, 139, 142–43
Centipede grass. See *Eremochloa*
Ceratiola, 40, 42, 189m
Ceratonia, 48
Cercidium, 51
Cercis, 48, 51, 192m
Cercocarpus, XV, 81–83, 203m
Chaff flower. See *Achyranthes*; *Alternanthera*
Chamaecyparis, XIII, 7–9, 178m
Chamaedorea, XV, 23–24, 184m
Chamartemisia (syn.). See *Tanacetum*
Chamomile. See *Anthemis*
Chasmanthium, 106–9, 113–14, 116–17, 215m
Chenopodiaceae, 133, 135, 152, 154–58
Chenopodium, XI, XIV, 133, 152, 153, 157–58, 229–31m
Cherry. See *Prunus*
Cherry palm. See *Pseudophoenix*
Chess. See *Bromus*
Chestnut. See *Castanea*
Chicory. See *Cichorium*
Chicory tribe. See Lactuceae
Chilopsis, 31, 187m
Chinquapin. See *Castanopsis*
Chionanthus, 73, 75, 77–78, 200–201m
Chlorideae, 106
Chloridoid subfamily. See Chloridoideae
Chloridoideae, 106
Chloris, 105–6, 109–11, 113–14, 117, 215m
Chokecherry. See *Prunus*
Chorisiva (syn.). See *Iva*
Chrysanthemum, 139, 142–43, 149
Chrysothamnus, 141
Cichorium, 138–39, 142–43, 148, 225m
Cigar tree. See *Catalpa*
Cinna, 106–10, 114, 117, 215m

Cinnamomum, 60, 67, 198m
Cirsium, 139, 143
Cistaceae, 36–37
X *Citrofortunella*, 85
Citrus, 84–86, 206m
Citrus family. See Rutaceae
Classification, phylogenetic, 1–3
Clearweed. See *Pilea*
Clematis, 171–73, 234m
Cliffrose. See *Cowania*
Clover. See *Trifolium*
Clover, sweet. See *Melilotus*
Club-moss. See *Lycopodium*
Clusiaceae, 37–38
Coca's solution preparation, 238
Cocklebur. See *Xanthium*
Cockscomb. See *Celosia*
Cocksfoot grass. See *Dactylis*
Coconut. See *Cocos*
Cocos, XV, 22–24, 184m
Coltsfoot. See *Tussilago*
Compositae (syn.). See Asteraceae
Composite family. See Asteraceae
Comptonia, XIV, 66, 68–69, 72, 200m
Coneflower. See *Rudbeckia*
Conifers. See Gymnosperms
Coral bean. See *Erythrina*
Cordgrass. See *Spartina*
Corema, 40, 190m
Corispermum, 152, 155, 229m
Cork tree. See *Phellodendron*
Corkwood, Florida. See *Leitneria*
Corkwood family. See Leitneriaceae
Corn. See *Zea*
Cornaceae, 38–39
Cornus, XV, 38–39, 188–89m
Coryloideae, 25
Corylus, XIV, 23–27, 29–30, 68, 185m, 187m
Cotinus, 15, 183m
Cottonweed. See *Froelichia*
Cottonwood. See *Populus*
Couchgrass. See *Agropyron*
Cowania, 81–83, 204m
Crabgrass. See *Digitaria*
Crabwood. See *Ateramnus*
Crataegus, XVI, 82–83, 203–4m
Creosote bush. See *Larrea*
Crossostephium (syn.). See *Artemisia*
Croton, 159
Crowberry. See *Empetrum*
Crowberry family. See Empetraceae
Cruciferae (syn.). See Brassicaceae
Cudrania, XIV, 66, 70, 199m
Cupressaceae, 7–9
Cupressus, XIII, 7–9, 178m
Currant. See *Ribes*
Cyclachaena (syn.). See *Iva*
Cycloloma, 152, 155, 231m
Cynareae, 139–40
Cynodon, 105–11, 113–14, 117, 215m
Cynosurus, 106–10, 112, 114, 117, 215m
Cyperaceae, 121–22
Cyperus, VIII, 121–22, 221m
Cypress. See *Cupressus*
Cypress, false. See *Chamaecyparis*
Cypress family. See Cupressaceae
Cytochrome c, 149

INDEX

Dactylis, VII, XIII, 106–11, 113–14, 117, 119, 215m
Daisy. See *Aster*
Daisy, English. See *Bellis*
Daisy, ox-eye. See *Chrysanthemum*
Daisy tribe or family. See Astereae; Asteraceae
Dallis grass. See *Paspalum*
Dandelion. See *Taraxacum*
Darnel. See *Lolium*
Date palm. See *Phoenix*
Daucus, XV, 135–36, 224m
Dermatitis, 15, 173
Deschampsia, 106–10, 114, 117, 216m
Desert-willow. See *Chilopsis*
Desmanthus, 45, 48, 50, 191m
Diarrhena, 106–7, 109, 117, 216m
Dicoria, 139–40, 142–43, 228m
Dicraurus, 129, 133
Digitaria, 106–11, 114, 117, 216m
Discoid tribes (Asteraceae), 139
Distichlis, 106–11, 114, 117, 216m
Dog fennel. See *Anthemis*
Dogtail. See *Cynosurus*
Dogwood. See *Cornus*
Dogwood family. See Cornaceae
Dondia (syn.). See *Suaeda*
Douglas fir. See *Pseudotsuga*
Dropseed. See *Sporobolus*

Echinochloa, 106–11, 113–14, 117, 216m
Elaeagnaceae, 39–41
Elaeagnus, 39–41, 189m
Elaeis, 22
Elderberry. See *Sambucus*
Elephant grass. See *Pennisetum*
Elm. See *Ulmus*
Elm, water. See *Planera*
Elm family. See Ulmaceae
Elymus, 106–10, 14, 117, 216m
Empetraceae, 40, 42
Empetrum, 40, 190m
Encelia, 139–40, 143
English ivy. See *Hedera*
Epilobium, 159, 163–64, 232m
Equisetum, 129–31, 223m
Eragrostideae, 106
Eragrostis, 106–10, 114, 117, 216m
Eremochloa, 106, 109, 111, 114
Erica, 40, 42, 190m
Ericaceae, 40, 42–44
Erigeron, 139, 141, 143
Eriogonum, 169–70, 234m
Erythrina, 192m
Eschscholzia, XV, 162, 164–65, 232m
Eucalypt. See *Eucalyptus*
Eucalyptus, XVI, 69, 71, 73–74, 200m
Euonymus, 36–37, 188m
Eupatorieae, 139–40, 149
Eupatorium, XVI, 139, 141–43, 146, 228m
Euphorbiaceae, 45–47, 158–60
Euphrosyne, 142
Eurotia, XIV, 152, 155, 157, 231m
Evening-primrose family. See Onagraceae
Everglades palm. See *Acoelorraphe*
Exine, 5

Fagaceae, 30, 51–57
Fagopyrum, 169, 234m
Fagus, III, XVI, 30, 51–54, 192–93m
Fan palm. See *Trachycarpus*
Feathergrass. See *Stipa*
Fennel. See *Foeniculum*
Ferns and fern-allies. See Pteridophytes
Fescue. See *Festuca*
Festuca, VII, XIII, 106–11, 113–14, 118, 216–17m
Feverfew. See *Parthenium*
Filbert. See *Corylus*
Filipendula, 82
Fimbristylis, 121–22, 221m
Fingergrass. See *Chloris*
Fir. See *Abies*
Fireweed. See *Epilobium*
Fleabane. See *Erigeron*
Florists' stevia. See *Piqueria*
Floss flower. See *Ageratum*
Flowering and aeropollen data map, for North America, 4
Flowering plants. See Angiosperms
Flowering times, 3–4; reference volumes for, 3
Foeniculum, 224m
Forsythia, 73, 75–78, 200m
Fortunella, 86
Fountaingrass. See *Pennisetum*
Foxtail. See *Alopecurus*
Franseria (syn.). See *Ambrosia*
Fraxinus, VI, XV, 73, 75–77, 79, 201–2m
Fringe tree. See *Chionanthus*
Froelichia, XIV, 129, 132–34, 224m
Frostweed. See *Halianthemum*

Gaertneria (syn.). See *Ambrosia*
Galleta. See *Hilaria*
Gamagrass. See *Tripsacum*
Garrya, XVI, 57–59, 196m
Garryaceae, 57–59
Ginkgo, XV, 11–13, 182m
Ginkgo family. See Ginkgoaceae
Ginkgoaceae, 11–13
Ginseng family. See Araliaceae
Gleditsia, 51, 192m
Glycerin jelly preparation, 237
Golden bells. See *Forsythia*
Golden-heather. See *Hudsonia*
Golden-rain tree. See *Koelreuteria*
Goldenrod. See *Solidago*
Gomphrena, 129, 133, 224m
Gooseberry. See *Ribes*
Goosefoot. See *Chenopodium*
Goosefoot family. See Chenopodiaceae
Grama. See *Bouteloua*
Gramineae (syn.). See Poaceae
Grape. See *Vitis*
Grape family. See Vitaceae
Grass: classification, 105–6; flowering, 106–11; pollen aerobiology, 110–11; pollen allergenicity, 113–14; pollen morphology, 112–18
Grass family. See Poaceae
Grayia, 152, 155, 231m

Greasewood. See *Larrea*; *Sarcobatus*
Grindelia, 139, 143
Groundsel tribe. See Senecioneae
Guayule. See *Parthenium*
Gum tree. See *Eucalyptus*
Guttiferae (syn.). See Clusiaceae
Gymnocladus, 50, 51, 192m
Gymnosperms, 7–12

Hackberry. See *Celtis*
Hairgrass. See *Deschampsia*
Hairgrass, crested. See *Koeleria*
Halogeton, 152, 155, 231m
Haloragaceae, 159, 163
Hamamelidaceae, 58, 60–61
Hamamelis, 58, 60–61, 196–97m
Hammerwort. See *Parietaria*
Hardiness zones, map of, 3, 177
Haw. See *Crataegus*
Hawthorn. See *Crataegus*
Hazelnut. See *Corylus*
Heath. See *Erica*
Heath family. See Ericaceae
Heather. See *Calluna*
Hedera, 19, 21–22, 184m
Helenium, 139, 142–43
Heliantheae, 139–40, 149
Helianthemum, 36
Helianthus, 139, 141, 143
Hemp family. See Cannabaceae
Hickory. See *Carya*
Hilaria, 106, 109–10, 114, 118, 217m
Hippocastanaceae, 58, 62
Hippophae, 40, 189m
Holcus, 106–8, 110, 113–14, 118, 217m
Holly. See *Ilex*
Holly family. See Aquifoliaceae
Holodiscus, 81–82, 84, 204m
Honeysuckle. See *Lonicera*
Honeysuckle family. See Caprifoliaceae
Hop hornbeam. See *Ostrya*
Hop tree. See *Ptelea*
Hops. See *Humulus*
Hop-sage. See *Grayia*
Hordeum, VII, 106–10, 112–14, 118, 217m
Hornbeam. See *Carpinus*
Horse chestnut. See *Aesculus*
Horse chestnut family. See Hippocastanaceae
Horsetail. See *Equisetum*
Hudsonia, 36–37, 188m
Humulus, XI, XIV, 150–53, 229m
Hydrangea, 87, 93, 209m
Hymenoclea, 139, 141–43, 145, 149, 228m
Hypericum, 37–38, 188m
Hypersensitivity: antibody-mediated (type I), 5–6; delayed (type IV), 17
Hystrix, 106–9, 113–14, 118, 217m

IgE, 5, 150
Ilex, 17, 21, 184m
Indian grass. See *Sorghastrum*
Inhalant allergy, important Angiosperms in, 2
Intine, 5

INDEX

Inuleae, 140
Iodine bush. See *Allenrolfea*
Iresine, 129, 133–35, 224m
Ironweed tribe. See Vernonieae
Iva, X, XVI, 137, 139–43, 145, 149, 228–29m

Japanese ivy. See *Parthenocissus*
Japanese lawngrass. See *Zoysia*
Jasmine. See *Jasminum*
Jasminum, 73, 75, 77, 201m
Johnson grass. See *Sorghum*
Jojoba. See *Simmondsia*
Juglandaceae, 58–60, 63–65
Juglans, III, XIV, 58–60, 63–65, 197–98m
Juncaceae, 121, 123
Juncaginaceae, 123–24
Juncus, 123, 221m
June grass. See *Poa*
June grass, western. See *Koeleria*
Juniper. See *Juniperus*
Juniperus, XIII, 7–9, 178–79m

Kalmia, XIII, 40, 42–44, 190m
Kentucky coffee tree. See *Gymnocladus*
Knapweed. See *Centaurea*
Knotgrass. See *Paspalum*
Knotweed. See *Polygonus*
Kochia, XV, 152, 154–55, 157–58, 229m, 231m
Koeleria, 106–7, 109–10, 114, 118, 217m
Koelreuteria, 87, 92, 209m

Lactuceae, 136, 139–40
Lamb's quarters. See *Chenopodium*
Laportea, 173–75, 235m
Larch. See *Larix*; *Pseudolarix*
Larix, XIII, 9, 179m
Larrea, XVI, 98, 102–3, 212m
Lauraceae, 60, 62, 66–67
Laurel, California. See *Umbellularia*
Laurel family. See Lauraceae
Lead tree. See *Leucaena*
Lechea, 36
Ledum, 42, 190m
Legume family. See Fabaceae
Leguminosae (syn.). See Fabaceae
Leitneria, 62, 68, 199m
Leitneriaceae, 62, 68
Leucaena, 45, 50, 191m
Leucanthemum (syn.). See *Chrysanthemum*
Leuciva (syn.). See *Iva*
Libocedrus, 7, 9, 179m
Light microscopy (LM), 237
Lignum vitae family. See Zygophyllaceae
Ligustrum, XVI, 73, 75–78, 201m
Lilac. See *Syringa*
Linden. See *Tilia*
Linden family. See Tiliaceae
Lindera, 60, 66–67, 198m
Liquidambar, XV, 58, 60–61, 197m
Liriodendron, 62, 64, 68, 199m
Lithocarpus, 51, 53–54, 56, 193m
Littorella, 164, 166, 233m

Locust, black. See *Robinia*
Locust, honey. See *Gleditsia*
Lolium, VII, XIII, 106–10, 112–14, 118–19, 217m
Lonicera, 31, 33–34, 187m
Lovegrass. See *Eragrostis*
Lucerne. See *Medicago*
Luzula, VIII, XIII, 123, 221m
Lycopodium, XIII, 129–31, 223m

Maclura, XIV, 64, 66, 69–70, 199m
Madrone. See *Arbutus*
Magnolia, XV, 62, 64, 68, 199m
Magnolia family. See Magnoliaceae
Magnoliaceae, 62, 64, 68
Mahonia (syn.). See *Berberis*
Maidenhair tree. See *Ginkgo*
Maize. See *Zea*
Maloideae, 82
Malus, 82, 203–4m
Mangifera, 15, 17, 20, 183m
Mango. See *Mangifera*
Mangosteen family. See Clusiaceae
Maple. See *Acer*
Maple family. See Aceraceae
Marigold. See *Tagetes*
Marijuana. See *Cannabis*
Marsh-elder. See *Iva*
Matricaria, 142
Matthiola, 229m
Mayweed tribe. See Anthemideae
Meadow-rue. See *Thalictrum*
Medicago, 159, 161–62, 232m
Medick. See *Medicago*
Melaleuca, 69, 71, 75, 200m
Melica, 106–7, 109, 114, 118, 217m
Melicgrass. See *Melica*
Melilotus, XVI, 159, 161–62, 232m
Mercurialis, 158–60, 232m
Mercury. See *Mercurialis*
Mercury, three-seeded. See *Acalypha*
Mesquite. See *Prosopis*
Methodology: for pollen grain preparation, 237; for photography, 237–38; for pollen extract preparation, 238
Milfoil family. See Haloragaceae
Millet, broomcorn. See *Panicum*
Millet, common. See *Panicum*
Mimosa, 45, 191m
Mimosa. See *Mimosa*; Mimosoideae
Mimosa, prairie. See *Desmanthus*
Mimosa tree. See *Albizia*
Mimosoideae, 45, 48–50
Mistletoe. See *Phoradendron*
Mistletoe, dwarf. See *Arceuthobium*
Mistletoe family. See Viscaceae
Mock-orange. See *Philadelphus*
Momisia (syn.). See *Celtis*
Monolepis, 152, 155, 231m
Moraceae, 64, 66, 69–70
Morus, IV, XIV, 64, 66, 69–70, 199–200m
Mountain ash. See *Sorbus*
Mountain laurel. See *Kalmia*
Mountain mahogany. See *Cercocarpus*
Mounting pollen, for permanent slides, 237
Muehlenbeckia, 169, 234m

Mugwort. See *Artemisia*
Muhlenbergia, 106–9, 114, 118, 217m
Muhly. See *Muhlenbergia*
Mulberry, paper. See *Broussonetia*
Mulberry family. See Moraceae
Mustard. See *Brassica*
Mustard family. See Brassicaceae
Mutisieae, 140
Myrica, V, XIV, 66, 68–69, 71–72, 200m
Myricaceae, 62, 66, 68–69, 71–72
Myriophyllum, XIV, 159, 163, 232m
Myrtaceae, 69, 71, 73–74
Myrtle family. See Myrtaceae

Napier grass. See *Pennisetum*
Nectandra, 60, 198m
Nectandro, Jamaica. See *Nectandra*
Needlegrass. See *Stipa*
Nettle. See *Urtica*
Nettle, false. See *Boehmeria*
Nettle, wood. See *Laportea*
Nettle family. See Urticaceae
Nexine, 5
Nothofagus, 51
Nyssa, XVI, 71, 73, 75, 200m
Nyssaceae, 71, 73, 75

Oak. See *Quercus*
Oak, tanbark. See *Lithocarpus*
Oatgrass, tall. See *Arrhenatherum*
Oats. See *Avena*
Oats, sea. See *Chasmanthium*; *Uniola*
Ocean spray. See *Holodiscus*
Oil palm. See *Elaeis*
Olea, XVI, 73, 75–77, 79, 201m
Oleaceae, 73, 75–79
Oleaster. See *Elaeagnus*
Oleaster family. See Elaeagnaceae
Olive. See *Olea*
Olive, American. See *Osmanthus*
Olive, Chinese. See *Osmanthus*
Olive, Russian. See *Elaeagnus*
Olive, wild. See *Elaeagnus*
Olive family. See Oleaceae
Onagraceae, 159, 163–64
Orach. See *Atriplex*
Orchard grass. See *Dactylis*
Oryza, 105–6, 108–10, 114, 118, 217m
Oryzeae, 106
Oryzopsis, 106, 109–10, 118, 217m
Osage orange. See *Maclura*
Osmanthus, 77, 201m
Ostrya, XIV, 23, 25–27, 29–30, 185m, 187m
Oxycedrus (syn.). See *Juniperus*
Oxyria, 169, 234m
Oxytenia (syn.). See *Iva*
Oxytheca, 169

Palm family. See Arecaceae
Palmae (syn.). See Arecaceae
Palmetto. See *Sabal*
Palo-verde. See *Cercidium*
Paniceae, 106
Panicoid subfamily. See Panicoideae
Panicoideae, 106
Papaver, 162

INDEX

Papaveraceae, 159, 162, 164–65
Papilionoideae (syn.). See Faboideae
Parietaria, 173
Parietarieae, 173
Parlor palm. See *Chamaedorea*
Parthenice, 139–40, 142–43, 146, 229m
Parthenium, 139, 142–43, 146, 149, 229m
Parthenocissus, XVI, 98, 101–2, 212m
Partridge pea. See *Cassia*
Paspalum, 105–11, 113–14, 119, 218m
Pear. See *Pyrus*
Pecan. See *Carya*
Pellitory. See *Parietaria*
Pennisetum, 106, 110–11, 113–14
Pepper tree. See *Schinus*
Persea, 60, 62, 67, 198–99m
Phalaris, 106–10, 113–14, 119, 218m
Phellodendron, XVI, 84–86, 206m
Philadelphus, 87, 92–93, 209–10m
Phleum, VIII, 106–11, 113–15, 118–19, 218m
Phoenix, 22–24, 184m
Phoradendron, 98
Phragmites, 106–11, 114, 119, 218m
Picea, 9–10, 179–80m
Picrothamnus (syn.). See *Artemisia*
Pigweed. See *Amaranthus*
Pigweed, Russian. See *Axyris*
Pilea, XIV, 173–76, 235m
Pinaceae, 9–11
Pine. See *Pinus*
Pine family. See Pinaceae
Pinus, XIII, 9–10, 179–81m
Pinweed. See *Lechea*
Piqueria, 139, 149
Pistachio. See *Pistacia*
Pistacia, XV, 15, 17, 20, 183m
Pithecellobium, 45, 49, 191–92m
Planer tree. See *Planera*
Planera, 92, 94, 96, 99, 211m
Planetree. See *Platanus*
Planetree family. See Platanaceae
Plant names, 1
Plantaginaceae, 164, 166–67
Plantago, XII, XIV, 164, 166–67, 233m
Plantain. See *Plantago*
Plantain family. See Plantaginaceae
Platanaceae, 77, 80–81
Platanus, XV, 77, 80–81, 202m
Platycarya, 60
Platycladus, 178m
Plum. See *Prunus*
Poa, VIII, 106–12, 114, 119, 218–19m
Poaceae, 3, 105–21
Poeae, 106
Poison ivy. See *Toxicodendron*
Poison oak. See *Toxicodendron*
Poison sumac. See *Toxicodendron*
Pollen: aerobiology, 4–5; allergens, 5; morphology, 5; wall, 5
Pollination, 4
Pollinosis, 5
Polygonaceae, 167–70
Polygonella, 169
Polygonum, XVI, 168–70, 234m
Pondweed. See *Potamogeton*

Pondweed family. See Potamogetonaceae
Pooid subfamily. See Pooideae
Pooideae, 106
Poplar. See *Populus*
Poppy, California. See *Eschscholzia*
Poppy family. See Papaveraceae
Populus, V, XIII, 85, 87–90, 207m
Potamogeton, XIII, 125, 222m
Potamogetonaceae, 124–25
Potentilla, 82
Poverty-grass. See *Corema*
Poverty weed. See *Iva*; *Monolepis*
Powderpuff. See *Calliandra*
Prick/scratch test concentration, 238
Prickly ash. See *Zanthoxylum*
Privet. See *Ligustrum*
Prosopis, 45, 48, 50, 192m
Prunoideae, 82
Prunus, XVI, 82, 84, 203m, 205m
Pseudolarix, 179m
Pseudophoenix, 23, 185m
Pseudotsuga, 9, 181m
Ptelea, 84, 206m
Pteridium, 129–31, 223m
Pteridophytes, 129–31
Pterocarya, XIV, 60, 63, 65, 197m
Purpletop. See *Tridens*
Pyrethrum. See *Chrysanthemum*
Pyrus, 82, 203m

Quackgrass. See *Agropyron*
Quaking grass. See *Briza*
Quassia family. See Simaroubaceae
Queen Anne's lace. See *Daucus*
Quercus, II, III, XV, 3, 30, 51–53, 55–57, 193–96m

Radiate tribes (Asteraceae), 139
Raffia palm. See *Raphia*
Ragweed. See *Ambrosia*
Ragweed, false (*Franseria*). See *Ambrosia*
Ragwort. See *Senecio*
Ranunculaceae, 171–73
Ranunculus, 171, 173
Rape. See *Brassica*
Raphia, 21
Redbud. See *Cercis*
Red-sage. See *Kochia*
Redtop. See *Agrostis*
Redwood. See *Sequoia*
Reed, common. See *Phragmites*
Reed, giant. See *Arundo*
Reedgrass. See *Calamagrostis*
Rescuegrass. See *Bromus*
Rhamnaceae, 80–81
Rhapidophyllum, 23, 185m
Rheum, 169, 234m
Rhododendron, 40, 44, 190m
Rhubarb. See *Rheum*
Rhus, XV, 17, 19–20, 183m
Rhynchelytrum, 106, 109, 111, 114, 119, 219m
Ribes, 87, 92–93, 209–10m
Rice. See *Oryza*
Rice, wild. See *Zizania*
Ricegrass. See *Oryzopsis*
Richweed. See *Pilea*

Ricinus, XV, 45–47, 190m
Robinia, 192m
Rockrose. See *Helianthemum*
Rockrose family. See Cistaceae
Rosa, 81–82, 84, 203m, 205m
Rosaceae, 80–84
Rose. See *Rosa*
Rose family. See Rosaceae
Rosemary. See *Ceratiola*
Rosinweed. See *Grindelia*
Rosoideae, 81
Roystonea, 23–24, 184–85m
Rudbeckia, 139, 141–43
Rumex, XVI, 167–70, 234m
Rush. See *Juncus*
Rush family. See Juncaceae
Russian thistle. See *Salsola*
Rutaceae, 82, 84–86
Rye. See *Secale*
Rye, wild. See *Elymus*
Ryegrass. See *Lolium*

Sabal, 21, 23, 185m
Sabina (syn.). See *Juniperus*
Saccharum, 106, 111, 113–14, 119, 219m
Sage, black. See *Artemisia*
Sage, white. See *Artemisia*
Sagebrush. See *Artemisia*
St. Augustine grass. See *Stenotaphrum*
St. John's wort. See *Hypericum*
Salicaceae, 85, 87–91
Salicornia, XV, 152, 154–55, 157, 231m
Salix, V, XV, 85, 87–89, 91, 207–9m
Salsola, XII, XV, 152, 154–55, 157–58, 229m
Salsoloideae, 152
Salt cedar. See *Tamarix*
Saltbush. See *Atriplex*
Saltgrass. See *Distichlis*
Saltwort. See *Salicornia*
Sambucus, XV, 31, 33–34, 187–88m
Sanguisorba, 81, 84, 203m
Sapindaceae, 87, 92
Sapindus, 87, 92, 209m
Sapium, 45, 47, 190m
Sarcobatus, XV, 152, 154–55, 231m
Sassafras, XIII, 60, 62, 66–67, 199m
Saw palmetto. See *Serenoa*
Saxifragaceae, 87, 92–93
Saxifrage family. See Saxifragaceae
Scanning electron microscopy (SEM), 237
Scheuchzeria, 123–24, 222m
Schinus, XV, 17, 19–20, 183m
Schizachyrium, 106–11, 114, 119, 219m
Scirpus, VIII, XIII, 121–22, 221m
Sea buckthorn. See *Hippophae*
Searush. See *Juncus*
Secale, 105–11, 113–14, 119, 219m
Sedge. See *Carex*
Sedge family. See Cyperaceae
Seepweed. See *Suaeda*
Senecio, 139, 142–43
Senecioneae, 139–40
Sequoia, XIII, 11–12, 182m

INDEX

Serenoa, XV, 23–24, 185m
Serviceberry. See *Amelanchier*
Setaria, 106–11, 114, 119–20, 219m
Sexine, 5
She-oak. See *Casuarina*
Shepherdia, 40–41, 189m
Silk tree. See *Albizia*
Silk-tassel family. See Garryaceae
Silverberry. See *Elaeagnus*
Simaroubaceae, 87, 89, 93–94
Simmondsia, 30–32, 187m
Sitanion, 106, 110, 114, 120, 219m
Skin tests, 6
Smartweed. See *Polygonum*
Smokebush. See *Cotinus*
Smoketree. See *Cotinus*
Smother weed. See *Bassia*
Snakeroot, white. See *Eupatorium*
Sneezeweed. See *Helenium*
Snowball. See *Viburnum*
Snowberry. See *Symphoricarpos*
Soapberry. See *Sapindus*
Soapberry family. See Sapindaceae
Solidago, XVI, 137, 139, 141–43, 147, 149, 229m
Sonchus, 139, 143
Sorbus, 82, 84, 203m, 206m
Sorghastrum, 106–9, 114, 120, 219m
Sorghum, VIII, XIII, 106–12, 114, 120, 219m
Sorrel. See *Rumex*
Sorrel, mountain. See *Oxyria*
Sparganiaceae, 127
Sparganium, XIII, 127, 222m
Spartina, 105–10, 114, 120, 219m
Speargrass. See *Poa*
Sphaeromeria (syn.). See *Tanacetum*
Spicebush. See *Lindera*
Spike heath. See *Bruckenthalia*
Spinacea, 152
Spinach. See *spinacea*
Spindle-tree. See *Euonymus*
Spindle-tree family. See Celastraceae
Spiraea, 81–82, 203m, 206m
Spiraeoideae, 81
Spore: morphology, 129; allergenicity, 129
Sporoboleae, 106
Sporobolus, 106–11, 114, 120, 220m
Sporosis, 129
Sporulation, 129
Spruce. See *Picea*
Spurge family. See Euphorbiaceae
Squirrel grass. See *Sitanion*
Staining pollen, 237
Stenotaphrum, 106, 109, 111, 120, 220m
Stipa, 106, 109–10, 114, 120, 220m
Stipeae, 106
Stock. See *Matthiola*
Suaeda, XV, 152, 154–55, 157, 321–22m
Sudan grass. See *Sorghum*
Sugar beet. See *Beta*
Sugarcane. See *Saccharum*
Sumac. See *Rhus*
Summer-cypress. See *Kochia*
Sump-weed. See *Iva*
Sunflower. See *Helianthus*

Sunflower tribe. See Heliantheae
Sweet fern. See *Comptonia*
Sweet gum. See *Liquidambar*
Switchgrass. See *Panicum*
Sycamore. See *Platanus*
Symphoricarpos, 31
Syntherisma (syn.). See *Digitaria*
Syringa, XVI, 73, 75, 77, 79, 201m

Tagetes, 139, 142–43, 147, 225m
Tallow tree. See *Sapium*
Tamarack. See *Larix*
Tamaricaceae, 89–90, 95
Tamarisk. See *Tamarix*
Tamarisk family. See Tamaricaceae
Tamarix, XV, 89–90, 95, 210m
Tanacetum, XVI, 139, 142–43, 147, 225m
Tanicum, 106–11, 113–14, 119, 218m
Tansy. See *Tanacetum*
Taraxacum, XVI, 139, 141–43, 148–49, 225m
Tarragon. See *Artemisia*
Taxaceae, 10–11
Taxodiaceae, 11–12
Taxodium, XIII, 11–12, 182m
Taxus, XIII, 10–11, 181m
Thalictrum, XIV, 171–73, 234–35m
Thatch palm. See *Thrinax*
Thistle. See *Cirsium*
Thistle, plumeless. See *Carduus*
Thistle, sow. See *Sonchus*
Thistle tribe. See Cynareae
Thorn. See *Crataegus*
Thrinax, 23, 185m
Thuja, 7, 9, 179m
Tidestromia, 129, 132–33, 135, 224m
Tilia, XVI, 90, 95–96, 210m
Tiliaceae, 90, 95–96
Timothy. See *Phleum*
Tobacco brush. See *Ceanothus*
Torreya, 11, 181m
Toxicodendron, 17, 19, 183–84m
Trachycarpus, 23, 184m
Tree-of-heaven. See *Ailanthus*
Trema, 92, 94–96, 98–99, 173, 211m
Tricholaena (in part) (syn.). See *Rhynchelytrum*
Tridens, 106–9, 111, 113–14, 120, 220m
Trifolium, 159, 161–62, 232m
Triglochin, 123–24, 222m
Tripsacum, 106–9, 111, 113–14, 120, 220m
Trisetum, 106–10, 114, 120, 220m
Triticeae, 106
Triticum, 105–10, 114, 120, 220m
Trumpet vine. See *Campsis*
Tsuga, XIII, 9–10, 181m
Tulip tree. See *Liriodendron*
Tumbleweed. See *Amaranthus*
Tupelo. See *Nyssa*
Tupelo family. See Nyssaceae
Turnip. See *Brassica*
Tussilago, 139, 143
Typha, XIII, 125–27, 222m
Typhaceae, 125–27

Ulmaceae, 90, 92–100

Ulmeae, 96
Ulmus, VI, XIV, 92–98, 100, 210–11m
Umbelliferae (syn.). See Apiaceae
Umbellularia, 60, 62, 66–67, 199m
Umbrella-sedge. See *Cyperus*
Uniola (in part) (syn.). See *Chasmanthium*
Urtica, XII, XIV, 173–74, 176, 235–36m
Urticaceae, 173–76
Urticeae, 173

Vaccinium, 40, 42, 44, 190m
Vasey grass. See *Paspalum*
Velvetgrass. See *Holcus*
Vernalgrass. See *Anthoxanthum*
Vernonia, 139, 143
Vernonieae, 139–40
Vetch. See *Vicia*
Viburnum, 31, 33–34, 187–88m
Vicia, 159, 161, 232m
Viguiera, 139, 149
Virginia creeper. See *Parthenocissus*
Viscaceae, 98, 101
Vitaceae, 98, 101–2
Vitis, 98, 101–2, 212m

Walnut. See *Juglans*
Walnut family. See Juglandaceae
Washington palm. See *Washingtonia*
Washingtonia, 23, 184–85m
Water-milfoil. See *Myriophyllum*
Waterhemp. See *Amaranthus*
Wattle. See *Acacia*
Wax-myrtle. See *Myrica*
Wax-myrtle family. See Myricaceae
Wedgescale. See *Atriplex*
Weigela, 31, 33, 187m
Wheat. See *Triticum*
Wheatgrass. See *Agropyron*
Willow. See *Salix*
Willow family. See Salicaceae
Wingnut. See *Pterocarya*
Winter fat. See *Eurotia*
Wire plant. See *Muehlenbeckia*
Witch hazel family. See Hamamelidaceae
Witchgrass. See *Panicum*
Woodreed grass. See *Cinna*
Woodrush. See *Luzula*
Wormseed. See *Chenopodium*
Wormwood. See *Artemisia*

Xanthium, X, XVI, 137, 139, 141–43, 145, 225m

Yarrow. See *Achillea*
Yew. See *Taxus*
Yew family. See Taxaceae

Zanthoxylum, 84–86, 206–7m
Zea, 105–11, 113–14, 120, 220m
Zelkova, 92, 96
Zizania, 106–9, 114, 120, 220m
Zoysia, 106, 111, 114, 120–21, 220m
Zoysieae, 106
Zygophyllaceae, 98, 102–3

The Johns Hopkins University Press

AIRBORNE AND ALLERGENIC POLLEN OF NORTH AMERICA

This book was composed in Palatino Linotron
by EPS Group, Inc., from a design
by Gerard A. Valerio.
It was printed on 70-lb. Paloma matte paper
and bound by the Maple Press Company.